NUCLEASES

Molecular Biology and Applications

Nawin C. Mishra

Department of Biological Sciences
University of South Carolina

WILEY-INTERSCIENCE
A JOHN WILEY & SONS, INC., PUBLICATION

Published by John Wiley & Sons, Inc., Hoboken, New Jersey.
Published simultaneously in Canada.

Library of Congress Cataloging-in-Publication Data:

Mishra, N. C. (Nawin C.)
Nucleases : molecular biology and applications / Nawin C. Mishra.
p. cm.
Includes bibliographical references and index.
ISBN 0-471-39461-0 (alk. paper)
1. Nucleases.
[DNIM: 1. Ribonucleases–physiology. 2. Deoxyribonucleases–physiology.
3. Deoxyribonucleases–therapeutic use.
4. Genetic Techniques. 5. Ribonucleases–therapeutic use. QU 136 M678n 2002] I. Title.
QP609.N78 M573 2002
572′.757–dc21 2002001354

Printed in the United States of America

10 9 8 7 6 5 4 3 2 1

Dedicated to the memory, of E. L. Tatum, and my parents

CONTENTS

PREFACE

Nucleases are usually considered to be a group of enzymes capable of hydrolyzing nucleic acids by breaking the phosphodiester bonds. However, recent advances in the fields of biochemistry, genetics, and molecular biology have established that nucleases occupy a central position in all aspects of the nucleic acid metabolism in every organism—particularly in view of the fact that they act on DNA and RNA, the molecules of inheritance.

Furthermore, nucleases are proven to be very useful and important tools in modern biological studies, crucial for the development of recombinant DNA technology and reverse genetics and the undertaking of grand projects such as the human genome project. Nucleases are essential for the molecular cloning of genes, leading to the identification and the characterization of genes responsible for several diseases and their possible alleviation by gene therapy and development of designer drugs. Nucleases are important for the development of biotechnology, industry, and forensic science.

Initially, I came to know about these enzymes from their brief description in biochemistry texts during my graduate school days in the early 1960s. However, my realization about the possible biological roles of the nucleases surfaced soon after my discussion with the late Professor Moses Kunitz in the late 1960s at the Rockefeller University when I went to his laboratory to borrow some enzymes (other than nucleases). Because of my interest in certain aspects of DNA transactions, particularly the process of DNA recombination from my early graduate days, the discussions regarding the nucleases with professor Kunitz always remained indelibly etched in my mind. Later I became involved with nucleases in my own research and in teaching a graduate course on enzymes involved in the different aspects of DNA replication, repair, and recombination. During that period, I found the information about these enzymes scattered throughout the literature in the form of research papers or specialized treatise. I felt a need for a book concerning the various aspects of nucleases. The main aim of this book is to introduce the properties and biological roles of nucleases to newcomers in the field and to provide the basis of their possible application in certain aspects of science, commerce, and industry. I very much hope that it will serve this purpose. In addition, my hope is that this book will provide a firsthand knowledge about nucleases to a large group of individuals in molecular biology, medicine, and biotechnology using these enzymes on a daily basis despite their knowledge of nucleases being mostly limited to the need of the use of nuclease-free buffers in their research kits.

I am thankful to a large number of people who have helped me in this endeavor. First, I wish to express my appreciation to my wife, Purnima, to acknowledge the fact that this work was made possible by her constant interest and devotion. I express my sincere thanks to Drs. Philip Hanawalt, Murray Fraser, M. Jayram, H. Hayes, Alex Almasan, and Brian Odom for reading or discussing different versions of certain chapters of this book. I also thank Drs. Rollin and Magda Hotchkiss, Steve Threlkeld, David Perkins, Stuart Linn, and Mike Felder for their interest, encouragement, and support. This book has been written entirely by me with the help of a personal computer. Therefore, I am solely responsible for its content and particularly for all errors in it. However, I am thankful to Candace Keel and Clint Cook for their assistance in word processing and to Virginia Parrott for correcting my grammar at an earlier stage of writing.

I would further like to acknowledge my sincere thanks to Dr. Darla Henderson and Ms. Amy Romano, both of John Wiley & Sons, Inc. Without their initiative, cooperation, and perseverance, this work would not have been possible. I also thank Ms. Kellsee Chu of John Wiley & Sons, Inc. for her sincere efforts ensuring the timely production of this book and a pleasant experience for me.

I would also like to thank my son, Prakash, for his interest in this work and for his continuous support. Finally, I would like to acknowledge the grant support from the U.S. Department of Energy which made the work reported from my laboratory possible.

Nawin C. Mishra

University of South Carolina
Columbia, South Carolina

LIST OF NOBEL PRIZE WINNERS FOR THEIR RESEARCH WORK WITH NUCLEASES

1968
R. W. Holley, Cornell University
For his work on the structure of tRNA. This work would not have possible without the use of ribonucleases.

1972
Christian B. Anfinsen, NIH
For his work on ribonuclease, especially concerning the connection between amino acid sequence and biological activity.
Stanford Moore, The Rockefeller University
William Stein, The Rockefeller University
For their work on the understanding of the connection between chemical structure and catalytic activity of the active center of the ribonuclease.

1978
W. Arber (Switzerland)
Daniel Nathans, John Hopkins University
Hamilton O. Smith, John Hopkins University
For their work on recombinant DNA and restriction endonuclease technology

1980
Paul Berg, Stanford University
For his work with particular reference to recombinant DNA and technology of gene manipulation

1984
Robert Bruce Merrifield, The Rockefeller University
For his work on the development of solid-matrix method for the chemical synthesis of proteins using ribonuclease.

1989
Sydney Altman, Yale University
Thomas Robert Cech, University of Colorado
For their work on catalytic RNA

1993
R. Roberts, New England Biolabs, Inc.
Phillip Sharp, MIT
For their work on RNA splicing and split gene structure.

ABOUT THE AUTHOR

Nawin C. Mishra is a Professor of Genetics in the Department of Biological Sciences at the University of South Carolina. He initiated the gene transfer experiments in fungi while he was a member of the laboratory of Dr. E. L. Tatum at the Rockefeller University (1967–1973). He has worked on various aspects of gene transfer and on the characterization of nuclease and DNA polymerase mutants with altered exonuclease activity in eukaryotes. He received his Ph.D in Genetics from McMaster University in 1967 and his Postdoctoral training with the late Nobel Laureate Professor E. L. Tatum at the Rockefeller University, supported by a postdoctoral fellowship from the Jane Coffin Childs Memorial Fund for Medical Research of the Yale University. In 1973 he joined the Molecular Biology faculty of the University of South Carolina. He has been invited to present his work in Australia, Europe, Russia, China, Japan, Thailand, and India. He has served as a scientific consultant to the FAO of the United Nations. He also served as the Chairman of the Program Committee of the Genetics Society of America (1977–1979) and as a member of the review panel of the Human Genome Project of the United States Department of Energy. He is a Fellow of the American Association for the Advancement of Science since his election to this organization in 1986 for his original contribution to the study of gene transfer in fungi.

1

INTRODUCTION

The genetic blueprint of all organisms consists of nucleic acids. In nature this choice is based on the stability of nucleic acid molecules due to the presence of phosphodiester linkages, which are the strongest among all chemical bonds (Westheimer, 1987). The stability of nucleic acid is a very important feature in maintaining the integrity of the genetic material. However, during the life of an organism the stability of phosphodiester linkages is compromised at times to facilitate certain other important processes such as the removal of damaged DNA and its repair in order to restore the accuracy of the genetic blueprint. Such breakage of phospodiester bonds in DNA chains is also allowed to provide the recombination of genes in a chromosome or the salvage of genetic material in a cell for reutilization of nucleotides or their components or for their final disposal in a cell destined for apoptosis during normal development of a multicellular organism as complex as human. Thus, it is not surprising that all living systems contain a group of enzymes capable of hydrolyzing the phosphodiester linkages present in nucleic acids; these enzymes are called *nucleases*.

The facts that these phosphodiester linkages are the most stable among all chemical bonds found in biological molecules and that nucleases can hydrolyze such phosphodiester linkages make them the most unique of all enzymes. No other groups of enzymes influence the physiology of an organism in such diverse ways as the nucleases. The fact that nucleases catalyze an array of biochemical reactions, all involving just one chemical bond, namely, the phosphodiester linkage, has no parallel in the biochemistry of enzymes. A number of nucleases possess other enzymatic functions in addition to their principal property of catalyzing the hydrolysis of phophodiester linkages. Many nucleases possess other associated catalytic activities such as DNA polymerase, ligase, helicase, or kinase and other functions such as repressor.

Initially, nucleases were considered to play only a degradative role in the salvage pathway of nucleic acid metabolism because of their association with the pancreas (Kunitz, 1940). However, they are now shown to play a multitude of important roles in different aspects of basic genetic mechanisms. These include their participation in (a) the processes of DNA replication, repair, recombination, and mutagenesis (Clark, 1971; Hanawalt et al., 1979; Kornberg and Baker, 1992), (b) the control of gene expression by determining the nature of transcript (Sharp, 1981) and its turnover and (c) transposition and other programmed gene rearrangements (Borst and Greaves, 1987). They are also a part of the host defense mechanism against alien nucleic acid molecules (Luria and Human, 1952). Nucleases play a great role in the mechanisms of immune systems in mammals by controlling the assembling of immunoglobulin genes, their allelic exclusion, and class switching and in the determination of the membrane-bound or secreted forms of the immunoglobulins, leading to immunoglobulin diversity (Sakano et al., 1980; Tonegawa 1983, 1985; Yancopoulos and Alt, 1985, 1988; Yancopoulos et al., 1986) and antigenic variation (Myler et al., 1984; Borst and Cross, 1982). Nucleases also play an important role during programmed cell death in the development of multicellular organisms, including human.

Thus the study of nucleases has been very useful from both the conceptual and technical points of view. The pancreatic ribonuclease was the first enzyme whose entire amino acid sequence was determined (Moore and Stein, 1973; Anfinsen, 1963). The amino acid sequence of the bovine pancreatic ribonuclease is presented in Figure 1.1. Knowledge of the amino acid sequence of the pancreatic ribonuclease was crucial in the confirmation of the idea that the secondary and tertiary structures of a protein are controlled by its primary structure (Anfinsen, 1964). Determination of the structure of the ribonuclease led to the development of an entirely new technology of protein synthesis and protein engineering (Gutte and Merrifield, 1971; Merrifield, 1986; Gutte, 1992). Nucleases have been studied extensively in order to seek answers to the question of the mechanism of enzyme catalysis and polypeptide folding (Anfinsen, 1964; 1973; Tucker et al., 1979; Cotton et al., 1979; Shortle, 1983; Botstein and Shortle, 1985; Kippen et al., 1994). Ribonucleases were used in the complete sequencing of the first tRNA molecule which led to the understanding of its role in the mechanism of genetic coding (Holley, 1965). Use of these nucleases led to the elucidation of the nucleosomal organization of eukaryotic chromosomes (Hewish and Burgoyne, 1973). Above all, the discovery of a new class of nucleases called *restriction endonucleases* (Smith and Wilcox, 1970) resulted in the development of recombinant DNA technology (Jackson et al., 1972), DNA sequencing methodology (Maxam and Gilbert, 1977; Sanger et al., 1977), and new methods for genetic mapping (Southern, 1975, 1982; Kan and Dozy, 1978; Botstein et al., 1980; White and Lalouel, 1988) and the extensive mapping of human chromosomes (Donis-Keller et al., 1987). Furthermore, the use of restriction endonucleases has been crucial in the development of a new branch of genetics called *reverse genetics* (Ruddle 1984; Orkin, 1986). The methods of reverse genetics have been very useful in the understanding of the molecular basis of several human diseases—for example, Huntington's chorea, cystic

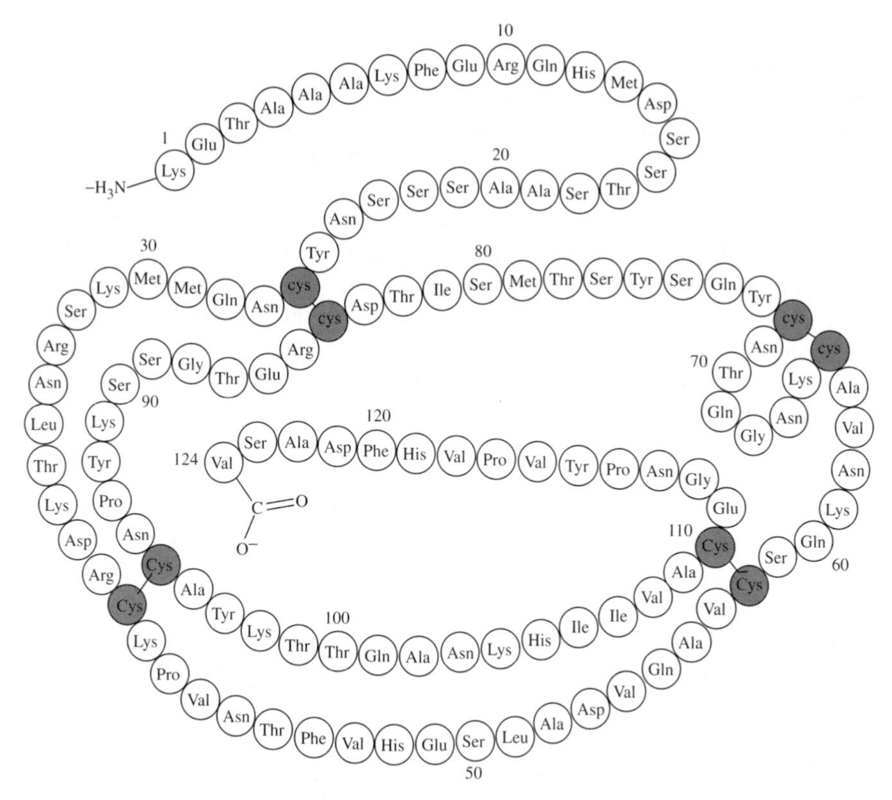

Figure 1.1. Amino acid sequence of the bovine pancreatic ribonuclease. Black circles represent cysteine residues forming disulfide bonds. (From *Biochemistry* by L. Stryer, copyright W. H. Freeman and Company, 1982, used with permission.)

fibrosis, retinoblastoma, arthritis, and many others (White, 1986; Gusella, 1986; Caskey, 1987; Davies, 1991; Mark, 1987; Monaco and Kunkel, 1987; Woron and Thompson, 1988; Martin, 1987; Buxton, 1992)—which may pave the way for the possible alleviation of certain inherited human diseases by gene therapy (Weatherall, 1982; Anderson, 1992; Ailliton, 1985; Nichols, 1988; Mulligan, 1993). All these new methodologies of genetics in which nucleases play critical roles hold great promise for the future of mankind through the development of science, technology, and commerce (Mishra, 1985).

I. HISTORICAL PERSPECTIVES

The enzymatic breakdown of nucleic acid was first observed in the early twentieth century (Araki, 1903). The enzymes involved in this process were named *nucleases* (Iwanoff, 1903). The fact that nucleases are indeed phosphodiesterases was, however, established much later (Brown and Todd, 1955; Hilmoe et al., 1961;

Laskowski, 1967; Westheimer, 1987; Gerlt 1992, 1993). Soon two groups of nucleases designated as ribonuclease (RNase) and deoxyribonuclease (DNase), capable of hydrolyzing ribonucleic acids and deoxyribonucleic acids, respectively, were described by Kunitz (1940, 1950). This specificity of nucleases toward the sugar moiety (i.e., deoxyribose or ribose) of nucleic acid was considered the most important criterion for the classification of nucleases (Kunitz, 1940, 1950). Studies of the properties of crystallized nucleases (Kunitz, 1940, 1950) laid down the foundation for the understanding of the biochemical nature, genetic control, and physiological role of these enzymes. Early studies of nucleases and their chemical modifications provide an excellent model for understanding the nature of the biochemical reaction mechanisms and the role of catalytic sites that facilitate substrate binding. These conclusions were later confirmed by x-ray crystallography. Analysis of the crystal structures of a number of nucleases such as *Escherichia coli* DNA polymerase I, RNaseH, endonuclease III, topoisomerase I, Hin invertase, T_4 endonuclease DNaseI, a number of recombinases and structure-selective nucleases pancreatic RNases, and several other RNases has provided insight into the structure and function of enzyme proteins and their specificity of action (Joyce and Steitz, 1987; Suck et al., 1984; Nakamura et al., 1991; Morikawa et al., 1992; Sevcik et al., 1990; Lima et al., 1994).

The study of nucleases, particularly the mechanistic view of the interaction of their active sites with their substrates, has provided a better understanding of the mechanisms of enzyme catalysis (Gerlt, 1992, 1993). Nucleases must possess the ability not only to recognize and bind with a DNA sequence, but also to hydrolyze it. Nucleases differ from other DNA recognizing and binding proteins such as transcription factors; the latter usually possess leucine zipper or zinc finger or helix-turn-helix motifs. Nucleases lack such motifs and therefore must recognize and bind with DNA by other methods. Certain restriction endonucleases (see Chapter 4) may undergo mutation to produce the mutant forms of the enzyme which can recognize a DNA sequence and bind with it but are unable to hydrolyze it.

II. PROTEIN, RNA, DNA, AND OTHER MOLECULES AS NUCLEASES

Enzymes are biocatalysts that facilitate chemical reactions in biological systems. Enzymes are traditionally known to be proteins. However, proteins are no longer the only molecules that have a catalytic function. In recent years a number of RNA and DNA molecules have been shown to possess such functions. Thus, it is no surprise that a number of nucleases are RNA or DNA even though a majority of nucleases are made up of proteins. In addition to proteins and RNAs or DNAs, a number of other smaller molecules (such as bleomycin and phenanthroline and others) have been shown to possess nucleolytic property. It is proposed here that the term *enzyme* or *proteinzyme* should be used to indicate all biocatalysts that are proteins. This nomenclature would be consistent with the fact that a majority of molecules with biocatalytic function are proteins. The RNA and DNA molecules with catalytic activities have been designated as *ribozyme* or *RNAzyme* (Kruger et al., 1982) and *deoxyribozyme* or *DNAzyme* (Breaker and Joyce, 1994; Breaker, 2000; Wilson and Szostak, 1999) respectively. Certain small molecules and chemical reagents with enzyme-like catalytic functions are called *chemzymes* (Corey and Reichard, 1989; Waldrop, 1989). The occurrence and the role of DNAzymes and chemzymes in biological systems remain to be established. An understanding of the divergent nature of the biocatalysts has made it possible to design synthetic or semisynthetic nucleases that can be targeted to a specific nucleic acid sequence. These nucleases are called *artificial* or *designer* or *chemical nucleases* (Sigman and Chen, 1990). The ability to design such nucleases may revolutionize the field of genome mapping (Ebright et al., 1990; Oakley and Dervan, 1990). Nucleases as protein enzymes are described in the different chapters of this book. The nature and role of ribozymes, deoxyribozymes, and certain chemzymes as nucleases are discussed in Chapter 9. All protein enzymes and ribozymes are encoded by genes or DNA molecules; therefore the DNA molecules are their genotypes and they represent phenotypes, whereas in the case of DNAzyme, both the genotypes and phenotypes are represented by the same molecule.

III. NATURE OF ENZYMATIC REACTIONS CATALYZED BY NUCLEASES

Nucleases are, in essence, phosphoesterases that hydrolyze the internucleotide linkage in a nucleic acid molecule. There are different kinds of phosphoesterases. Phosphomonoesterases do not act on internucleotide linkage but cleave the terminal phosphate from a nucleotide chain. Among the three classes of phosphodiesterases, the first group of enzymes acts on phosphodiester bonds not involving internucleotide linkage. The enzymes responsible for the breakdown of cAMP and cGMP are examples of such phosphodiesterases. The second group of enzymes acts on different types of phosphodiester bonds, including internucleotide linkages; the snake venom phosphodiesterases belong to this group of enzymes. The third group specifically acts on internucleotide phosphodiester linkage(s); protein and ribozyme nucleases belong to this group of phosphodiesterases. Both protein and ribozyme nucleases act by hydrolysis of phosphodiester linkages in the nucleic acid chain. This, however, is not true of the chemzyme nucleases; they cause cleavage by disruption of the sugar moiety via oxidation or by alkylation of bases (Sigman and Chen, 1990).

Most enzymatic reactions are reversible in nature. Both the synthesis and degradation of micromolecules are usually carried out by the same enzymes. However, the degradation and biosynthesis of macromolecules in all biological systems are carried out by distinct sets of enzymes (Kornberg, 1974). Nucleases that cleave internucleotide linkages during the degradation of nucleotides cannot reform the internucleotide linkage leading to the biosynthesis of nucleic acid molecules. Instead, the internucleotide linkages are either (a) formed by DNA or RNA polymerases during the synthesis of specific nucleic acid molecules or (b) joined by the enzyme ligase. However, there are exceptions to this rule. A number of DNA polymerases carry nuclease activity in different parts of the same polypeptide. Also, a class of nucleases called *topoisomerases* has been shown to combine both properties; that is, they can hydrolyze the internucleotide linkages transiently and then rejoin them as discussed in Chapter 6. The action of different enzymes that act on nucleic acids are illustrated in Figure 1.2. Nucleases and phosphodiesterases

1. Polymerases

   a. DNA Polymerase

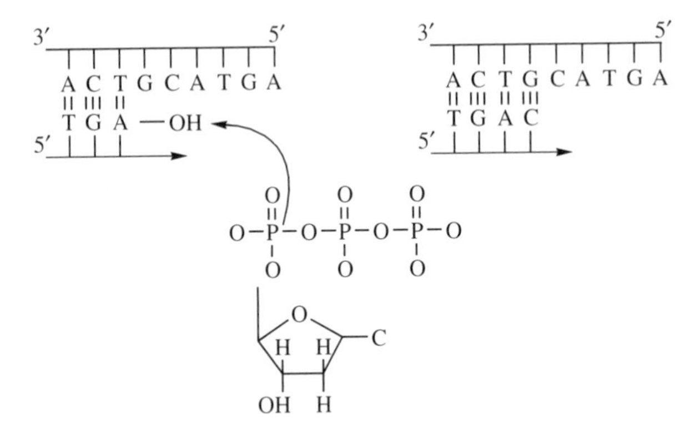

   b. RNA Polymerase

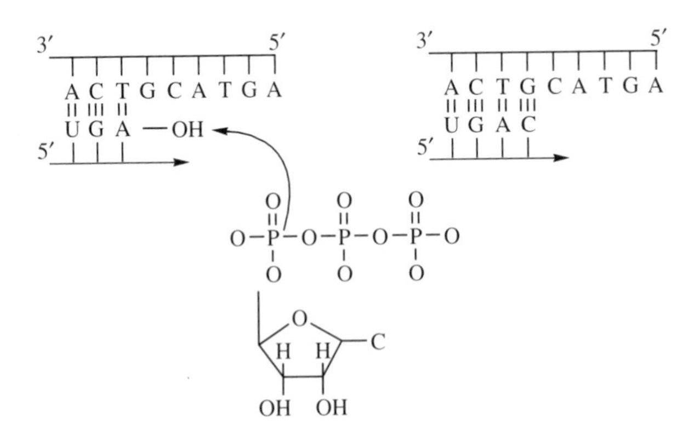

   c. Reverse Transcriptase

Figure 1.2 Action of different enzymes on nucleic acids.

2. Ligase

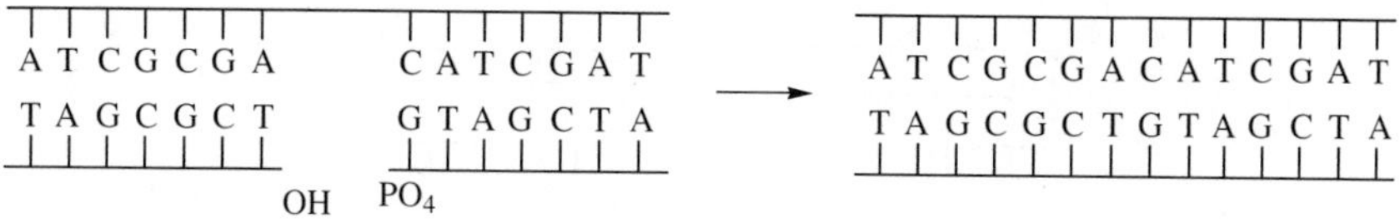

3. Terminal Transferase

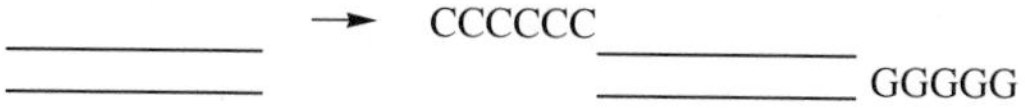

4. Phosphatase and Polynucleotide Kinase

a. Phosphatase

b. Polynucleotide Kinase

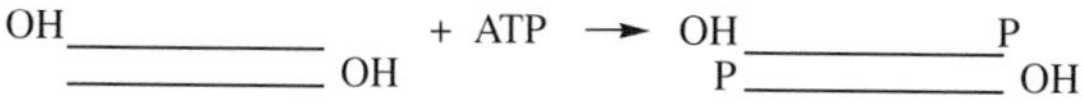

5. Nucleases

a. DNase

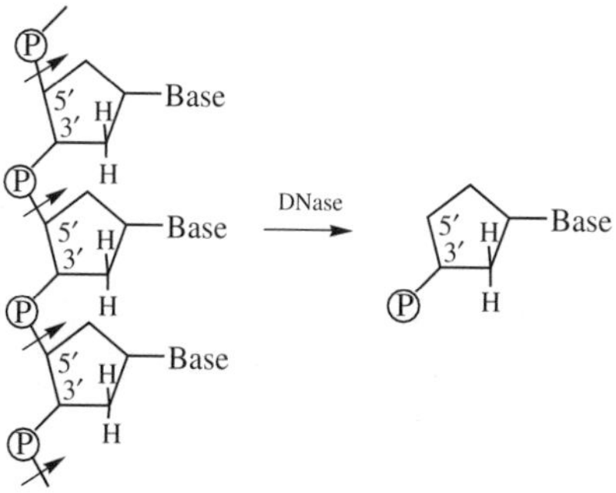

b. Restriction Endonuclease

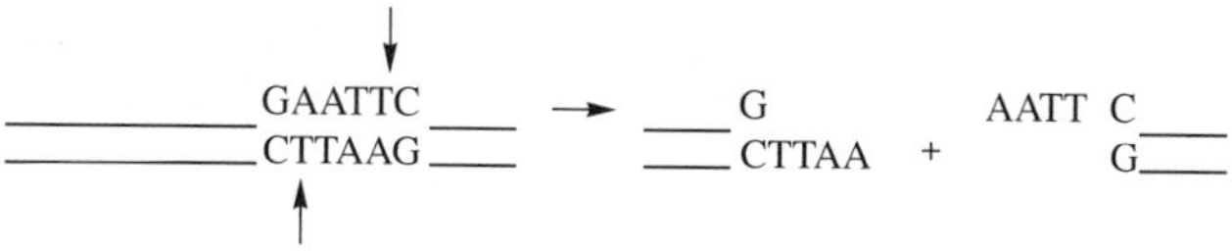

Figure 1.2 (*continued*)

c. Damage-Specific Nuclease

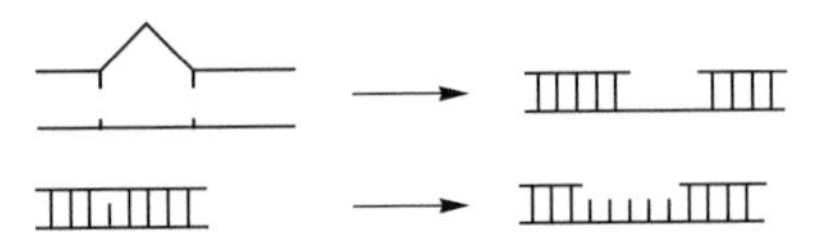

d. Topoisomerase

(i) Relaxed DNA —Gyrase→ Supercoiled DNA

(ii) Supercoiled DNA —Topoisomerase I→ Relaxed DNA

e. Recombinase

A B → A b

a b a B

f. RNase

P 5′ 3′ H Purine OH

P 5′ 3′ H Pyrimidine OH —RNase→ 5′ 3′ H Base P OH

P 5′ 3′ H Purine or pyrimidine P OH

g. RNA Splicing Enzyme

A I B C III G I V J E II H IV —Splicing enzyme→ I II V A B C G I J E H II IV

h. RNaseH

DNA RNA hybrids

DNA RNA —RNase H→ DNA ribonucleotides

Figure 1.2 (*continued*)

i. Nonspecific Nucleases or Phosphodiesterases

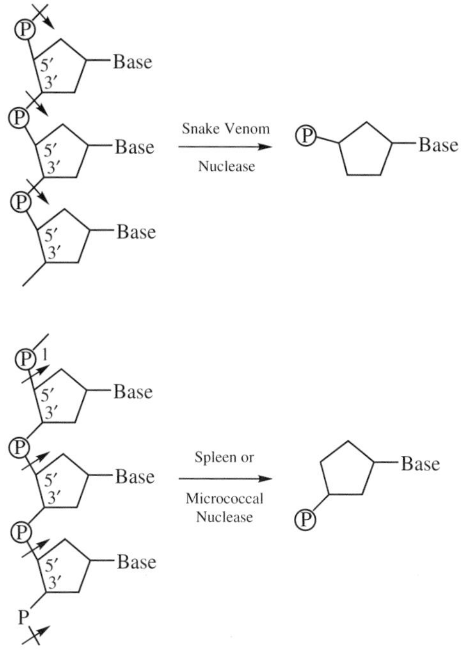

P 5′ 3′ Base
P 5′ 3′ Base
P 5′ 3′ Base

Snake Venom Nuclease → P Base

P 1 5′ 3′ Base
P 5′ 3′ Base
P 5′ 3′ Base
P

Spleen or Micrococcal Nuclease → Base P

6. DNA Methylase

A — T → A ($-CH_3$) — T

C — G → C ($-CH_3$) — G

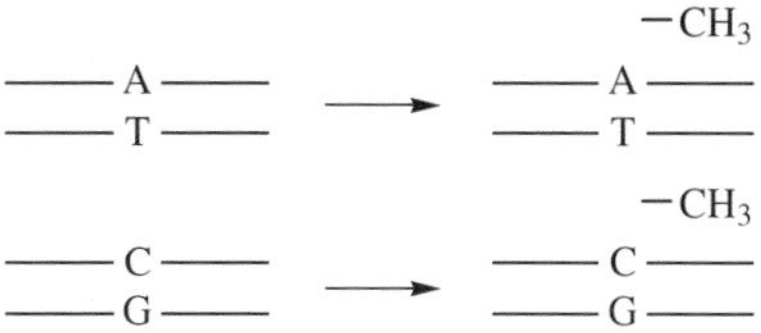

Figure 1.2 (*continued*)

cleave the bond between phosphorus and oxygen in the internucleotide linkage of nucleic acid (Hilmoe et al., 1961) as shown in Figure 1.3.

Protein nucleases are characterized by their processive or distributive mode of action. A processive enzyme remains engaged with the substrate molecules, resulting in the hydrolysis of successive internucleotide linkages before it leaves to act on another nucleic acid molecule. In contrast, a distributive nuclease will usually hydrolyze only a few internucleotide linkages before it falls off the substrate molecule and will then be free to engage with the same or another nucleic acid molecule.

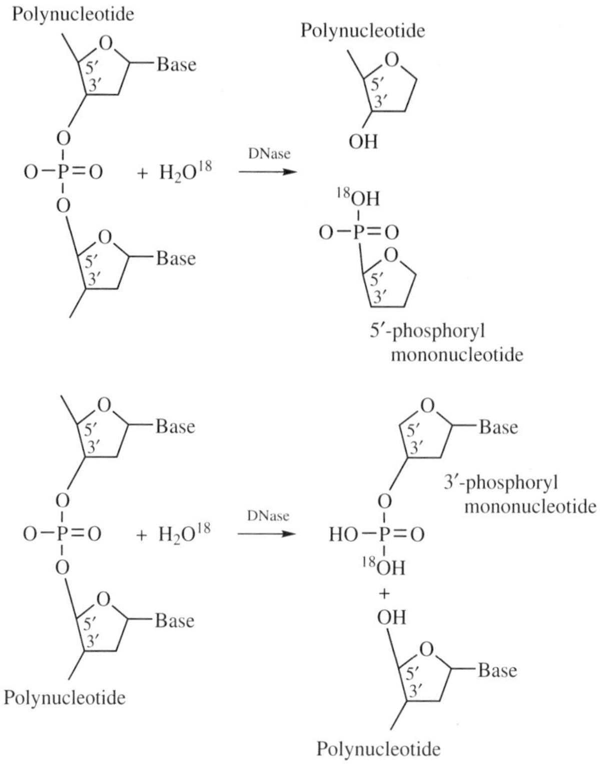

Figure 1.3. Nuclease acts on bond in P–O.

IV. CLASSIFICATION

Kunitz (1940) classified nucleases into two major groups, ribonucleases and deoxyribonucleases, based on the specificity of their nucleolytic attack toward the sugar moiety of the different nucleic acid molecules. Even though snake venom phosphodiesterase hydrolyzed both ribonucleotides and deoxyribonucleotides (Schmidt, 1955), the classification of nucleases into DNase and RNase by Kunitz was considered valid at that time. This was partly because of the nonspecific nature of snake venom phosphodiesterase that could attack phosphodiester linkages other than those found in nucleic acid. The classification of nucleases into DNase and RNase, however, appeared somewhat inappropriate with the discovery of micrococcal nuclease and snake venom phosphodiesterases that attacked both DNA and RNA (Laskowski, 1982). Soon a new class of sugar-nonspecific nucleases was added to the list of nucleases. This method of classification led to a trend in which the terms DNase and RNase were used for the sake of convenience. Later, several new groups were added as dictated by new evidence.

In view of these difficulties with the classification of nucleases, Bernard (1969) and Laskowski (1959, 1982) introduced the idea of consensus criteria (instead of absolute criteria) for the classification of nucleases. These include (a) the nature of substrate hydrolyzed (DNA, RNA), (b) the type of nucleolytic attack (exonuclease and endonuclease), (c) the nature of the nucleolytic products (i.e., mono- or oligonucleotides terminated by a 3′- and 5′-phosphate group), and (d) the nature of the bonds hydrolyzed. Additional criteria, such as the nature of the substrate DNA (mismatch, damaged or topological isomers), site-specificity, structure-selectivity, and functional ability to restructure DNA molecules (i.e., to facilitate genetic recombination), were added to this list at a later stage (Laskowski, 1982; Linn, 1982a; Lieber, 1997 and Suck, 1998; Mishra, 1995). This set of consensus criteria for the classification of nucleases seems compatible with the recent advances made in the understanding of nucleases as a whole. This eliminates many of the problems in the classification of nucleases, and it also provides an easy method for the accommodation of an enzyme with exceptional properties into a particular group.

However, this system is still riddled with difficulties, by the fact that the pancreatic DNase can attack a polynucleotide chain containing a mixture of both ribo- and deoxyribonucleotides (Pruch and Laskowski, 1980). Furthermore, exonucleases were initially considered to be enzymes that sequentially removed mononucleotides from a free terminus. This group was later expanded to include enzymes that attacked internal phosphodiester linkages but required a free terminus for the identification of polynucleotide chain to be hydrolyzed (Frankel and Richardson, 1971). This view was based on the observation that a covalently closed circular DNA could be broken by an endonuclease, but not by an exonuclease, since the latter required a free terminus for the recognition of the substrate. However, recently it has been shown that some exonucleases are capable of opening a supercoiled DNA 10,000 times faster than a relaxed DNA (Pritchard et al., 1977), suggesting that an exonuclease is, in essence, an "exophilic" nuclease that prefers an open terminus (Laskowski, 1982). Furthermore, certain nucleases such as micrococcal nuclease can attack a nucleic acid molecule either endo- or exonucleolitically. The properties of these nucleases and their possible roles in the basic genetic mechanisms that occur during the life of an organism are discussed in the different chapters of this book. The major classification groups of nucleases are based on a set of consensus criteria. Some of the other major criteria considered during the classification of nuclease as discussed by Laskowski (1982) are described below.

A. Nature of Substrates

Nucleases differ significantly in the nature of the substrates that they can hydrolyze. This is implicit in the idea of the consensus criteria discussed above. Nucleases hydrolyze the phophodiester linkages in DNA and RNA. Both of these molecules can exist in different helical structures. DNA can exist as a right-handed or a left-handed helical structure (Figure 1.4). The biologically active form of DNA found in the living systems is the right-handed form of DNA called B-DNA. The DNA whose structure was first elucidated by the Watson–Crick model of DNA is the B-DNA. This DNA has 10 base pairs per turn. The B-DNA depending on hydration or ionic environment may transform into several isomeric forms such as the

B′, C, C′, C″, D, E, and T forms. The phage T2 DNA occurs in the B form under conditions of relative high humidity but transform directly into T-DNA. The other right-handed DNA helix is the A-DNA, which possesses 11 base pairs per turn—in contrast to the naturally occurring B-DNA, which passesses 10 base pairs per turn. In addition, DNA can exist as a left-handed helix with 12 base pairs per turn; this left-handed DNA is called a Z-DNA. There is no evidence for the existence of Z-DNA in nature, but alternating purine–pyrimidine occurring as tandem repeats of dinucleotide pair CA/AG found in intergenic DNA and in introns and rarely in the coding DNA may assume Z-DNA form. Most nucleases act on B-DNA. Only one naturally occurring nuclease such as Mung bean nuclease has been found to make cuts in the intergenic DNA, suggesting its ability to use Z-DNA as substrate. A number of chemical nucleases can utilize Z-DNA as substrate. The structure of biologically active forms of DNA is presented in Figure 1.4.

Most RNAs occur as single-stranded structures, but these may include a certain double-helical domain that is very characteristic of tRNA ribosomal RNA and many mRNA. Double-stranded RNAs assume a helical structure in two isomeric

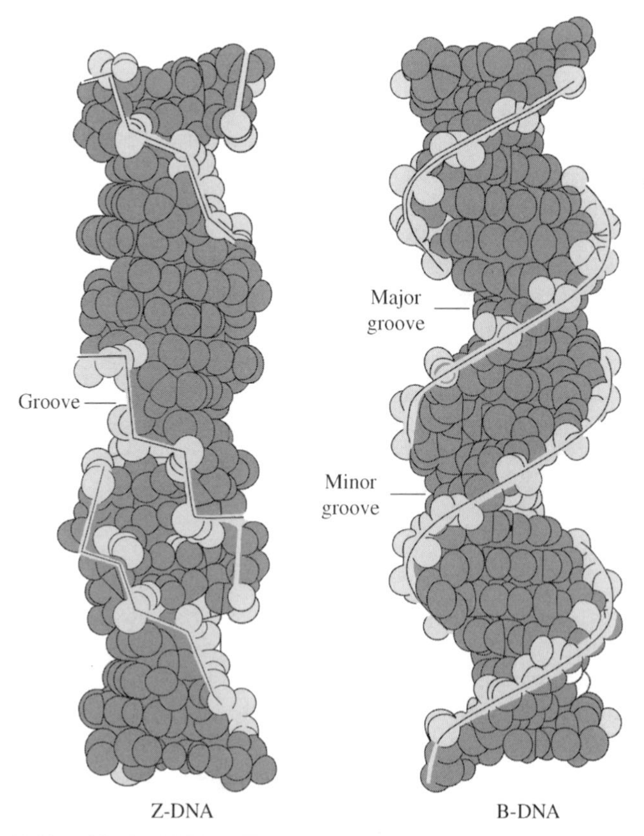

Figure 1.4. Z- (*left*) and B- (*right*) DNA. The dark lines connect phosphate groups. (Reproduced with permission from the *Annual Review of Biochemistry*, Volume 53, Copyright 1984 by Annual Reviews, Inc.)

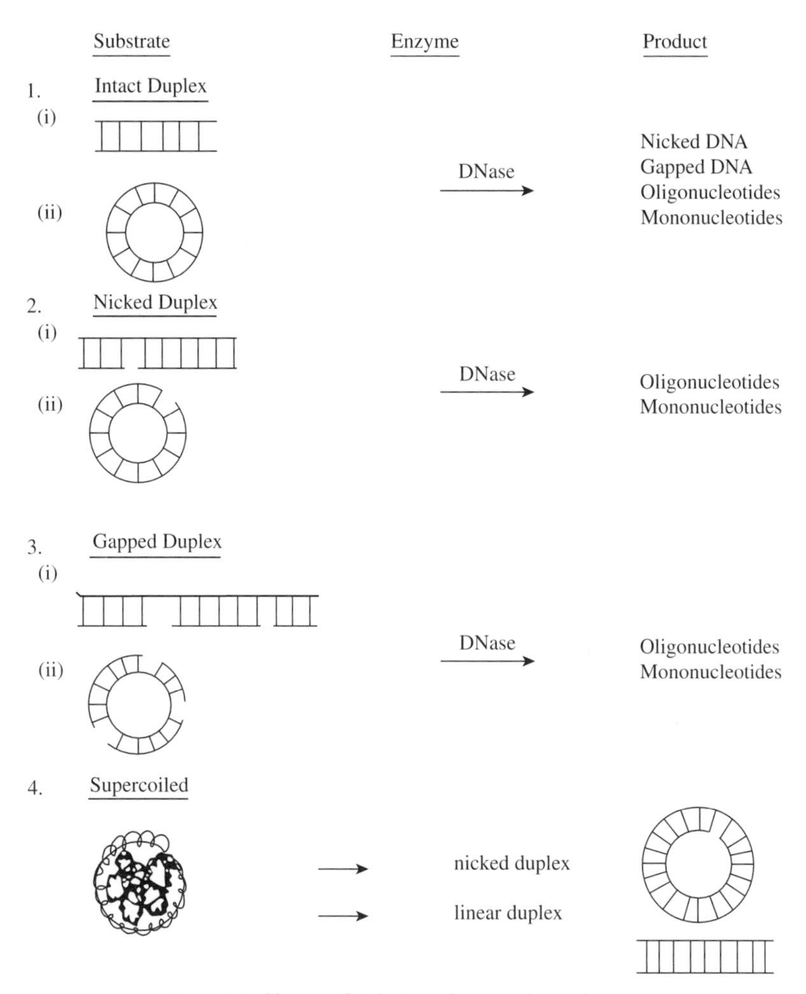

Figure 1.5. Nature of substrate for protein nucleases.

forms called A-RNA and A′-RNA depending on the ionic concentration of the media. At low salt concentration, the RNA helix exists as A-RNA with 11 base pairs per turn; however, at high salt concentration the A-RNA may assume the A′-RNA form with 12 base pairs per turn. Both forms of the double-stranded RNA show features typical of Watson–Crick base pairs; however, certain homopolymers may yield triple-stranded RNA structure simultaneously showing Watson–Crick and Hoogsteen base pairing. Different forms of DNA and RNA substrates are shown in Figure 1.5.

In addition to DNA and RNA, small DNA pieces with a hairpin structure containing a stem with a fluorophore-quencher pair and a loop with sites for cleavage by nucleases are used as the substrates to probe the nuclease activity of unknown molecules. The nuclease activity of the unknown molecules can be measured by the amount of fluorescence released due to the separation of quencher from the fluorophore after the digestion of the DNA substrate (Figure 1.6).

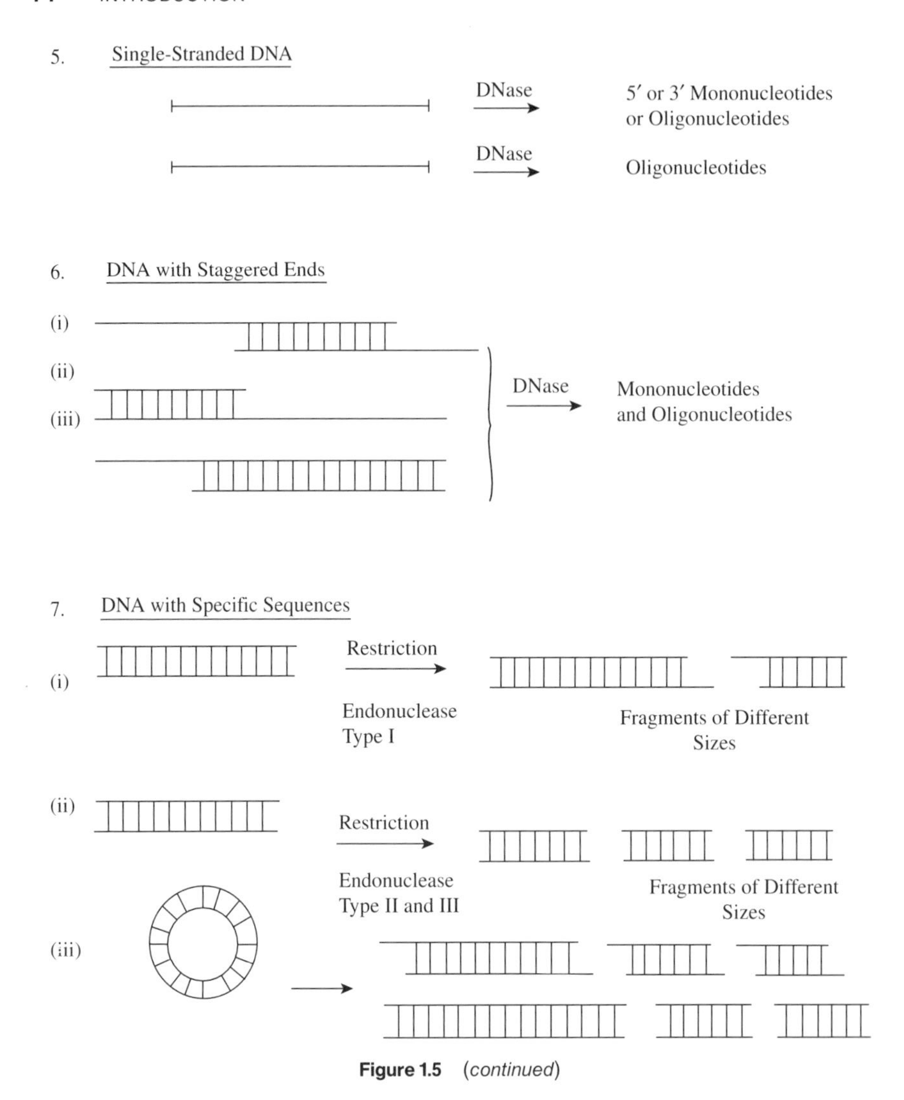

Figure 1.5 (*continued*)

Peptide nucleic acids (PNAs), a class of artificially synthesized oligomers containing a neutral peptide-like backbone instead of ribose phosphate backbone, mimic nucleic acids. PNAs can hybridize to complementary DNA/RNA with higher affinity or specificity than the corresponding oligonucleotides. PNAs can be recognized by different DNA polymerases as substrate (Lutz et al., 1999). However, PNAs are never recognized as substrates by nucleases even though binding of a complementary PNA oligomer to DNA may target a single-strand-specific nuclease such as S1 nuclease in a sequence-specific manner to DNA, leading to a double-strand DNA cleavage (Demidov et al., 1993).

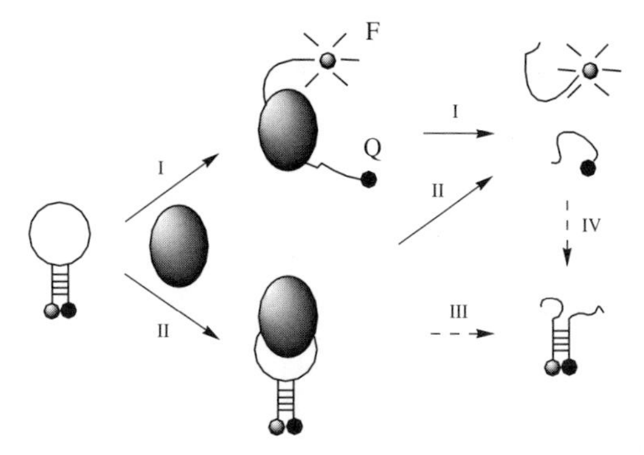

Figure 1.6. Schematic representation of the fluorescence mechanism of the molecular beacon during cleavage by single-stand-specific DNA nuclease (indicated by solid arrows). The solid arrows indicate two paths (I and II) leading to fluorescence enhancement during digestion. The dashed arrows represent two possible processes (III and IV) in which no fluorescence enhancement is produced. Only the first cut is shown here. Even though the nuclease may keep on cutting one single strand many times, only the first cut contributes to the fluorescence signal increase. The ball represent the nuclease. MB, E, and Q represent molecular beacon, fluorophore, and quencher, respectively. Here the fluorophore and quencher are tetramethylrhodamine (TAMRA) AND 4-(4′-dimethylaminophenylazo) benzoic acid (DABCYL), respectively. (From Nucleic Acid Research, reproduced with permission.)

B. Mode of Attack

Nucleases differ in their modes of attack. Enzymes may attack either from the 3′ end or from the 5′ end of the nucleic acid molecule; some nucleases can attack from either end. The mode of nucleolytic attack is presented in Figure 1.7.

Nucleases hydrolyze a wide spectrum of substrates with different modes of attack and produce either mono- or oligonucleotides as products. However, the nucleolytic products are always either 3′- or 5′-phosphorylated. The same nuclease is not known to produce both 3′- and 5′-phosphorylated products, regardless of which direction the enzyme may traverse the substrate. Thus, this is the only criterion which can be strictly applied to the classification of all nucleases without exception. Recently, this criterion has been a point of major consideration in the classification of a large number of protein nucleases (Linn, 1982a). However, this criterion is not applicable to topoisomerases, recombinases and damage-specific nucleases. The system of nuclease classification devised by Linn (1982a) represents the most suitable system of classification. The classification of protein nucleases is presented in Table 1.1.

Presumably the active site of the nuclease may use either the 3′ or 5′ carbon in the internucleotide linkage as a guide to make the incision; it can hydrolyze the P–O bond adjacent to 3′ carbon or 5′ carbon of the sugar moieties involved in the internucleotide linkage. The fact that the topography of the region encompassing the 3′ or 5′ carbon of the sugar molecule attached to a particular base is significantly different may explain why a particular nuclease produces either a 5′ or 3′ mononucleotide but never both. The enzyme must have evolved to recognize distinctly

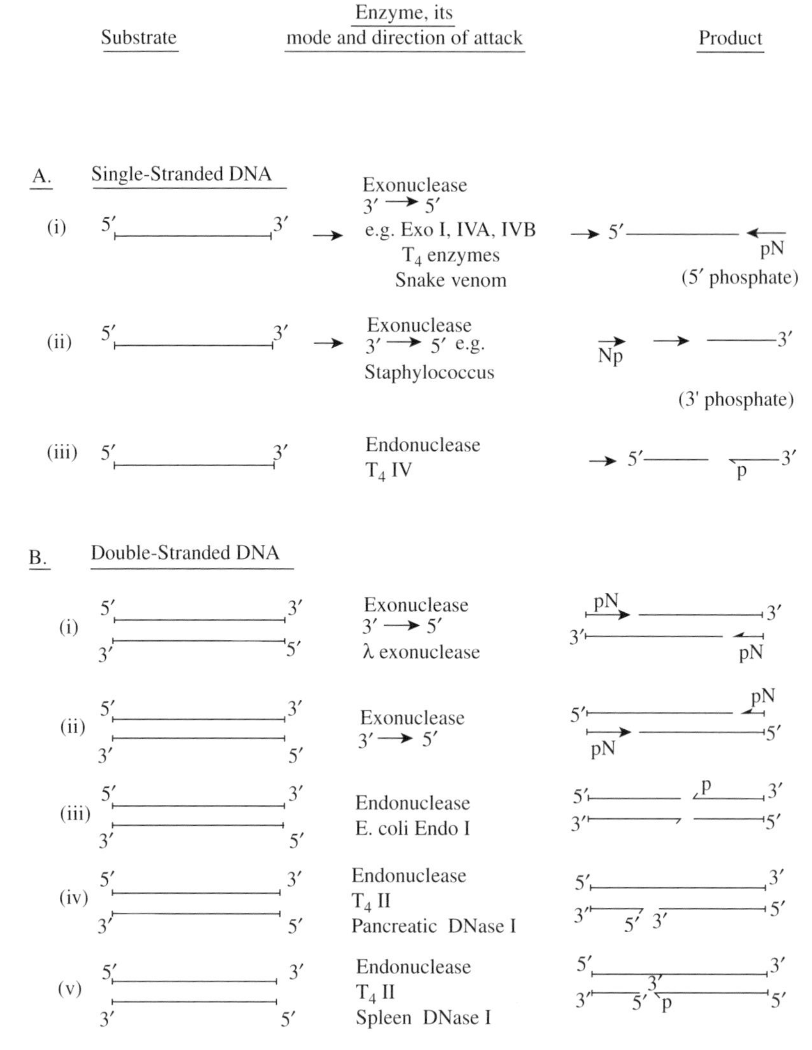

Figure 1.7. Mode of nucleolytic attack.

these two regions of the sugar moiety encompassing the internucleotide phosphodiester linkage of a nucleic acid chain. There is some evidence that recognition of oxygen atoms adjacent to the phosphorus atom may be used as a guide by a nuclease before it cleaves the phosphodiester bond. This view is based on the fact that nucleotides containing sulfur in place of oxygen in the α phosphate are good substrates for polymerization (by DNA polymerase I) or for ligation (by DNA ligase) but not for removal from a polynucleotide chain by an exonuclease (Putney et al., 1981; Kunkel et al., 1981a,b).

Table 1.1 Classification of protein nucleases[a]

Sugar-Specific Nucleases		Sugar-Non-Specific Nucleases
Ribonucleases	**Deoxyribonucleases**	
I. *Exoribonuclease*	I. *General deoxyribonucleases*	I. *Exonucleases*
a. 3′-Phosphomonoester producer	A. Exodeoxyribonucleases	a. 3′-Phosphomonoester producer
b. 5′-Phosphomonoester producer	a. 3′-Phosphomonoester producer	b. 5′-Phosphomonoester producer
II. *Endoribonuclease*	b. 5′-Phosphomonoester producer	II. *Endonuclease*
a. 3′-Phosphomonoester producer	B. Endodeoxyribonuclease	a. 3′-Phosphomonoester producer
b. 5′-Phosphomonoester producer	a. 3′-Phosphomonoester producer	b. 5′-Phosphomonoester producer
	b. 5′-Phosphomonoester producer	
	II. *Restriction endonucleases*	
	a. Type I	
	b. Type II	
	c. Type III	
	III. *Damage-specific deoxyribonucleases*	
	IV. *Topoisomerases*	
	Topoisomerases I	
	Topoisomerases II	
	V. *Recombinase*	

[a] This classification is based on a consensus of criteria as reflected in earlier classification (Linn, 1982).

C. Site-Specificity and Structure-Selectivity

A variety of nucleases such as restriction endonucleases, certain recombinases, and a number of ribonucleases are site-specific. They recognize a sequence of (two or more) nucleotides and make a cut either within these sequences or outside of the recognition sequence. Certain other nucleases such as micrococcal DNase and the topoisomerases also make site-specific cleavages (Horz and Altenburger, 1981). Site-specific cleavage is a characteristic property of several RNases which were used in the sequence analysis of RNA (Holley, 1965). However, the apparent site-specificity of certain RNase was found to be dependent on the size, stereospecificity, or secondary structure of the substrate RNA molecules (Penswick and Holley, 1965). Certain nucleases, although not specific to nucleotide sequences, are specific to distortions in the DNA structure caused by a damage in the DNA structure or by mismatches in complementary base pairing. Certain nucleases are specific to cruciform structures or to bends in DNA molecules. However, none are known to make distinctions among the different forms of DNA such as B-DNA and Z-DNA and thus are unable to make cleavages in a specific DNA form. There seems to be an exception to this rule because mung bean nuclease can make a cleavage at the beginning or end of a gene by recognizing a short stretch of Z-DNA, presumably present at sites preceding and following a gene (McCutchen et al., 1984).

In contrast to those nucleases that make cleavage at a particular site with a specific nucleotide sequence, a number of nucleases can instead recognize particular structural characteristics of their corresponding DNA substrates (Suck, 1998). Such structural features for recognition and subsequent action by these structure-specific nucleases may include (a) the strandedness of the substrate nucleic acid molecules such as nucleases P1 and S1, (b) helical parameters (i.e., width and/or flexibility) of DNA grooves such as DNase I, (c) distortion in DNA helical structure due to thymine dimer or abasic sites such as exonuclease III and HAP1, and (d) certain specialized DNA structures including a flap DNA recognized by viral, bacterial, and eukaryotic FEN-1 or a four-way junction of Holliday structure by T4 endonuclease VII or RuvC or yeast enzyme.

Both site-specific or structure-selective nucleases show little sequence homology but contain certain structural motifs that allow such sequence or structural selectivity.

V. METHODS FOR THE STUDY OF NUCLEASES

A. Methods for the Assay of the Enzymatic Activity

Several methods for the assay of nucleases are available. These include:

1. Viscocity measurements: This method (Laskowski and Siedel, 1945) is based on the decrease in the viscosity of a nucleic acid polymer in a solution upon its digestion with nucleases.
2. Spectrophotometric method: This method measures changes in hyperchromacity of nucleic acid after enzyme digestion (Kunitz, 1950; Privat de

Garilhe and Laskowski, 1956). The increase in the optical density of DNA molecules in solution increase upon nuclease digestion because of the release of the nucleotides that absorb more UV light. Such chromic shift is also seen during the process of denaturation and denaturation of DNA molecules in solution.

3. Increase in the amount of inorganic phosphate: Phosphates are released due to phosphatase activity of certain nucleases.
4. Increase in the amount of acid soluble (or decrease in the amount of acid insoluble) radioactivity due to mono- or oligonucleotides released after enzyme digestion of radioactively labeled nucleic acid molecules as substrate (Roth and Milstein, 1952). In the case of damage-specific nuclease, radioactivity can be released only when a radioactively labeled damaged DNA is used as substrate.
5. The release of radioactivity from ^{3}H-labeled DNA substrate, bound (either directly or through anti-DNA antibodies) to plastic depression plates after incubation with nuclease preparation, has been used to measure nuclease activity. This method has proven to be very accurate for nucleases that attack native DNA, single-stranded DNA, and damaged DNA.
6. A rapid and sensitive assay for endonucleolytic activity of DNase is based on the fact that nitrocellulose filters can retain only large fragments of denatured DNA. In this method, radioactively labeled denatured DNA is treated with enzyme and then passed through nitrocellulose filters. The decrease in retention of radioactivity by the filter is correlated to enzyme activity (Eron and McAuslan, 1966; Geiduschek and Daniels, 1965).
7. A change in the size or conformation of the nuclease-treated DNA as determined by their mobility in agarose gels after electrophoresis (Wang, 1971). This method is useful for the assay of the activity of restriction endonucleases, topoisomerases, and recombinase. Histochemical and immunological methods are also available for the *in situ* demonstration of the localization of nucleases (Sierakowska and Shugar, 1977).
8. Several quantitative assays of nucleases are based on the digestion of nucleic acids incorporated into the medium on which an organism secreting nuclease has been grown or when nucleases are added as a disc overlay. Nuclease activity is then visualized as a "halo" around the growing colony or around the disc overlay containing the enzyme either after the precipitation of undigested nucleic acid with HCl or due to the change in color of metachromatic dyes bound to the DNA. The use of several dyes such as Toluidine blue and green has been described for the detection of nuclease activity (Lacks et al., 1974). Changes in color due to the release of dye as a result of the digestion of nucleic acid can be visualized as a halo.
9. The activity of nuclease can also be detected on an electrophoretogram. The DNA polyacrylamide gel is appropriately incubated to facilitate the digestion of DNAs by nucleases, and then the gel is stained to visualize DNA. In

such a gel, the nuclease activity is visualized as a band without a stain. These methods of visualization of nuclease activity can be very useful in screening mutants deficient in nuclease activity or in identification of different protein components with nuclease activity, in a cell-free extract resolved by electrophoretic analysis. This aspect is discussed further in the section on genetics of nucleases.

10. Enhancement of a fluorescence signal due to separation of a fluorophore and a quencher (see Figure 1.6) in a DNA probe used as a substrate for digestion by different nucleases such as restriction endonucleases and other DNases and chemzymes has been developed to measure the activity of nucleases (Li et al., 2000; Biggins et al., 2000).
11. Flow cytometry has been developed to measure DNA cleavage by the structure-specific nuclease FEN-1. This technique is very fast: It can measure the enzymatic cleavage in approximately 300 millisecond (Nolan et al., 1996).

The use of a specific method is also dictated by the nature of the enzyme being assayed and the nature of information one wishes to obtain (Schein, 2001). Viscosity and hyperchromacity measurements can readily provide some insight into the mode of the nucleolytic attack by a specific nuclease. Snake venom phosphodiesterases can cause a significant change in the hyperchromacity of calf thymus DNA without making any significant change in the viscosity of the DNA. This suggests that the snake venom enzyme is an exonuclease; in contrast, pancreatic DNaseI can cause significant drop in viscosity of the calf thymus DNA without producing significant changes in the hyperchromacity. This suggests that pancreatic DNaseI is essentially an endonuclease. The micrococcal nuclease, which can produce significant changes in the hyperchromacity and viscosity of DNA, is now known to possess both endo- and exonucleolytic modes of attack. The nature of the products of nuclease digestion has been usually determined by sucrose gradient analysis or by electrophoresis. However, at times, these methods do not provide the complete information regarding the nature of the nuclease digestion. For example, when examined by these methods, a sample of DNA of various lengths usually appears as a smear without providing any information about the nature of the smaller molecule. A better picture of the nature of the nuclease digestion products—such as their sizes, nicked or relaxed, and/or strandedness—can be determined by the application of the atomic force microscopy because of the unique ability of this methodology to visualize individual molecules present in the pool of the nuclease digests (Umemura et al., 2000).

B. Methods for the Study and Characterization of Nucleases

A large number of nucleases have been characterized with respect to their biochemical and molecular properties, both as enzymes and as the genes encoding them, using methods of classical biochemistry and genetics and that of molecular biology.

Traditionally, the study of nucleases includes their biochemical purification and characterization of their enzymatic properties such as substrate specificity, K_m and K_i, and other such parameters. Purification may involve several steps such as precipitation with ammonium sulfate and column chromatography, leading to the separation of proteins based on their net electrical (cation/anion) charge, molecular size, and affinity to a particular matrix. Nucleases purified to homogeneity, as ordinarily revealed to contain a single band of protein upon polyacrylamide gel electrophoresis, may further be subjected to crystallization and the determination of three-dimensional (3-D) structure via x-ray and/or NMR analyses leading to establish the active site and the role of the different N- and C-termini peptide segments in determining the enzymatic and other properties of the nuclease. Pancreatic RNase was the first nuclease crystallized (Kunitz, 1940) and its 3-D structure determined. Nucleases are relatively thermostable and have been the favorite of biochemist for the analysis of the structure and function of the proteins. Essentially, purification and crystallization and 3-D structure analysis have been the usual methods to understand the structuture and function of ribozyme nucleases as well. The Hammerhead ribozyme is a typical example of such nucleases other than protein nucleases.

The methods of classical biochemical genetics (Beadle and Tatum, 1941) involving comparision of the properties of the wild-type and mutant enzymes have been routinely used for understanding the structure and function of nucleases as well. Such analysis involves the identification of mutants in the natural population of an organism and/or creation of mutants in the laboratory for the comparision of their biochemical and structural properties. Such analysis has been further aided by the application of *in vitro* mutagenesis to create mutation at a specific site in the gene encoding the nucleases resulting in an enzyme with a site-specific change which is then used to evaluate the role of specific amino acid (or nucleotide in the case of ribozyme) in the enzymatic and other role of the nucleases. Study of nucleases has been further facilitated by the application of the modern methodologies of genomics, proteomics, and bioinformatics, leading to the comparision of their nucleotide and inferred amino acid sequences and their 3-D structure and the identification of different enzymatic and regulatory motifs helpful in the understanding of the multifunctional nature of several nucleases and finally the process of evolution of these nuclease and the role of nuclease in the evolution of the organisms possessing them.

VI. GENETICS OF NUCLEASES AND BIOLOGICAL ROLES

The fact that nucleases were initially obtained from animal pancreas led to the conclusion regarding their possible degradative role. This view was further supported by their occurrence as a lysosomal component in different cells. However, the fact that the cell possesses a variety of nucleases in different cellular components, besides those present in the lysosome, suggests biological roles of nucleases in

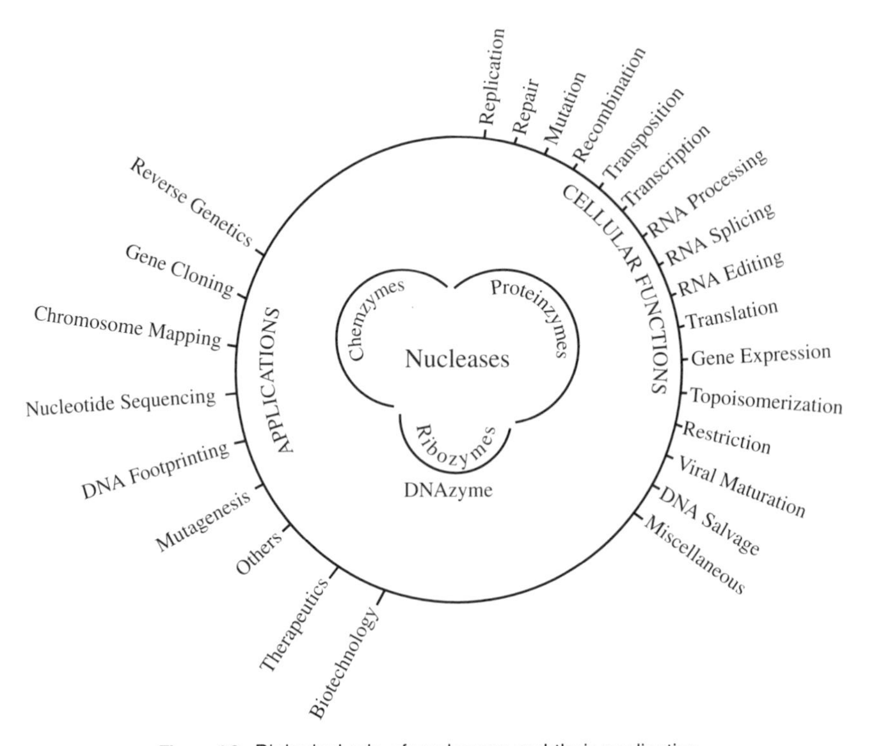

Figure 1.8. Biological role of nucleases and their application.

addition to the degradative role. Nucleases have been shown to participate in a variety of basic cellular and genetic processes.

Within the last few years, the role of nucleases in basic genetic mechanisms (as implied in Figure 1.8) has been elucidated to a great extent. This has been made possible through the genetic analysis of nucleases. A comparison of wild-type and mutant enzymes has been very useful in dissecting the role that a protein may play in controlling a specific biological function. Such an approach, first initiated by Beadle and Tatum (1941), has become crucial for investigating the biochemical, or molecular, basis of any biological component and its function. An alternative method, involving characterization of the gene and its product following the cloning of a specific gene, has also become available due to the development of recombinant DNA technology and *in vitro* site-directed mutagenesis (Jackson et al., 1972; Barany, 1985; Smith, 1985; Clarke et al., 1988; Botstein and Shortle, 1985; Shortle et al., 1981). Such methods of molecular genetics have been utilized in the characterization of micrococcal nuclease (Shortle, 1983). Both methods have been utilized in the study of nucleases.

An assessment of the role that nucleases play in several genetic processes has been made possible by the isolation and characterization of specific genetic mutants. A majority of these mutants have been obtained by the variety of both general and novel methods. Some of these are summarized here. First, mutants with a possible defect in specific biological processes have been obtained and then characterized to uncover the specific role of different nucleases. Mutants defective in DNA replication, repair, and recombination belong to this group of mutants (see Clark, 1971, 1973; Hanawalt et al., 1979; Kornberg, 1980; Thompson et al., 1980). Second, mutants defective for particular nucleases were characterized in depth to assess their roles in biological processes (Weiss, 1981). Such nuclease defective mutants have been obtained by "brute force" methods (Milcarek and Weiss, 1972; Yajko and Weiss, 1975). Weiss and his colleagues (1972, 1975) developed a method to screen large numbers of *Escherichia coli* in order to detect a mutant colony deficient in nuclease. Likewise, yeast nuclease mutants that are defective in splicing of tRNA were identified by the brute force screening of a large number of putative mutants (Winey et al., 1989). Third, nuclease mutants have been isolated using their resistance to particular inhibitors; several topoisomerase mutants have been obtained by this method (Cozzarelli, 1980; Andoh et al., 1987). Fourth, a number of nuclease mutants have been obtained by certain novel methods such as (a) the isolation of *Neurospora* and yeast nuclease mutants by their inability to utilize nucleic acid as the sole phosphorus source in a defined medium (Ishikawa et al., 1969; Hasunuma and Ishikawa, 1972; Forsthoefel and Mishra, 1983; Mishra and Forsthoefel, 1983); (b) the identification of nuclease mutants by the lack of "halo" around the growing colony on a DNA–or RNA–agar medium after treatment with trichloroacetic acid or perchloric acid (Badman, 1972; Käfer and Fraser, 1979; Fraser et al., 1980); (c) the isolation of certain nuclease mutants of *E. coli* by their ability to survive in the presence of toluene as bactericidal agent; mutants unable to degrade DNA and RNA from the bacterial cell were able to survive the bactericidal effects of toluene and were identified after staining with basic dyes; (d) isolation of nuclease mutant by the change in the color of the medium containing metachromatic dye as previously mentioned; a number of mutants of *Diplococcus* and *Staphylococcus* have been obtained by this method (Shortle, 1983); (e) isolation of certain nuclease mutants by their altered repressor function; certain recombinases such as yeast Flp and δ resolvase have been known to possess repressor function in addition to their nuclease activity (Lebreton et al., 1988); and (f) the isolation of nuclease mutants following *in vitro* mutagenesis (Shortle, 1983; Botstein and Shortle, 1985). A number of staphylococcal nuclease mutants have been isolated after *in vitro* mutagenesis of the cloned nuclease gene. The mutant colonies were identified by the color change in the medium containing Toluidine blue (Shortle, 1983). Finally, a large number of nuclease mutants in phage, bacteria, fungi, and animal cells have been isolated by their sensitivity to UV light, x-ray, chemical mutagens, and carcinogens.

As depicted in Figure 1.8, there is hardly any basic biological process involving nucleic acids which is independent of participation by nucleases. As a matter of fact, one could learn a good deal about the molecular biology of nucleic acid structure, organization, transactions, and expression just by knowing the various properties of the different nucleases. Some of the properties of nucleases and their possible biological roles have been described earlier (Privat de Garilhe, 1967;

Davidson, 1972; Boyer, 1981, 1982; Linn and Roberts, 1982; Linn et al., 1993). Most of these descriptions are specialized, and at least some of them are now out of date in view of the developments of the last few years. In addition, there has been a change in the emphasis on the nature of approach to study the properties of the enzymes in general, due to the arrival of the techniques of molecular biology and reverse genetics. The new methodology has a tendency to ignore the details of the hard-core biochemistry of protein (Kornberg, 1987). Instead, the methodology of molecular biology relies on deducing the biochemical properties of an enzyme from its cloned DNA sequences. The molecular biological approach has certain advantages over the methods of hard-core biochemistry. First, it is much faster. Second, it can provide a better insight into the functional aspects of the enzyme that cannot be obtained readily, via classical approach of the hard-core biochemistry. The identification and characterization of a number of recombinases and other nucleases would not have been possible either without the molecular cloning and the amplified expression of the genes encoding these enzymes, at times via *in vitro* transcription and translation or without characterization of gene and enzyme after genetic complementation of a defective function in the mutant by the cloned wild-type gene.

An attempt is made in this book to provide the reader with an introduction to the properties and biological roles of nucleases, discuss their contribution to the emergence of modern biology, and, above all, elucidate the role of nucleases in evolution via their participation in various aspects of DNA transaction such as DNA replication, repair, recombination (including transposition), and transcription. Nucleases are the only common group of enzymes that participate in all these genetic processes by acting directly or in conjunction to other proteins, or in response to environmental (or SOS) factors. In addition, nucleases are triggered during the apoptosis of cells and are therefore crucial for controlling the life and death of cells and in causing cancer. In addition, nucleases are the basis for a number of devastating human diseases such as blindness or Kalazar in tropical countries caused by parasites like trypanosomes and leschmania. Nucleases also control the unfolded protein response (UPR) which is the cause of several other devastating human diseases (such as cystic fibrosis, amphysema, and osteoporesis imperfecta). On the other side of the world of ribonucleases, DNase has been used to improve the lung function of a large number of cystic fibrosis patients, and certain RNases such as onconase is under stage III of trial for treatment of cancer. Above all, nucleases have been instrumental in the development of recombinant DNA-based biotechnology.

VII. APPLICATIONS OF NUCLEASES

There are several major considerations underlying the application of nucleases. The first is their utility in (a) the construction of recombinant DNA molecules crucial for gene-cloning, (b) the generation of RFLP useful for identification of genes controlling diseases, and (c) the mapping of genes on chromosomes and utilization in forensic science. The second is the fact that certain nucleases are the target of anti-exitcancer drugs such as camptothecin, which interferes with resolution of the DNA-topoisomerase intermediate in cancer cells, leading to cell death; thus the understanding of nucleases structure and function can become the basis for drug development; likewise the manipulation of certain recombinases such as cre-lox could be utilized for temporal control of the expression of certain genes, as already seen to occur in nature by various microorganisms. The third is their ability to control the expression of certain genes by destroying certain mRNA; ribozymes have been exclusive tools for this approach called *antisense strategy*, which is being developed to treat certain diseases via control of the expression of certain genes. The fourth nucleases are known for their notoriety in digesting nucleic acid molecules during the purification of the latter for scientific or commercial and other investigative purposes; therefore, much of the application strategy is directed in protecting the digestion of the nucleic acids. The fifth is about the strategy of developing nucleases that completely removes nucleic acids in the cell extract in order to facilitate the preparation of proteins by reducing the viscosity of the cell extracts or to remove the viscous material in the lungs of patients suffering from the genetic disorder cystic fibrosis due to the abundance of nucleic acid in the bacterial cellular debris. The sixth basis for the application of nucleases includes breakdown of oligonucleotides by nucleases for the making of certain additives used as food flavoring. Finally, nucleases are used for diagnostic purposes as an indicator of environmental pollutants or infectants.

The fact that nucleases can be obtained in the purified form in a large amount has led to their utilization in science, medicine, and biotechnology. In 1970, the discovery of restriction enzyme opened the avenue for the cloning of genes and led to the coming of biotechnology based on recombinant DNA technology. Nucleases have been utilized as the major tool in developing the methodologies for recombinant DNA technology and molecular cloning of genes. Genentech, one of the first biotechnology companies founded in 1976 based on recombinant DNA technology, developed the procedure for producing human insulin; this was the first drug produced by recombinant DNA technology that was approved by the US Food and Drug Administration (FDA) and later marketed by Eli Lilly Company. In 1985 Genentech became the first company that developed human growth hormone via gene cloning and marketed this hormone for the treatment of children with genetic growth disorders. In 1995 the complete genome sequence of the first organism, *Hemophilus influenzae*, was obtained. This was followed by deciphering a number of other organisms, including yeast and fruit flies. Finally, in the spring of 2000 the announcement of the completion of the entire sequence of human genome was made possible; none of these were feasible without the application of recombinant DNA technology and gene cloning utilizing nucleases. Nucleases have been used for the diagnostic purposes in identifying the genes controlling human and other diseases and for the treatment of certain diseases. The gene controlling Huntington's disease was the first human disease gene that was cloned by positional cloning utilizing nucleases. The genetic disorder of the severe combined immunodeficiency disease (SCID) was the first human genetic disease that was treated by gene therapy using the cloned human adenosine deaminase (ADA) gene; the SCID patients are defective for the ADA gene. Nucleases find their application in forensic science by their utilization in DNA fingerprinting/DNA profiling. Industrial use of nucleases involves making of certain sweetener and flavoring materials. Nucleases have great potential to be used in medicine and agriculture via the application of recombinant DNA technology and of course in tackling many environmental problems.

2

RIBONUCLEASE

Among nucleases, ribonuclease was the first one to be described. A heat-stable enzyme from the pancreas capable of digesting yeast ribonucleic acid was isolated and purified, later crystallized, and named as ribonuclease (Kunitz, 1940). A number of treatises have been written on ribonucleases (Bernard, 1969; Davidson, 1972; Egami and Nakamura, 1969; Linn and Roberts, 1982). With certain exceptions, ribonucleases are small and heat-stable proteins that usually make-site-specific cleavage. The ribonucleases may act on a single-stranded or double-stranded RNA, or on the RNA present in a DNA–RNA hybrid. It may act as an endonuclease or as an exonuclease producing 5′ or 3′ phosphomonoester(s). Also, nucleases may occur within a cell or be secreted as extracellular enzymes. In addition to the salvage of cellular or extracellular RNAs, ribonucleases perform certain vital functions in cellular biochemistry by their participation in DNA replication, transcription, and RNA processing, splicing, and editing and in the control of the translation of the message by determining the turnover of the mRNA. These properties of ribonucleases are discussed in Chapter 11.

I. GENERAL RIBONUCLEASES

A general description of certain important ribonucleases is presented in this chapter. A list of different RNases is presented in Table 2.1.

A. Microbial Ribonucleases

A number of bacterial and fungal ribonucleases have been very well characterized. These are low-molecular-weight enzymes that are specific for guanyl residues in RNA molecules. These enzymes have been purified to homogeneity. Their amino

Table 2.1 Properties of certain ribonucleases

Source	Specificity	Molecular Weight	Amino Acid	Homology
Bacterial				
RNaseBa (*Bacillus amyloliquefaciens*)	$G > A >$ pyrimidine	12 kDa	110 AA	Close
RNaseBi (*Bacillus intermedian*)			109 AA	Close
RNaseSa (*Streptomyces aureofaciens*	G	10.5 kDA		
RNaseSt (*Streptomyces erythraeus*)	G		101 AA	Close
			101 AA	Close
Fungal				
RNaseT1 (*Aspergillus oryzea*)	G	11 kDa	104 AA	Close
RNaseC2 (*Aspergillus clavatus*)	G	11 kDa	102 AA	Close
RNaseMS (*Aspergillus saltori*)	$G \gg A \gg C > U$		106 AA	Close
RNasePb1 (*Penicillus brevicompactum*)	G	11 kDa		
RNaseU1 (*Ustilago sphaerogena*)				
RNaseU2 (*Ustilago sphaerogena*)	G		112 AA	
	Purine			
Mammalian				
RNaseA (bovine pancreas)	Pyrimidine		124 AA	Close
RNase HPR (human pancreas)			127 AA	Close
RNaseH NSR (human urine HLR human liver)		13 kDa	—	Complete
RNaseK1 (human kidney)		14 kDa	—	Complete
RNaseK2		—	—	
MBP				
ECP		13 kDa	67% homology to EDN	
EDN		21 kDa	complete homology to liver	
		18 kDa	enzyme	
Angiogenin		14 kDa	123 AA; 35% homology to RNaseA	

acid sequences and their three-dimensional structures at high resolution have been determined (Hill et al., 1983; Sevcik et al., 1990; Mauguen et al., 1982; Takahashi, 1978). Some of the properties of different ribonucleases and their relation to enzyme function and regulation (Nicholson, 1999) are summarized in Table 2.1. Understanding the amino acid sequence analysis and the three-dimensional structure of the microbial RNase supports a two-step mechanism for the hydrolysis of RNA by these enzymes. The first step involves the process of transesterification to generate a cyclic 2′,3′-phosphate intermediate that is hydrolyzed to 3′-phosphate in the second step. This mechanism for catalysis by RNase is supported by results of different lines of investigation including NMR studies and replacement of amino acids at specific sites by protein engineering (Ikehara et al., 1986; Nishikawa et al., 1987). The reaction mechanism of the mammalian RNaseA also involves the above two-step process. Also, studies with different fragments of barnase (RNase from *Bacillus amyloliquefaciens*) provide insight into the mechanism of protein folding; results of studies with barnase fragments show that the enzyme folds via association of the independently folded parts, thus supporting the role of nucleation in enzyme folding (Kippen et al., 1994). Barnase consists of 110 aminio acids and is most of the time complexed with an 89-amino-acid inhibitor protein called *barstar*. The three-dimensional (3-D) structure of the complex between barnase and barstar proteins has been known. Free barnase is toxic to cells; therefore, under physiological conditions its binding to barstar is favored to prevent its toxic effect.

Application of Barnase and Barstar in Agriculture. Transgenic corn plants containing barnase gene are pollen-sterile, whereas such corn plant containing genes for both barnase and barstar are pollen-fertile. Such transgenic corn plants have been utilized to produce hybrid seeds. This technology of transgenic plants has been of great use in agricultural practice.

1. RNaseT1. This is an endoribonuclease purified from Taka-diastase, an enzymatic preparation of *Aspergillus oryzae*. It specifically hydrolyzes Gpx bonds. The enzyme has been purified to homogeneity with a molecular weight of 11 kDa (Egami and Nakamura, 1969). Its amino acid sequence is completely known (Takahashi, 1978). It contains 104 amino acids with two cysteine bonds; His 40 and His 92 are required for the catalytic activity of the enzyme. A comparison of its properties with respect to RNaseA is presented in Table 2.1.

Chemically synthesized RNaseT1 gene and its mutant forms have been cloned in *E. coli*; the overproduced enzyme, when analyzed for catalytic function, showed that an enzyme containing the wild-type Gly 71–Ser 72–Pro 73 amino acid sequence was functional, whereas an enzyme containing mutant sequence Pro 71–Gly 72–Ser 73 was not functional (Ikehara et al., 1986).

2. RNaseT2. This is a group of nonspecific secretory endoribonucleases of wide spread occurrence. This enzyme has been found to occur in every organism that has been investigated. It is found extracellularly or in the endomembrane system. The RNaseT2 of oryzae consists of 239 amino acids with amino acids Asn 15, Asn 76, and Asn 239 as glycosylation sites. It is characterized by the presence of two stretches of amino acids, FTIHGLWP and KHGTC, at the active site; these sequences have been conserved in the family of RNaseT2 in all organisms (Kawata et al., 1988). RNaseT2 has very little homology of amino acid sequences with RNaseT1 except for some amino acid in the sites comprising the substrate recognition and binding. The RNase T2 has been vey well characterized from a number of organisms, and the gene encoding this protein has been cloned and overexpressed. The *E. coli* enzyme RNase1 belonging to this family has been very well characterized, but it has a long hydrophilic segment near the amino terminus; the crystal structure of the bacterial enzyme has been determined. RNaseT2 may have a protective function to the cells elevated in response to wounding and pathogen invasion; certain viral T2RNases suppress host immune function whereas plant S-RNases, also belonging to this group of ribonucleases, prevent self-fertilization. In yeast this enzyme is activated in response to stress such as changes in the osmolarity of the growth medium and/or heat shock.

B. Mammalian Ribonucleases

Bovine pancreatic RNase is the most well-charcterized of all nucleases. This is the first enzyme that has been characterized with respect to amino acid sequencing, protein crystallographic, and protein folding studies because of its abundance and stability (Raines, 1998).

Sierakowska and Shugar (1977) have classified mammalian ribonucleases into two groups: secretory and nonsecretory ribonucleases. The secretory ribonucleases occur mostly in secretory organs. They have a pH optimum of 8 and show a high preference for poly C. They are usually both heat- and acid-stable. The nonsecretory ribonucleases occur more predominately in liver, lung, spleen, and leukocytes, show a lower preference for poly C, and act most effectively at pH 6.5–7.0. Pancreatic ribonuclease is an example of secretory enzyme, whereas the liver ribonuclease is an example of nonsecretory enzyme. Urinary ribonucleases are both secretory and nonsecretory types. Some of the properties of these ribonucleases are summarized in Table 2.2.

Table 2.2 Classification and properties of mammalian ribonucleases

	Secretory	Nonsecretory
Occurrence	Secretory organs, such as pancreas	Other organs, such as liver, spleen, kidney
pH optimum	8.5	6.5–7
Poly C as preferred substrate	Most active	Less active
Uridine 3′-(α-naphthyl) phosphate	Used as substrate	Not used as substrate
Heat-stable	Heat-stable	Less heat-stable
Acid-stable	Yes	Yes
Examples	Pancreatic ribonuclease	Liver ribonuclease

Pancreatic ribonucleases from 40 different organisms have been characterized with respect to their amino acid sequences and other physical and chemical properties. Among these ribonucleases, pancreatic bovine and human ribonucleases, human liver ribonucleases, human urinary ribonucleases, and a family of basic proteins with ribonuclease activity have been very well characterized. The basic proteins such as eosinophil granule major basic protein (MBP), eosinophil cationic protein (ECP), and eosinophil-derived neurotoxin, initially identified to possess bactericidal and/or helminthotoxic and neurotoxic activities, have now been shown to possess ribonuclease activity. The N termini of these groups of proteins show a 26% identity to the sequence of human pancreatic ribonuclease. The complete cDNA-derived amino acid sequence of EDN was found identical to a human urinary ribonuclease of the nonsecretory type (HNSR). The ECP showed a 16% sequence identity to EDN and to HNSR. The ribonuclease activity, as well as the cytotoxic effect of EDN, was abolished by the commercial ribonuclease inhibitor RNasin (Promega®, Madison, Wisconsin) or by a competitive substrate such as yeast ribonucleic acid (Molina et al., 1988). The properties of some of these ribonucleases are discussed in Table 2.2.

1. Bovine Pancreatic RNase. The crystallized bovine pancreatic RNase of Kunitz was found to contain RNaseA, RNaseB, RNaseC, RNaseC1, and RNaseD when chromatographed on ICR-50 (Plummer and Hirs, 1963). The different RNases are derived from RNaseA and are the products of the same gene; any differences are due to post-translational modifications as shown in Table 2.3. The bovine kidney ribonuclease has been very well characterized; this enzyme possesses properties similar to those of bovine pancreatic ribonuclease (Niwata et al., 1985).

2. RNaseA. This enzyme is an endonuclease that cleaves single-stranded RNA at the 3′ end of pyrimidine residues producing 3′-phosphorylated mono- or oligonucleotides. RNaseA has been most commonly used for degrading contaminating RNA molecules during the preparation of (plasmid) DNA and for the mapping single nucleotide mismatch in DNA:RNA hybrids by the ability of this enzyme to make cleavage in the RNA strand at the site of mismatch.

Table 2.3 Comparison of the structure of different bovine pancreatic RNase

RNaseA[a]	No carbohydrate moiety
RNaseB[a]	Glycoprotein containing N-acetyl mannose and glucosamine
RNaseC[a]	Sialo glycoprotein
RNaseD[a]	

[a]All forms of the pancreatic RNase have the same amino acid sequence; therefore, they are encoded by the same gene. The carbohydrate moieties are attached by an aspartoamidohexose linkage (Salnikow et al., 1970).

Ribonuclease A is the most well-described nuclease and the first protein whose entire amino acid sequence was determined (Smyth et al., 1963). It is also the first protein whose chemical synthesis was achieved by solid-phase synthesis (Gutte and Merrifield, 1971) and by solution methods (Yagima and Fujii, 1980). The methods of enzyme preparation using affinity chromatography and the physical, chemical, and catalytic properties of RNaseA have been reviewed (Richards and Wyckoff, 1971; Blackburn and Moore, 1982). Bovine pancreatic ribonuclease consists of 124 amino acids (see Figure 1.1). It contains four disulfide bonds that stabilize the protein.

Ribonuclease has a 25-amino-acid extension on its N-terminus which represents the signal peptide of this secreting protein. The ribonuclease has four glycosylation sites at the asparagine residue at positions 21, 34, 62, and 76. The RNaseB is derived from RNaseA by removal of the signal peptide due to digestion between amino acids 20 and 21 by subtilisin. The RNaseB is also called RNaseS. In RNaseA, His 119, His 12, and Lys 41 are crucial for enzymatic activities. Analysis of crystal structure of the enzyme confirms this conclusion by showing the proximity of these amino acids. In addition, Phen 120, Thr 45, and Val 30 form the groove in which pyrimidines can fit in and are attacked by the catalytic center of the enzyme. The role of the histidine residues in determining the catalytic activity of the enzymes of RNase family has been established. RNaseA exists as a dimer; the N-terminal portion of each subunit is shared by peptic domain swapping during the formation of a dimer; such sharing of peptide, via swapping of domain, is not seen anywhere else in the protein structure.

3. Human Pancreatic Ribonuclease (HPR). This enzyme has been purified to homogeneity and was found to resemble the bovine pancreatic ribonuclease in its properties (Weickman et al., 1981). The entire amino sequence of the human enzyme has been determined (Beintema et al., 1984). It differs at 37 positions from bovine pancreatic ribonuclease, and it has three more amino acids at the C terminus. In addition, the human pancreatic ribonuclease contains many positively charged residues at the N terminus and negatively charged residues at the C terminus. Antibodies prepared against pure human pancreatic RNases inhibited RNase activities from brain, kidney, serum, and urine up to 60–80%. However, the RNase activities from spleen (8%) and liver (30%) were not significantly inhibited, suggesting the presence of another species of RNase in these organs.

Human pancreatic ribonuclease has become very important because of possible therapeutical value of several RNases such as the tumoricidal activity of onconase or antiviral activity of other RNases (Ardelt et al., 1991 Rybak et al., 1992; Schein, 1997; Domachowske et al., 1998; I. Lee et al., 2000; Iordanov et al., 2000). It is therefore crucial to produce this enzyme in large quantity. In order to achieve this goal, human gene encoding this enzyme has been cloned and expressed in *E. coli*. It has been shown that pro50 residue is very crucial for the recovery of human enzyme. It is also important to change the amino acid at position R4, K6, Q9, D16, and S17 to the corresponding amino acids found in the Bovine seminal ribonuclease in order to enhance the heat stability of the human enzyme (Canals et al., 1999).

4. Human Nonsecretory Ribonuclease (HNSR). This enzyme found in human urine is identical to liver, spleen, or EDN enzymes in amino acid sequence and shows identity at about 30% of amino acid positions with those of the secreted mammalian ribonucleases (such as pancreatic ribonuclease). The active site residues (histidine 12, histidine 119, and lysine 41) and all eight half-cysteine residues are identical (Beintema et al., 1988). The liver enzyme has been purified to homogeneity and characterized in detail (Sorrentino et al., 1988). The liver enzyme has a molecular weight of 19.5 kDa. Urinary enzymes have also been characterized after purification to homogeneity (Morita et al., 1986). The human nonsecretory ribonucleases such as the human liver and urine enzyme are immunologically distinct from the human pancreatic enzyme.

However, two serologically distinct enzymes have been identified from human kidney; these are RNaseK1, and RNaseK2 (Niwata et al., 1985). RNaseK1 has a molecular weight of 13 kDa and a pH optimum of 8.5; this enzyme is immunologically similar to bovine pancreatic RNaseA. Whereas RNaseK2 has a molecular weight of 14.6 kDa and a pH optimum of 6.5, this enzyme is similar to human urinary enzyme.

5. Human Major Basic Protein (MBP), Eosinophil Cationic Protein (ECP), and Eosinophil-Derived Neurotoxin (EDN). Human eosinophil granules contain highly cationic protein. These include MBP (13 kDa), ECP (17–21 kDa), and EDN (18 kDa); of these, MBP and ECP have microbicidal activity, whereas EDN is toxic to helminths, protozoa, and mammalian cells (McGrogan et al., 1988). All three proteins possess RNase activity. The amino acid sequence analyses showed homology to human liver enzyme (Slifman et al., 1986; Gleich et al., 1986) and human pancreatic ribonuclease, including the four cystine residues and lysine 41 at the active site. The RNase activity of EDN is comparable to that of the bovine pancreatic RNase (Slifman et al., 1986). Moreover, the EDN and ECP RNase activities were found to be immunologically distinct (Slifman et al., 1986). Amino acid sequence analysis showed 67% identity between the two proteins. However, EDN was found to be completely identical to human liver enzyme in amino acid sequences.

a. MBP. Human MBP has been purified to homogeneity, and its NH-terminal amino acid sequence has been determined. An oligonucleotide probe corresponding to NH-terminus His-Asn-Phe-Asn-Ile peptide was used to obtain a full-length cDNA encoding MBP (McGrogan et al., 1988). The encoded NH-terminal sequence of this mRNA was found to be identical with the reported NH terminal amino acid sequence of the MBP (Barker et al., 1988). The sequence analysis of the full-length cDNA suggests that MBP is synthesized as a 23-kDa precursor that is later cleaved to yield a 14-kDa MBP (Barker et al., 1988). However, there is no evidence regarding the RNase activity of MBP.

b. ECP. Human ECP belongs to a family of immunologically identical extremely basic proteins (Olsson et al., 1986) with isoelectric points higher than pH 11. The major ECP has a molecular weight of 21 kDa and contains 21 arginine residues and is a zinc-containing protein. ECP undergoes changes in its size subsequent to its synthesis in bone marrow cells.

c. EDN. Human EDN has a molecular weight of 18.6 and 20.1 kDa. EDN shows 67% identity with ECP, 29% identity with HPR, and 100% identity with HNSR and HLR (Slifman et al., 1986; Gullberg et al., 1986) in their amino acid sequence. Recently, a 725-bp cDNA clone for EDN has been obtained (Rosenberg et al., 1989) which contains the complete amino acid sequence of EDN and is identical to that of HNSR.

6. Angiogenin. Angiogenin is a 14.4-kDa protein purified from human cell extract which induces angiogenesis *in vivo* (Fett et al., 1985). It shows sequence homology to pancreatic ribonuclease (Strydom et al., 1985) and possesses ribonuclease activity. It consists of 123 amino acids with three disulfide bonds that link half-cystinyl residues at position 26–89, 39–92, and 57–107. Its amino acid sequences shows a 35% homology to the pancreatic ribonuclease. The similarities are especially obvious around the major activities residues such as His 12, Lys 41, and His 119 of ribonuclease including three out of four disulfide linkages (Shapiro et al., 1987). The gene for human angiogenin has been cloned, and the amino acid sequence encoded by the cloned gene completely matches the amino acid sequence of the purified angiogenin (Kurachi et al., 1985). Mutation of aspartic acid 116 enhances both the RNase activity and angiogenic property of angiogenin (Harper and Vallee, 1985). Angiogenin contains a cell binding site in addition to the catalytic site. The cell binding site that involves amino acids 60–68 and ASN 109 facilitates the internalization of angiogenin to specific endothelial cells leading to cell proliferation during angiogenesis (Moroinu and Riordan, 1994). However, the RNase activity is required for the angiogenic activity. Angiogenin is shown to inhibit biosynthesis by nucleolytic digestion of ribosomal RNA. It would be of further interest to examine the interaction of thalidomide with angiogenin because thalidomide seems to prevent angiogenesis during fetal development.

Human pancreatic RNase (RNase1) , EDN (RNase2), ECP (RNase3), RNase4, angiogenin (RNase5), and RNase6 belong to the superfamily of RNaseA and share more than 50% of amino acid homology. Among RNaseA, onconase is the best characterized and is being evaluated as therapeutic agent with tumoricidal properties. The crystal structure of RNaseA and its inhibitor protein called RI has been determined (Kobe and Deisenhofer, 1996). RNaseA is a kidney-shaped protein. The structure of RI contains 15 leucine-rich motifs and several cysteine residues. The structure of the complex shows that inhibition of RNase activity is achieved by preventing the access of active site by the inhibitior protein (RI). The RNaseA inhibitor protein is a 50-kDa cellular protein. It only interacts with the RNase superfamily of enzymes except onconase. The RNase inhibitor protein does not inhibit the RNase activity of sialic acid-binding proteins.

7. Interferon-Induced Mammalian Ribonuclease. RNaseL, a 2,5pAn-dependent ribonuclease, has been characterized from human and other mammalian cells treated with interferon. This enzyme has been called RNaseL or RNaseF. The molecular size of the enzyme in SDS gel has been found to be 75 kDa. The enzyme has an optimal activity at 50–150 mM KCl in the presence of M^{2+}. The enzyme usually cleaves RNA at U-A and U-U. This ribonuclease is one of the mediators of the action of interferon induced by dsRNA in mammalian cells. Induction of interferon by dsDNA leads to activation of 2,5pAn synthetase, which synthesizes 2,5pAn from ATP. In the presence of 2,5pAn the RNaseL is activated which digests single-stranded RNA (such as mRNA, ribosomal RNA, and viral RNA). The RNaseL is dependent on 2,5pAn for its activity. The RNaseL becomes inactive when 2,5pAn is removed from the reaction mixture. The activity of the RNaseL is one of the main pathways through which interferon exerts its antiviral action (Floyd-Smith et al., 1982; Pestka et al., 1987; Kumar et al., 1988). The human gene encoding RNaseL enzyme has been cloned and located to human chromosome 1p25. Molecular characterization of the cloned gene and its overexpression has led to the identification of a 2 to 5-Å-binding domain. The N-terminal half of RNaseL contains ankyrin repeats with two P-loop motifs and a cysteine-rich region presumably involved in protein interaction, whereas the sites for substrate binding and the nuclease activity are confined to the C-terminal half of this protein. The ankyrin repeats are involved in 2 to 5-Å binding and also function as a repressor that regulates the ribonuclease activity of the enzyme. In addition, the ankyrin repeats are involved in protein–protein interaction, and a dimer of RNasel polypeptide possesses the enzymatic activity (Dong and Silverman, 1997). Unlike yeast endoribonuclease IREI, the mammalian RNaseL does not show protein kinase activity. However, RNaseL contains sequences on the N terminus with homology to protein kinase, which is characteristic of yeast endoribonuclease IREI from which the RnaseL might have evolved. RNaseL may exist as a 37-kDa polypeptide, which may result from the proteolytic degradation of a 75 to 80-kDa polypeptide; both forms may possess the nuclease activities but show other differences besides their molecular size. RNaseL is inhibited by an inhibitor protein RLI. Its exact mechanism of inhibition is not yet known.

Mutation in RNaseL has been implicated in several diseases, particularly those caused by viral infection. RNaseL have been also found to modulate the kinetics of the muscle differentiation by causing the selective degradation of MyoD mRNA (Bisbal et al., 2000).

8. Human RNase with a Possible Role in Tumor Suppression. A human RNase with homology to R2/Th/ Styler RNase family has been identified upon the YAC clones of human chromosome 6 encompassing the region 6q26-27 (Trubia et al., 1997). Allelic loss of this segment of human chromosome 6 has been associated with a number of malignancies. The molecular analysis of this human segment showed the presence of a RNase sequence found in plant (styler), fungus (R2/Th), and fruit fly (Hime et al., 1995). The styler RNase in plants is known to control the self-incompatibility (McClure et al., 1989). The human RNase homologous to plant styler and fungal R2/Th showed the presence of two distinct sequences of five amino acids such as HGLWP and KHGTC at the catalytic site of these RNases (Kurihara et al., 1992). The crystal structure of the fungal enzyme has been determined. The level of this RNase in humans is reduced in malignant cells, suggesting its role in causing malignancy by controlling a certain aspect involved in the loss of heterozygosity (LOH) tumor suppressor genes associated with several cancers. The plant S-RNase is described below in the section on plant ribonucleases.

C. Plant Ribonucleases

A number of plant ribonucleases have been described. Their amino acid sequence similarity among diverse groups and their physiological roles in self-incompatibility, phosphate remobilization, mRNA decay, and plant defense mechanisms have been established (Green, 1994).

Plant S-RNases. Among the plant ribonucleases, S-RNases are one of the very important group of RNases. These control self-incompatibility in plants and help in maintaining hybrid vigor and genetic diversity. They have been very well characterized from different groups of plants, particularly Solanaceae (the tobacco family). They are styler glycoproteins that reject pollens from the same plant. The S-RNases are produced in large amounts (i.e., millimolar amount) and digest the RNA of the pollen and thus ward off the process of fertilization by self-pollination. The tobacco S-RNase is a glycoprotein of 32 kDa. The gene encoding this S-RNase from tobacco have been cloned and characterized. The S-RNases differ in their amino acid composition in different plants which provide the basis for self-incompatibility. However, these RNases possess sequences FTIHGLWP and KHGTC which possess similarity to amino acid sequences found at the active site of *Aspergillus orizae* RNaseT2 (McClure et al., 1989). As mentioned above, a homolog of this protein has been found to exist in different organisms (including humans) where this group of RNase is involved in controlling developmental stages, including cancer.

D. Evolution of Ribonucleases

The amino acid sequence data suggest that human nonsecretory ribonuclease belongs to the same ribonuclease family as the mammalian secretory ribonucleases, turtle pancreatic ribonuclease, and human angiogenin. The data show that a gene duplication occurred in an ancient vertebrate ancestor; one branch led to nonsecretory ribonuclease while another branch led to a second duplication with one line leading to the secretory ribonuclease (in mammals) and a second line leading to pancreatic ribonuclease in turtles and to angiogenin in humans (Beintema et al., 1988).

II. RIBONUCLEASE INVOLVED IN RNA PROCESSING (TRIMMING, SPLICING, AND EDITING)

A number of enzymes may act on primary transcript leading to formation of a mature functional RNA; this process involves several steps and has been collectively called *RNA processing*. Besides addition of nucleotides to the 5′ end (capping) or to the 3′ end (poly A or CCA addition) of the RNA molecule, along with modification of canonical nucleotides, this usually involves changes in the size and nature of the transcript via a cleavage or trimming of RNA by endo- and exoribonucleases. In addition, the transcripts may undergo removal of noncoding RNA segment (introns) and rejoining of coding RNA segments (exons); this process is called *splicing*. This involves action by enzymes(s) possessing both ribonuclease and ligase activities and is analogous to the process of recombination (Knapp et al., 1978; Greer et al., 1983; Deutscher, 1984; Greer, 1986). Furthermore, a transcript may undergo a process of RNA editing involving insertion, deletion, or substitution of a nucleotide, producing an RNA molecule that differs from a genetically coded structure. Ribonuclease plays an important role in the process of RNA editing (Benne, 1988).

All functional tRNA and rRNA are products of RNA processing; only a few prokaryotic mRNA undergo RNA processing. Certain viral mRNA, eukaryotic mRNA, and certain prokaryotic and eukaryotic tRNA and rRNA are the products of RNA splicing. A number of RNAs in different organisms, including at least two RNAs encoding different proteins in humans, have been now shown to undergo the process of RNA editing (Benne, 1988). Trypanosome mtDNA transcripts have been shown to undergo RNA editing (Lamond, 1986). In plants, the chloroplast and mitochondrial mRNAs usually undergo the process of RNA editing (Simpson, 1990; Araya et al., 1994). RNA editing is considered to be an ancient feature during the process of evolution (Landweber and Gilbert, 1994).

Thus, a number of ribonucleases play a significant role in this important RNA transaction involving the production of mature RNA molecules. The prokaryotic endoribonucleases and exoribonucleases involved in cleavage and trimming of RNA have been very well characterized. Their role *in vivo* has been established by the study of genetics and biochemistry of mutants and inferred from the cleavage of RNA by a purified enzyme *in vitro* (Garber and Altman, 1979; Abelson, 1979; Gegenheimer and Apirion, 1981; Engelke et al., 1985; Goldfarb and Daniel, 1980). Some of the RNases involved in this process are described below, and their properties are summarized in Table 2.4.

A. RNaseIII and RNaseIII-Like Enzymes

1. RNaseIII. This enzyme was first purified as an endoribonuclease which hydrolyzed high-molecular-weight dsRNA into an acid-soluble form (Robertson et al., 1968). The enzyme consists of two identical polypeptide subunits of 25 kDa. The *rnc* gene of *E. coli* encoding this enzyme is located at 55 min on the *E. coli* map (Kindler et al., 1973). The enzyme plays an important role in the processing of

Table 2.4 RNA-Processing Nucleases in Prokaryotes[a]

Enzyme	Molecular Weight	Substrate	Product(s)	Monovalent	Divalent	pH Optimum
RNaseIII	50 kDa (2 subunits, 25 kDa)	Double-stranded RNA fragments	Random, acid-soluble	0.08–0.2 M	0.1 mM Mn^{2+}; 1–10 mM Mg^{2+}	9.75
		Secondary cleavage of natural RNAs	Site-specific cleavages	0.004–0.08 M (inhibitory, >0.09 M)	1–10 mM Mg^{2+}	ND (probably 7.6–8.0)
		Precursor rRNAs (double-stranded regions)	Site-specific cleavages	0.15–0.3 M	1–10 mM Mg^{2+}	ND (probably 7.6–8.0)
RNaseD	38–40 kDa	tRNA-$CCACCC_{OH}$	tRNA-CCA_{OH} 3pC (nonprocessive exonuclease)	Nose (inhibitory)	5 mM Mg^{2+}; 5 mM Mn^{2+}; 5 mM Co^{2+}	9.1–9.5
RNaseE	~70 kDa	*E. coli* 9S rRNA	Pre-5S rRNA + 5′ and 3′ fragments	0.1–0.2 M	1 mM Mn^{2+}; 5 mM Mg^{2+}	7.6–8.0
RNaseF	ND	T4 p2 species 1 RNA	Species 1 RNA + a 3′ fragment	0.15–0.2 M	None	6.8–7.6
RNaseM5	ND (2 subunits; β, 17 kDa; α, ND)	*B. subrilis* pre-5S rRNA	5S rRNA + 5′ and 3′ fragments	0.05 M (K^+, NH_4^+)	1 mM Ca^{2+}; 5–10 mM Mg^{2+}; 10 mM Mn^{2+} (or 0.3 mM spermidine; 10 mM putrescine)	7.0–7.5
RNase P	Protein, ~17,500; RNA, ~360 bases	Pre-tRNAs	tRNA + a 5′ fragment	0.1 M (K^+, NH_4^+; inhibitory, 0.1 M Na^+)	1 mM Mn^{2+}; 5 mM Mg^{2+}	8.0

[a] For each processing ribonuclease, enzymatic properties were determined for the substrate cited under the Enzyme column. If properties are known for more than one substrate, the one listed is the one that predominates *in vivo* or most closely resembles an *in vivo* substrate. Monovalent cations refer to Na^+, K^+, or NH_4^+ unless otherwise noted. Reactions require only a single cation from those listed. RNases III, E, M5, and P cleave at the 5′ side of a phosphodlester bond; that is, they generate products bearing 5′-phosphoryl and 3′-hydroxyl groups. ND, not determined. From Gegenheimer and Apirion (1991), copyright *Microbiol. Rev.*, American Society for Microbiology, used with permission.

rRNA and mRNA precursors (Dunn and Studier, 1975). The DNA segment containing the *rnc* gene for RNaseIII has been cloned (March et al., 1985). It encoded a protein of 227 amino acid residues with RNaseIII activity. The long leader region of the gene contains a stable stem bulge-stem structure similar to the RNaseIII cleavage site, suggesting an autoregulated expression of the *rnc* gene (March et al., 1985). RNaseIII is responsible for the processing of *E. coli* precursor rRNA into 16S, 23S, and 5S RNA. In RNaseIII mutant, a 30S pre-rRNA molecule accumulates which can be cleaved *in vitro* by purified RNaseIII to yield RNA molecules slightly larger than the mature rRNAs (Dunn and Studier, 1975; Ginsburg and Steitz, 1975). The important features of this enzyme are listed in Table 2.4. The functions of RNaseIII and other bacterial RNase are reviewed by Nicholson (1999).

2. RNaseIII-like Enzymes. RNaseIII-like enzymes have been conserved in fungi, plants, worms, flies, and mammals and are known to participate in the process of RNA interference (RNAi). RNAi is a mechanism in which the introduction of a double-stranded RNA into a cell leads to silencing of a particular gene via destruction of mRNA upon transcription from this gene. At least two RNase enzymes are involved in this process. These include Dicer (so named because of the defective nature of this protein in *Drosophila* mutants called dicer) and RISC. Dicer has two RNaseIII motifs (Nicholson, 1999) and a helicase motif and PAZ domain, whereas RISC is multicomponent nuclease. Dicer and RISC are separable as soluble and insoluble components in the extract of *Drosophila* embryo cells (Yang et al., 2000; Bernstein et al., 2001). Dicer chops the ds-RNA into ~22-nucleotide piece sequences ; latter serve as a guide to multicomponent nuclease RISC to destroy specific mRNA leading to gene silencing (Montgomery and Fire, 1998; Sharp, 1999; Bass, 2000).

B. RNaseP

This *E. coli* enzyme is a ribonucleoprotein containing a catalytic RNA subunit (RNA MI) and a protein subunit (C5 protein). Both components are needed for *in vivo* activity (Guerrier-Takada et al., 1983). However, the MI RNA component is sufficient for *in vitro* cleavage activity of a precursor tRNA molecule to generate the correct 5′ terminus of the tRNA. Because this enzyme is essentially a ribozyme, it is discussed along with other ribozymes in Chapter 9. Genes *rnpA* and *rnpB* encode the polypeptide (C5) and RNA (M1) subunit of the ribozyme (Kole et al., 1980). The gene *rnpB* which encodes for the ribozyme has been isolated from different members of Enterobacteriaceae by its ability to restore the correct *in vivo* processing function in *E. coli* mutant that is temperature-sensitive for RNaseP activity (Lawrence et al., 1987).

C. RNaseE

This enzyme catalyzes the formation of precursor rRNA from larger RNA transcripts in *E. coli.* The enzyme has been purified and characterized using a 9S rRNA substrate (250 bases long) accumulated in a temperature-sensitive RNaseE (*rne*) mutant strain of *E. coli* (Apirion and Lassar, 1978; Ghora and Apirion, 1978, 1979; Misra and Apirion, 1978, 1979). The native enzyme has a molecular weight of 70 kDa, and its characteristics are presented in Table 2.4. The enzyme is encoded by gene *rne* located at 23 min on *E. coli* map. The *rne* gene has been cloned (Ray and Apirion, 1980). Recent studies show that RNaseE occurs as a high-molecular-weight protein (180 kDa) and is always associated with polynucleotide phosphorylase (PNPase) as evidenced by their co-sedimentation and co-immunoprecipitation. RNaseE is an endonuclease whereas PNPase is an exonuclease, and both functions are required for the processing of rRNA. Furthermore, RNaseK is a proteolytic product of RNaseE (Carpousis et al., 1994). RNaseE is required for the action of RNaseP (Ray and Apirion, 1980).

D. RNaseM5

This endoribonuclease catalyzes the production of 5SrRNA (116 bases long) in *Bacillus subtilis* from precursor rRNA molecules (240, 179, and 150 bases long) (Sogin et al., 1977). The enzyme activity can be fractionated into two components α and β. The β subunit possesses the substrate binding property, and seemingly the α component has the catalytic activity. This enzyme is similar to *E. coli* RNaseE, because both process precursors to 5S rRNA. The major properties of this enzyme are listed in Table 2.4.

E. RNaseD

This enzyme is an exonuclease and catalyzes the removal of extra nucleotides distal to CCA sequence in a precursor tRNA molecule to yield the mature 3′ end of tRNA. The characterization of this enzyme was facilitated by synthesis of tRNA precursor substrate such as tRNA-CCA CC $[C]_n$ by addition of 14 cytosine residue to a normal tRNA CCAH terminus of *E. coli* tRNA. Using this substrate, a tRNA maturation activity was identified as RNaseD enzyme (Cudny et al., 1981; Ghosh and Deutscher, 1978). The enzyme has been purified to homogeneity. It is a monomeric protein with molecular weight 38 kDa, and its properties are summarized in Table 2.4. Another *E. coli* exonuclease RNaseII resembles RNaseD; however, both are distinct enzymes because they are completely separated during purification. Moreover, they differ in their substrate specificity and processivity; polyadenylate, an excellent RNaseII substrate, is not attacked by RNaseD.

F. Eukaryotic RNA-Splicing Enzymes

In eukaryotes, all three kinds of RNA (tRNA, rRNA, and mRNA) are the product of the RNA processing. In each case, intron(s) are spliced out by the catalytic activity of a protein enzyme, or a ribozyme, or a ribozyme in conjunction with protein(s). Of these, the splicing of tRNA is facilitated by a protein endonuclease and a ligase. tRNA splicing endonucleases have been described from a variety of eukaryotes. Among these, the yeast splicing endonuclease is very well characterized. Enzymes involved in splicing of tRNA have been found to be localized in nucleus (De Robertis et al., 1981).

1. Yeast tRNA Splicing Endonuclease. The yeast enzyme is a membrane-bound protein (Peebles et al., 1979, 1983). Two unlinked genes *sen-1* and *sen-2* control the yeast tRNA splicing endonuclease activity. The *sen-1* mutant possesses a deficient endonuclease activity, whereas *sen-2* mutant has a temperature-sensitive endonuclease that is active at 25°C but not at 34°C. Endonucleases were characterized using Proline pretRNA$_{UGG}$ *in vitro* assay. At present, it is not known whether *sen-1* and *sen-2* genes encode the two subunits of the tRNA splicing endonucleases or whether *sen-1* is a structural gene that requires *sen-2* gene product for the endonuclease activity. A number of yeast mutants that are defective in RNA processing and accumulate pretRNA molecules within the cells have been isolated and characterized (Hopper et al., 1978; Hopper and Schultz, 1980; Atkinson et al., 1985).

The tRNA splicing endonucleases have also been described from animals and plants such as *Xenopus,* HeLa cells, and wheat germ. The wheat germ splicing endonuclease is highly specific for plant tRNAs. The length of the intron may vary in different organisms; the plants have a shorter intron than do yeasts and invertebrates. It seems that while yeast and vertebrate splicing enzymes can tolerate the structural variation, the plant enzymes cannot and they are highly specific to plant pretRNAs (Stange et al., 1988).

Using mutant pre-tRNA, it has been shown that there are two steps in the splicing endonuclease activity. The first step involves the binding of the protein enzyme to the substrate pre-tRNA, whereas the second step involves the cleavage of the pre-tRNA into two halves and intron by its splicing activity (Baldi et al., 1986). It is known that the yeast enzyme can recognize the substrate and the splice sites (Greer et al., 1987).

III. RIBONUCLEASE H

Ribonuclease H is a protein enzyme that specifically hydrolyzes the ribonucleotide moeity of the RNA–DNA hybrid molecules. The enzyme was first described by Stein and Hausen (1969). RNaseH activity has been described from a variety of organisms including bacterial virus and higher organisms, including humans. At least two biochemically distinct RNaseH have been found to exist (Crouch and Dirksen, 1982; Itaya, 1990). Typically, these are high-molecular-weight and low-molecular-weight RNaseH activities in different organisms investigated so far; these RNaseH enzymes differ in other biochemical properties. All RNaseH enzymes produce 5′-phosphate and 3′-hydroxyl ends. In general, the one group of RNaseH enzymes produces oligonucleotides (2–9 bases in length), while the other produces mono- or dinucleotides. All RNaseH enzymes are endonucleases (Krug and Berger, 1989). Comparisons of the amino acid sequences of the cloned RNaseH genes from different sources, their site-directed mutagenesis and the analysis of their crystal structures reveal that Asp 10, Glu 48, and Asp 70 (corresponding to amino acid position in *E. coli*) are crucial for the enzymatic activity (Itaya et al., 1991; Schatz et al., 1989; Yang et al., 1990; Nakamura et al., 1991; Katayanagi et al., 1990; Hizi et al., 1990; Davies et al., 1991).

A. *E. coli* RNaseH

E. coli has two enzymes, namely, RNaseHI and RNaseHII. The RNaseHI and HII are encoded by *rnhA* and *rnhB* genes located at 5.1 and 4.5 min on the *E. coli* genetic map. Both genes have been cloned and shown to code 155 amino acid (RNaseHI) and 213 amino acid (RNaseHII) proteins (Itaya, 1990). The *E. coli* RNaseHI showed almost complete homology to *Salmonella typhimurium* RNaseHII (Itaya et al., 1991). Besides RNaseH of *E. coli,* other enzymes such as DNA polymerase I and exonuclease III have been shown to degrade RNA from DNA–RNA hybrid molecules.

B. Retroviral Reverse Transcriptase RNaseH

All RNA tumor viruses contained RNaseH activity associated with their reverse transcriptase. Both activities are contained by the same polypeptide as shown by biochemical, genetic, and immunological analyses. A single polypeptide chain obtained from SDS-PAGE contained both activities. The viral mutants were found to show temperature-sensitive activities of both enzymes. Furthermore, the antibody prepared against reverse transcriptase activity was found to cross-react with RNaseH (Crouch and Dirksen, 1982). The RNaseH activity is contained in the C terminus of the protein. The RNaseH and the reverse transcriptase activity is contained in a single polypeptide of 90 kDa (encoding a protein of 895 amino acids). This polypeptide is called a β polypeptide. The enzyme is known to exist as a dimer of β and α subunits; the α subunit (65 kDa) is a derivative of β subunit after proteolytic digestion. However, the β and α subunits alone can function as RNaseH. The HIV-1 reverse transcription gene has been cloned (Schatz et al., 1989). It encodes for a protein of 560 AA of which the C-terminus 120 AA contains RNaseH activity. The 65 kDa polypeptide has been further obtained as 51p (N-terminus) and 15p (C-terminus) peptides. Analysis of crystal-structure, site-directed mutagenesis and *in vitro* complementation confirms that the presence of both 60p and 15p is required for RNaseH activity (Yang et al., 1990; Hostomsky et al., 1991).

C. Yeast RNaseH

Two RNaseH activities have been described from yeast: a high-molecular-weight RNaseHI (<90 kDa) and a low-molecular-weight RNaseHII (21 kDa, pI10). RNaseHI is a typical enzyme that requires divalent cation, reducing agent and is optimally active at alkaline pH. The yeast RNaseH has been cloned. The yeast enzymes showed 30% homology to *E. coli* enzyme (Itaya et al., 1991).

D. Human RNaseH

Human RNaseH (RNaseH1) is an 89-kDa protein that acts on a DNA–RNA hybrid and degrades RNA, producing 5′-phosphoryl and 3′-OH termini. It has absolute requirement for divalent cations such as Mg^{2+}, Mn^{2+}, and Co^{2+}. It has a pH optimum of 8–8.5 and a PI of 6.4. Unlike *E. coli* enzyme, the human enzyme can act on a heteroduplex with a minimum of one ribonucleotide in one of the strand. This property of human RNaseH suggests its role in the removal of genomic ribonucleotides during DNA replication (Eder and Walder, 1991). The C terminus of the human RNaseH is highly conserved containing the catalytic and substrate binding domains. The N terminus of human enzyme contains a dsRNA binding motif (Wu et al., 2001).

E. Other Eukaryotic RNaseH

Calf thymus contains two size classes of RNaseH: high molecular weight (70 kDa) and low molecular weight (30 kDa). The large RNaseH has a pl around pH 5 and consists of two polypeptides 32 kDa and 25 kDa. The antibody prepared against large RNaseH does not cross-react with the small RNaseH, but does cross-react with many other RNaseH such as those obtained from *Xenopus laevis* and HeLa cells. The large RNaseH is a nuclear enzyme. The large and small RNaseH have been found in other mammalian cells, including ascites cells.

F. Biological Function of RNaseH

RNaseH plays a greater role in the initiation and termination steps of DNA replication. It can provide RNA primers for the initiation of DNA and also suppresses initiation from any primer site other than *ori-C* in *E. coli*. It participates in the termination steps of DNA replication via the removal of RNA primers in the Okazaki fragments. RNaseH also participates in DNA repair and in SOS-induced mutagenesis. Retroviral RNaseH participates in the removal of RNA from the DNA–RNA hybrid during the replication of viral RNA via the process of reverse transcription. Once the RNA is digested by RNaseH, the dsDNA is synthesized. A comparison of certain important properties of different RNases (both protein enzymes and ribozymes) is presented in Table 2.5. RNase H can be used to degrade a specific

Table 2.5 Comparison of the properties of RNases

Type	Substrates	Products	Catalytic Groups
Pancreatic A	ssRNA (C,U)	5′-OH, 3′-PO_4	His, His, Lys
Microbial	ssRNA	5′-OH, 3′-PO_4	His, Glu, Arg
Bacterial (Ba)	(N)		
Fungal (Tl)	(G)		
Ribonuclease H	dsRNA:DNA	5′-PO_4, 3′-OH	Asp, Asp, Glu, Mg^{2+}
Ribozymes	ssRNA		
E. coli RNase P	pre-tRNA	5′-PO_4, 3′-OH	RNA, Mg^{2+}
Tetrahymena	pre-rRNA	5′-RNA, 3′-OH	

Source: Yang et al. (1990), copyright 1990 by Science, AAAS, used with permission.

mRNA, which has been complexed with a complementary oligonucleotide (Dash et al., 1987; Walder and Walder, 1988). Thus RNase H can be used for possible targeted degradation of several viral RNA, including HIV-causing human diseases.

IV. PROOFREADING ACTIVITY OF RNA POLYMERASE

Influenza virus RNA-directed RNA polymerase has been shown to possess proofreading RNase activity that removes misincorporation of nucleotide by RNA polymerase (Plotch et al., 1981; Ishikama et al., 1986). Similar RNase domains of bacterial RNase, such as barnase containing glutamic acid, arginine, and histidine, have been shown to occur in RNA polymerase II of *Drosophila* and yeast. This RNase domain may be involved in proofreading of misincorporated nucleotide during transcription by eukaryotic RNA pol II (Shirai and Go, 1991).

3

DEOXYRIBONUCLEASE

Deoxyribonucleases were first described from calf thymus and pancreas. The pancreatic deoxyribonuclease (DNaseI) was used by Avery et al. (1944) to establish the transforming ability of the donor DNA in pneumococcal transformation because the DNA treated with DNaseI was found to abolish the transforming ability of the donor DNA. Deoxyribonuclease was the second enzyme whose amino acid sequence was fully determined (Liao et al., 1973). In this chapter, the properties of the certain deoxyribonucleases are described. The descriptions of certain special groups of deoxyribonucleases are included in the subsequent chapters.

I. CLASSIFICATION OF ENZYMES

Deoxyribonuclease as described here includes all those enzymes that, in general, can hydrolyze phosphodiester bonds present in a molecule of deoxyribonucleic acid. Deoxyribonucleases with special properties, such as specificity for nucleotide sequence (restriction endonuclease), topology (topoisomerase), and damage in a DNA molecule (damage-specific nucleases), are described in subsequent chapters. Deoxyribonucleases described in this chapter are specific to deoxypentose sugar moiety in a nucleic acid chain. Initially, deoxyribonucleases were considered to belong to two groups, namely, endonuclease and exonuclease. Endonucleases usually attack within a DNA chain and produce oligonucleotides, whereas the exonucleases are considered to attack from one end of the DNA molecule, sequentially producing mono- and/or dinucleotides.

Frankel and Richardson (1971) redefined the term *exonuclease* in view of the fact that an exonuclease was unable to hydrolyze a covalently closed circular DNA (such as Simian virus genomic DNA). Thus, it was shown that an exonuclease required a free terminus for nucleolytic attack on internucleotide phosphodiester bonds in a nucleic acid molecule. An exonuclease was also shown to produce oligonucleotides besides mononucleotides. Furthermore, an exonuclease can hydrolyze a phosphodiester bond in a highly supercoiled circular DNA but not in a relaxed circular DNA (Pritchard et al., 1977). In view of this fact, it has been suggested that such exonucleases may be better described as "exophilic nuclease," to indicate the preference for the open termini. Such classification of deoxyribonucleases into endo- and exonucleases is further complicated by the discovery of nucleases with both exo- and endonucleolytic properties (for example, the enzyme exoV of *E. coli* encoded by RecBCD genes possesses both exo- and endonucleolytic functions).

These deoxyribonucleases may show preference for single- or double-stranded DNAs as substrate and may attack from the 5′ end (i.e., 5′ → 3′ attack) or from the 3′ end (i.e., 3′ → 5′ attack). As a principle, no matter what may be the direction or mode of attack, a particular deoxyribonuclease produces only phosphomonoesters, either 5′ monophosphoesters or 3′ monophosphoesters, never both. However, no exodeoxyribonuclease is known to produce 3′ monophosphoesters (Linn, 1982). The examples of these groups of deoxyribonuclease are included in Table 1.1.

Such classification of deoxyribonuclease is also applicable to sugar nonspecific nucleases because they can attack a nucleic acid molecule endonucleolytically or exonucleolytically, or exo–endonucleolytically. The distinction among these properties of exonuclease, endonuclease, and exo–endonuclease can be seen by the change in the viscosity of nucleic acid determined after digestion of DNA with a particular nuclease (Laskowski and Seidel, 1945). Phosphodiesterase from snake venom (a classical example of an exonuclease) has very little or no effect on the viscosity of DNA as substrate, whereas treatment of the same DNA with pancreatic DNaseI (a classical example of an endonuclease) caused drastic change in the viscosity of the DNA. The effect of micrococcal nuclease (now proven to be an endo–exonuclease) had an intermediate effect upon the viscosity of the substrate DNA. These enzymes can further be distinguished based on their ability to produce a double-strand breakage (cleavage) or a single-strand breakage (nick) in the DNA. It has been shown that DNaseI produced nicks four times more than the cleavage of DNA, whereas DNaseII produced both kinds of breaks with equal frequencies (Young and Sinshiemer, 1965).

Exonucleases prefer to attack DNA molecules with free ends. They may attack a DNA chain either from the 3′ end or from the 5′ end and always produce a 5′ phosphomononucleotide. The 3′ and 5′ mononucleotides produced as a result of the nucleolytic digestion of a substrate (DNA) by an exonuclease can be readily identified by their susceptibility to 3′- and 5′ nucleotidase. No exodeoxyribonuclease is known to yield 3′ phosphomononucleotide (Linn, 1982b). Exodeoxyribonucleases are known to be processive or distributive enzymes. The processivity of an exodeoxyribonuclease is defined by its ability to complete the nucleolytic degradation of a DNA chain before the enzyme can attack a new DNA chain. The processivity of an exodeoxyribonuclease may be altered as a result of genetic mutation. Enzymes unable to digest substrate molecules completely are designated distributive.

The processivity of an enzyme can be determined by different methods (Nossal and Singer, 1968; Thomas and Olivera, 1978). The enzyme is mixed with an excess of ^{3}H-labeled DNA chains and then after a few minutes (say 10–15 min) of enzymatic action, another substrate such as DNA chains labeled with ^{32}P is added. An inverse relationship is known to exist between the processivity of an enzyme and the amount of radioactivity released from the second substrate (i.e., ^{32}PDNA) added to the reaction mixture. The method described by Nossal and Singer (1968) is based on the observation that a distributive enzyme yields partially digested molecules whereas a processive enzyme yields a mixture of entirely digested and completely intact molecules when a homogeneous substrate preparation is subjected to digestion by exonuclease. In the method described by Thomas and Olivera (1978) a processive enzyme was identified by its ability to release radioactivity from a DNA substrate labeled with ^{3}H at the 3′ end and with ^{32}P at the 5′ end equally well when treated with an exonuclease capable of hydrolyzing the DNA from 3′ and 5′ ends. At times some nucleases may act as both a processive enzyme and a distributive enzyme. For example, *E. coli* exonuclease III acts as a processive enzyme at 23°C but as a distributive one at higher temperature (37°C).

The directionality of an exonuclease is indicated by whether it traverses a substrate from the 3′ end to the 5′ end or from the 5′ end to the 3′ end of a DNA chain. This can be demonstrated by the use of terminally labeled substrate. DNA can be labeled at the 3′ end by the use of radiolabeled dNTPs and DNA polymerase or at the 5′ end by the use of ^{32}P labeled ATP and polynucleotide kinase.

A. Deoxyribonucleases

They are principally divided into two groups: exonuclease and endonuclease. They are distinguished by their requirement for a free terminus; thus an exonuclease can act only on a linear DNA chain but not on a covalently closed circular DNA, whereas an endonuclease can act on all kinds of DNA molecules (closed, linear, or open circle). The exonuclease can attack a DNA chain from either the 3′ or 5′ end in a stepwise manner releasing mononucleotides, whereas the endonucleases attack internal phosphodiester bonds in a DNA chain almost randomly, producing DNA fragments of various lengths. The endonucleases are also characterized by their site specificity.

B. Endonucleases

Endodeoxyribonucleases attack the internal phosphodiester bonds within a DNA chain with or without free terminus. Most enzymes of this class possess site specificity; this characteristic of the enzyme has been realized only recently. The pancreatic DNaseI and spleen DNaseII make similar preferential site-specific nucleolytic attacks within a DNA chain. Restriction endonucleases, damage-specific endonucleases, and a number of topoisomerases (described in the following chapters) are indeed site-specific. Deoxyribonucleases produce either 5′ or 3′ phosphomonoesters as a result of the nucleolytic attack; for example, pancreatic DNaseI

Table 3.1 Comparison of the properties of exo- and endodeoxyribonucleases

Property	Endonuclease	Exonuclease
Free terminus	Not required for enzyme activity	Required for enzyme activity
Manner of nucleolytic attack	Random	Stepwise
Nature of product	(a) DNA fragments of different lengths	(a) Usually mononucleotides
	(b) Both 5′ and 3′ monoester formed	(b) Only 5′ mononucleotides, no 3′ monoester formed
Site specificity	Invariably all of them are site-specific	None
Processivity	Nonprocessive	Most exonucleases are processive

produces 5′-monophosphate esters, whereas spleen DNaseII produces 3′-monophosphate esters. Certain exonucleases also possess endonucleolytic function. Such enzymes may conveniently be grouped as endo–exonucleases to reflect their enzymatic abilities. The *E. coli* exonuclease V (recBCD gene product) is an endo–exonuclease. The properties of the two groups of deoxyribonucleases (exonucleases and endonucleases) are summarized in Table 3.1. Deoxyribonuclease may produce oligonucleotides as well as mononucleotides (Deutscher and Kornberg, 1969) or may produce dinucleotides exclusively (Trilling and Aposhian, 1968) or produce high-molecular-weight fragments initially (Friedman and Smith, 1972).

C. Exonuclease

Escherichia coli exodeoxyribonucleases are among the most well-described enzymes in terms of both the genetics and the biochemistry. The important features of exodeoxyribonucleases from different organisms are listed in Table 3.1. All exodeoxyribonucleases known so far are 5′-monoester formers. No such enzyme which forms 3′-monoesters has yet been described (Linn, 1982b). This suggests that all exodeoxyribonucleases, whether transversing the DNA chain from the 5′ or 3′ end or both, must be able to recognize the P–O bond adjacent to the 3′ carbon of the sugar moiety in the DNA chain and hydrolyze it, releasing 5′ phosphomonoesters. The fact that a 3′-OH group is generated as a result of the exodeoxyribonuclease action suggests the role of this group of enzyme in DNA replication, repair, and recombination because a 3′-OH group can be used for the initiation/elongation of DNA chain during DNA synthesis involved in these processes. Among the exodeoxyribonucleases, *E. coli* enzymes are the most well-characterized. Their biochemical genetic properties are listed in Table 3.1 and discussed in the following paragraphs.

II. PROPERTIES OF ENZYMES FROM DIFFERENT ORGANISMS

A. Bacterial Enzymes

1. Exonuclease I. This enzyme has a molecular weight of 70 kDa and is coded by the *Sbc*B gene, which maps at 43.6 min (near *his*) on the genetic map of *E. coli* (Bachman and Low, 1980). Deletion of *Sbc*B gene has no effect on the growth of the mutant; therefore, seemingly this gene is not essential for the bacteria. This fact, as well as the proximity of ScbB gene to phage P2 attachment site, suggests possible phage origin for this gene (Weiss, 1981). Mutants of *E. coli* exonuclease have been characterized (Yajko et al., 1974). Earlier *Sbc* mutations were identified as the ones capable of restoring the frequency of recombination in the recBCD (exonuclease V)-deficient mutants in the presence of the recF gene (Templin et al., 1972). *Xon* mutants that map at the same site as the *Sbc* mutation on the genetic map of *E. coli*, however, cannot restore recombinational abilities of recBCD mutants in the combination with the recF gene. These properties of *xon* and *Sbc* mutations suggest that *xon* is a polar mutation that abolishes both the exonuclease I and recombinational abilities of the gene. In addition, *xon*A mutants produce temperature-sensitive exonuclease I (Weiss, 1981). Exonuclease I shows a greater preference for single-stranded DNA. It attacks from the 3′-OH end of a DNA chain and produces 5′ mononucleotides in a stepwise manner, leaving the terminal 5′ dinucleotide intact. The role of exonuclease I in DNA recombination is discussed later in Chapter 11.

2. Exonuclease II. Exonuclease II is the DNA polymerase I associated with proofreading 3′–5′-exonucleolytic function that is discussed later.

3. Exonuclease III. Exonuclease III is a small protein of molecular weight 28 kDa. The monomeric protein shows four kinds of catalytic activity. These are (a) a 3′–5′-exodeoxyribonuclease activity (specific for double-stranded DNA), (b) RNase activity, (c) 3′ phosphatase, and (d) AP-endonuclease activity. This enzyme was first identified as the enzyme that enhances the ability of calf thymus DNA to act as a primer. This activity was possible because of the phosphatase activity of the enzyme. The association of exonuclease and phosphatase activities to the same protein was based on the facts that both activities were found to copurify and to undergo heat inactivation at the same rate. The protein with these two enzyme activities was also found to possess RNaseH activity (Keller and Crouch, 1972). Later it was shown that exonuclease III also possessed AP-endonuclease activity of an independently described enzyme called endonuclease II (Friedberg and Goldthwait, 1969; Veerley and Paquette, 1972). The fact that the two *E. coli* enzymes (i.e., exonuclease III and endonuclease II) were the same and encoded by a single gene is supported by the evidence that (a) a homogeneous monomeric protein of molecular weight 28 kDa possessed both enzyme activity and (b) independently isolated mutants were defective for both activities and possessed an identical mutant phenotype such as the temperature-sensitive enzyme activity. However, the AP-endonuclease activity of the *E. coli* exonuclease III has now been termed endonuclease IV, which constitutes 85% of the cellular AP-endonuclease activity. This is discussed later in Chapter 5 under the damage-specific deoxyribonucleases. Enzymes with exonuclease III activity have been described from *Hemophilus influenzae* and *Streptococcus pneumoniae* (Gunther and Goodgall, 1970; Clements et al., 1978). However, the *Streptococcus* enzyme lacks AP-endonuclease and RNaseH activities (Lacks and Greenberg, 1967).

The enzymatic properties of exonuclease III have been discussed by Weiss (1981). Some of the interesting features of this enzyme are as follows:

1. The exodeoxyribonuclease activity of the enzyme is processive at 23°C but distributive at 37°C.
2. In a DNA–RNA hybrid it selectively degrades RNA strand with a rate 10,000 times faster than that of DNA degradation releasing dinucleotides.
3. It can remove 3′-terminal phosphate and generate 3′-OH group and then act as an exonuclease. The 3′-phosphatase activity is somewhat specific for single-stranded DNA.
4. The enzyme has an AP-endonuclease activity on apurinic and apyrimidinic sites but with no demonstrable endonuclease activity with intact DNA as substrate.

Exonuclease III is encoded by the *xthA* gene, located at 38 min on the chromosomal map. Factors such as mitomycin C or mutation in UTPase gene which can cause the production of AP sites have no effect on the induction or production of exonuclease III. It is the mutation in the *xthA* gene which led to the conclusion that exonuclease III accounts for the 85% of all cellular AP-endonuclease activity. The remaining (i.e., 15%) AP endonuclease activity is due to endonuclease IV (Ljungquist et al., 1976; Ljungquist, 1977) whose properties are discussed later (see Chapter 5). Exonuclease V, an ATP-dependent enzyme, and exonuclease III account for almost all the exonuclease activity of *E. coli*. *E. coli xth* mutants show a hyper rec-phenotype and a slight sensitivity to alkalyting agents, and they are unable to tolerate a *dut* mutation (Weiss et al., 1978) but are normal with respect to their growth rate on UV-sensitivity or mutagenic activity (Milcarek and Weiss, 1972). These properties of *xth* mutant have been explained on the basis of the loss of AP-endonuclease activity (Weiss, 1981). However, this exonuclease seems to be a nonessential enzyme because an *xth* deletion mutant by itself or in combination with polA1 (i.e., deficient in 35 exonuclease activity) mutant or with *Sbc-B-recBCD* mutant (deficient in both exonuclease I and exonuclease V) has no obvious or detectable effect on the growth of the mutant. Furthermore, the overproduction of exonuclease III does not affect either the growth of the bacteria or the replication of the chimeric plasmid containing the cloned *xth* gene. Thus the exonucleolytic as well as the associated DNA-phosphatase and RNAaseH activity of exonuclease III are not crucial for the physiology of *E. coli*. The AP-endonuclease activity of the exonuclease III is, however, required for the survival of *E. coli* because a double mutant (i.e., *xth* and *dut*) deficient in both exonuclease III and UTPase is lethal. Such a double mutant can, however, become viable in the presence of *ung* mutation, because mutants deficient in uracil DNA glucosylase stop the formation of AP sites in DNA (Weiss et al., 1978). Thus the AP-endonuclease activity of exonuclease III is the only enzyme activity that seemingly plays a role in the bacterial physiology. The hyper-rec phenotype of the *xth* mutants may be due to a failure of the breakdown of the recombination intermediate in the absence of exonuclease III or due to availability of recombinogenic sites created by excessive AP sites.

4. Application of the Enzyme Exonuclease III. Exonuclease III has been used for the following:

1. Generation of 3′ OH in DNA which provides excellent primers for DNA synthesis
2. Controlled synchronous exonuclease degradation which is used for DNA sequencing
3. Production of localized mutation by the generation of small deletions in a cloned gene in combination with S1 nuclease
4. The identification of AP sites in DNA and distinction from the misincorporated ribonucleotide in a DNA chain
5. Mapping specific protein-binding sites on DNA because protein-bound sites are partially protected from nuclease digestion

5. Exonucleases IVA and IVB. These enzymes catalyze the hydrolysis of oligonucleotides to nucleoside 5′ monophosphates. They have a broad alkaline pH optimum (8.0–9.5) and require magnesium for activity and differ from exonuclease I in their ability to degrade dinucleotide and in greater heat stability. These enzymes have not yet been purified to any great extent (Jorgenson and Koerner, 1966).

6. Exonuclease V (RecBCD Enzyme). This *E. coli* enzyme has been extensively studied. It consists of three subunits of 140, 180, and 60 kDa encoded by *recB*, *recC*, and *recD* genes, respectively, which map at 60 min on the *E. coli* genetic map. Like exonuclease III, this enzyme has several functions. These include (a) an ATP-dependent exodeoxyribonuclease activity that acts processively from either end (5′ or 3′) yielding oligonucleotides, (b) a DNA-dependent ATPase, and (c) a duplex DNA unwinding activity (Demple and Linn, 1982). Exonuclease V has been demonstrated to participate in DNA recombination and repair and in the restriction of infecting phage DNA and other bacterial cellular activities that are based on a detailed analysis of recBCD mutants unable to carry out these functions as discussed later (see Chapter 11).

7. RecBCD (Exo V) from Other Organisms. The ExoV-like activity was first reported from *Micrococcus luteus* (Tsuda and Strauss, 1964). Since then, such enzyme activities have been reported from a number of other bacterial species. An ATP-dependent deoxyribonuclease activity on dsDNA has been purified from sea urchin (Gafurov et al., 1979) and bovine lymphocytes (Graw et al., 1981).

Furthermore, there is genetic evidence for the presence of an ATP-dependent DNase in yeast (Mahler and Bastos, 1974). Presence of an enzyme resembling the activity of the recBCD enzyme has been demonstrated in *Neurospora* (Chow and Fraser, 1979). Thus this enzyme is ubiquitously present in both prokaryotes and eukaryotes.

The properties of bacterial enzyme (ExoV) listed in Table 3.1 show a number of common characteristics. These include (1) an ATP-dependent exodeoxyribonuclease capable of degrading dsDNA processively into oligonucleotides, (2) a DNA-dependent ATPase, (3) a single-stranded DNA-specific exonuclease and an endonuclease activity, the latter accompanied by dsDNA unwinding activities, and (4) the multimeric nature of the enzyme with requirement for divalent cation for enzyme activity and an alkaline pH optimum. The exonuclease V resembles restriction endonucleases type I, gyrase (topoisomerase), and helicase in possessing ATP-dependent DNase and DNA-dependent ATPase activities. These properties of bacterial exonuclease V have been discussed elsewhere (Muskavitch and Linn, 1981; Muskavitch, 1982).

8. Exonuclease VI. This represents the 5′–3′-exonucleolytic activity of DNA polymerase I in *E. coli*. This enzyme is similar in its activity to mammalian DNase IV or Flap-endonuclease or FEN 1 enzyme (Lindahl et al., 1969; Robins et al., 1994) as discussed later in this chapter.

9. Exonuclease VII. This enzyme is an 88-kDa protein encoded by an *XSC* gene, which maps at 53 min on the *E. coli* genetic map. The *E. coli* enzyme and its mutant have been very well characterized (Chase and Richardson, 1974, 1977). It acts on single-stranded DNA or single-stranded termini processively from both ends producing oligonucleotides. The enzyme can act on a UV dimer from a nick incised by *Micrococcus luteus* endonuclease. Studies of temperature-sensitive and deletion mutants suggest that this enzyme is nonessential for the survival of the bacteria. However, *XSC* mutants are sensitive to UV light and to nalidixic acids and possesses hyper-rec phenotype. The double mutant *XSC-Pol Aex* is more hyper-rec and temperature-sensitive than either of the single mutants. *XSC* mutation also reduces the rate of thymidine dimer removal when introduced into multiply defective recBCD PolA ex mutant (Chase and Masker, 1977; Chase et al., 1979). Exonuclease VII has been used in combination with other nuclease to determine the location and nature of single-stranded regions in DNA molecules.

10. Exonucleases Associated with DNA Polymerases. *E. coli* DNA polymerase I possesses both the 3′–5′- and 5′–3′-exonuclease activity which corresponded to previously described exonuclease II and exonuclease VI activities. The 3′–5′-exonuclease activity is essentially the proofreading activity of the DNA pol I, whereas the 5′–3′-exonuclease activity is known to (a) remove the primer after the completion of DNA synthesis, (b) facilitate nick translation and remove thymine dimers, and (c) possess an RNaseH activity. The monomeric 109-kDa DNA pol I protein can be separated into two fragments either by proteolysis or by certain mutation in PolA gene located at 33 min on *E. coli* map. The 35-kDa fragment containing the 5′–3′-exonuclease activity and the 76-kDa fragment containing both the DNA polymerase and 3′–5′-exonuclease (proofreading) activities are the product of such proteolysis or of mutation in PolA genes. The 76-kDa fragment is also known as the *Klenow fragment* (Klenow and Henningsen, 1970). In the Klenow fragment the N-terminus domain possesses the proofreading activity whereas the C-terminus domain possesses the polymerizing activity. The existence of two distinct domains has been visualized by x-ray crystallography of the *E. coli* wild type and mutant DNA pol I (Derbyshire et al., 1988).

The *E. coli* DNA pol II, a 102-kDa enzyme encoded by the dinA gene (Bonner et al., 1990) located near the ara-D gene on the genetic map possesses only the 3′–5′-exonuclease (proofreading) activity but lacks the 5′–3′-exonuclease activity. The DNA pol III, a 140-kDa enzyme of *E. coli* encoded by *PolC* (*dnaE*) located at 4 min on the *E. coli* genetic map, possesses both the 3′–5′- and 5′–3′-exonucleolytic activities analogous to DNA pol I enzyme. However, the exonuclease activities of the DNA pol III holoenzyme are the properties of distinct proteins. The proofreading activity is encoded by the DNAQ gene. DNA polymerases of bacterial viruses (*T*2, *T*4, and *T* 7) also possess both 3′–5′- and 5′–3′-exonucleolytic activities.

Among lower eukaryotes, the yeast DNA polymerase has been shown to possess exonucleolytic function. In mammalian cells, the DNA polymerase α does not possess an exonucleolytic function per se, but both the DNA polymerases δ and ε possess 3′–5′-exonucleolytic (proofreading) activity. The eukaryotic DNA polymerase ε has been shown to be associated with a separate polypeptide carrying a 5′ → 3′-exonucleolytic function required for the removal of RNA primers from Okazaki fragments (Siegel et al., 1992).

11. Exonuclease VIII. This is a dimeric protein with an apparent molecular weight of 140 kDa which shows a 40-fold preference for duplex DNA. Unlike exoV, it lacks a requirement for ATP for exonucleolytic activity. It does not act from a gap or nick and requires a free 5′-OH group and acts in the direction of 5′–3′. It is encoded by the *recF* gene, which is expressed only in the *Sbc*A mutant; presumably the *Sbc*A gene product is a repressor of the *rec*F gene. The exonuclease VIII can overcome the recombinational deficiencies of *rec*BC mutants or of the λ phage mutants defective in λ exonuclease. This enzyme is responsible for the RecF pathway of recombination in *E. coli* as discussed later in Chapter 11. The evolutionary relation of this enzyme and its possible origin from a cryptic λ phage gene are discussed elsewhere (Weiss, 1981).

B. Endonucleases

1. Bacterial Enzymes

a. EndonucleaseI. *E. coli* endonuclease I is a small 12-kDa protein and is encoded by an *end* A gene located at 63 min on the genetic map of *E. coli*. The enzyme exists in periplasmic space and is easily released by osmotic shock (Nossal and

Heppel, 1966). The *end* A mutant has no obvious effect on the phenotypes of the bacteria. The enzyme appears during the early log phase of the growth and then disappears significantly toward the stationary phase. The enzyme attacks both strands of DNA simultaneously and degrades a duplex DNA into oligonucleotides. The duplex DNA is degraded first into oligonucleotides of 400 base pairs and later oligonucleotides of 7 base pairs upon complete digestion. Double-stranded RNA strongly inhibits the enzyme activity by forming a complex with endonuclease I. Such an enzyme–tRNA complex has, however, been found to cause exactly one random nick into a supercoiled circular DNA molecule (Goedel and Helsinki, 1970). The physiological implication of such a property of the enzyme remains unclear. The periplasmic location and other properties of the enzyme and the fact that end A mutation has no obvious effect on the bacteria suggest that endonuclease I essentially plays a degradative role during the growth of bacteria.

b. Endonuclease II. This enzyme, initially thought to act on alkylated and apurinic DNA, has now been shown to correspond to AP-endonuclease activity of exonuclease III (Weiss, 1976). No enzyme capable of acting on alkylated DNA alone has ever been shown to occur in *E. coli*. Thus endonuclease II is a nonexistent enzyme. Likewise, AP-endonuclease activity described as endonuclease VI also corresponds to exonuclease III. Thus both the endonuclease II and VI should be reassigned to AP-endonuclease activity of exonuclease III as proposed by Linn (1978, 1982a). The term endonuclease II has, however, been retained to include enzymes capable of acting on alkylated DNA alone if and when described in the future.

c. Endonuclease III. Endonuclease III is a 27-kDa protein specific for damaged DNA and will be described later in the subsequent chapter. The *n*th gene controlling the structure of this protein has been cloned and characterized (Asahara et al., 1989). The enzyme has been purified to homogeneity and shown to possess both AP-endonuclease and DNA–*N*-glucosylase activities. Another *E. coli* protein encoded by the Mut Y gene is homologous to endonuclease III in possessing these two enzyme activities (Tsai-Wu et al., 1992).

d. Endonuclease IV. This is a low-molecular-weight (33 kDa) AP endonuclease. No mutation in the gene controlling its structure has yet been known. The properties of this enzyme have been described later in Chapter 5.

e. Endonuclease V. This is also a damaged low-molecular-weight (27 kDa) DNA-specific enzyme. No mutant of this enzyme has yet been described. This enzyme has also been discussed in Chapter 5.

f. Endonuclease VI. This enzyme is synonymous with the AP-endonuclease activity of the exonuclease III as discussed earlier in this chapter and also in Chapter 5.

g. Endonuclease VII. This is a protein of apparent molecular weight 56 kDa which acts on damaged DNA and is discussed in Chapter 5.

Thus, except for endonuclease I, all other *E. coli* endonucleases are essentially damaged DNA-specific endonucleases. They are low-molecular-weight (27 kDa to 56 kDa) proteins. Genes controlling the structure of these enzymes are not known because no mutants have been described. The properties of these endonucleases are described in Chapter 5 on damage-specific nucleases.

h. Other Microbial Deoxyribonucleases. A number of deoxyribonucleases have been characterized from different bacterial and other microbial organisms. The enzymes similar to *E. coli* ExoV have been described from a number of organisms, including several bacteria and fungi like *Neurospora* and yeast.

2. Mammalian Deoxyribonuclease. The mammalian endodeoxyribonucleases are essentially of two kinds, namely, DNaseI and DNaseII. DNaseI is much more prevalent than DNaseII. DNaseI is characteristically found in the pancreas and in digestive tissues, whereas DNaseII has been found in the spleen, thymus, and other tissues. A large number of other deoxyribonucleases may include a variety of these deoxyribonucleases due to changes in the level and nature of glycosylation and/or proteolytic digestion. The identity of a large number of proteins can be established only by determining their amino acid sequences, their level of glycosylation, and antigenicity. Only two mammalian endodeoxyribonuclease such as DNaseI and DNaseII have been characterized with respect to these criteria.

a. Deoxyribonuclease I. This is the most well-characterized deoxyribonuclease. The bovine pancreatic DNase was first crystallized by Kunitz (1950) and later characterized by Moore (1981). This is a glycoprotein of 260 amino acids with two disulfide bonds (Liao et al., 1973; Moore, 1981; Blackburn and Moore, 1982; Suck et al., 1984). Chemical modification experiments suggest the role of tyrosine, serine, tryptophan, histidine, and carboxyl residues in the activity of the enzyme. Also, amino acids that modulate phospholrylation of mannose in this glycoprotein have been identified (Nishikawa et al., 1999). Like RNase, the folding of DNaseI is determined by the residues at the C-terminus. The enzyme is an endonuclease with an absolute requirement for divalent cations such as Mg^{2+} or Mn^{2+}. The enzyme makes a double-stranded break or a single-stranded nick in the presence or absence of Mg^{2+} and Ca^{2+}. The native DNA seems to be attacked in two different phases. The enzyme possesses site specificity during the initial phase leading to single-stranded break. The enzyme, however, loses specificity during the terminal phase and is subject to phenomena of autoretardation. The crystalline bovine DNases of Kunitz was found to separate into four different forms upon phosphocellulose chromatography (Salnikow et al., 1970; 1973). These multiple forms are caused by the replacement of histidine at the 118 position by proline in the primary structure of the protein and by the presence of a neutral or sialic acid containing residues in the carbohydrate side chains. These characteristics of bovine pancreatic DNase are presented in Table 3.2. The multiple forms of DNaseI have been found in all other mammals including humans and in tissues other than pancreas. Multiple forms of DNaseI are encoded by at least two different genes.

Table 3.2 Comparison of different bovine pancreatic DNaseI

Enzyme	Amino Acid at the 118 Position in the Primary Structure	Carbohydrate Moiety
DNaseI A	Histidine	*N*-Acetyl glucosamine and mannose
DNaseI B	Histidine	Galactose and sialic acid in addition to *N*-acetyl glucosamine
DNaseI C	Proline	*N*-Acetyl glucosamine and mannose
DNaseI D	Proline	Galactose and sialic acid in addition to *N*-acetyl-glucosamine and mannose

The DNaseI is inhibited by actin (Lazarides and Lindberg, 1974). DNaseI may control the expression of several genes *in vivo* because it has been shown that DNaseI, when introduced into mammalian cells via lyposomes, can cause neoplastic transformation (Zajac-Kaye and Tso, 1984).

b. Human DNaseI. The human DNaseI has been very well characterized from human pancreas, duodenal juice, urine, and serum. The human pancreatic DNaseI shows tremendous homology to bovine DNaseI, including the four cystine residues and the amino acids at the active sites. A comparison of the amino acid sequences of the bovine and human DNaseI showed significant homology. The gene for the human DNaseI has been cloned (Shak et al., 1990). At least five alleles of human DNaseI have been described (Lida et al., 1997), and some of the DNase phenotype appears to be associated with gastric carcinoma (Tsutsumi et al., 1998). The analysis of cloned human pancreatic DNaseI revealed that this human enzyme contained only 258 amino acids, and not 260 amino acids as inferred from previous chemical analysis of the amino acid sequence.

Bovine pancreatic DNaseI interacts with the minor groove of the substrate DNA. DNase has a low specificity but possesses homology to exonuclease III, an enzyme involved in repair of AP-DNA. This enzyme has been engineered to possess AP-DNA repair activity by adding the additional 12 amino acids present in the exonuclease III (Cals et al., 1998).

The human recombinant DNase has been used for therapeutic purposes to reduce the viscoelasticity of sputum and to improve the function of lungs in patients suffering from cystic fibrosis. The recombinant human DNase has also been obtained as DNaseI-Fc fusion protein with improved functions (Dwyer et al., 1999). The application of human DNaseI to treat cystic fibrosis patients has been discussed in Chapter 14 of this book.

c. RNase–DNase-Cross-Linked Bifunctional Enzyme. A bifunctional enzyme capable of degrading both DNA and RNA has been synthesized by the cross-linkage of bovine pancreatic DNaseI and RNaseA by Wang (1979). The coupling of RNase to DNase was achieved by thiol–disulfide interchange at pH 6.2 and 25°C for 90 min. The hybrid enzyme had 75% DNase and 40% RNase activity as compared to the parental enzymes. Furthermore, the RNA strand of a hybrid substrate (f1, DNA-^{3}HRNA) was rapidly attacked by the bifunctional enzyme but not by the parental RNase alone. Thus, the RNase activity of the bifunctional enzymes compares with that of the RNaseH but not with the RNaseA.

d. Deoxyribonuclease II. This deoxyribonuclease has been found throughout the animal kingdom (Laskowski, 1967). The enzyme from the calf thymus and from hog spleen has been purified to homogeneity and is very well characterized. The enzyme has a molecular weight of 38 kDa and an isoelectric point of pH 10.2. It is a glycoprotein containing glucosamine and one disulfide bond. The enzyme acts in two phases that include an initial rapid phase and a very slow second phase. The enzyme makes double-stranded breaks during the initial rapid phase of action, whereas the slow phase requires a minimum of two hits to cause a double-strand break. This mode of action is in sharp contrast to DNaseI, which on the average makes only one in four attacks causing a double-strand break. However, like DNaseI, DNaseII shows the phenomena of autoretardation. DNaseII is strongly inhibited by Mg^{2+} at concentrations higher than 1 mM but requires a higher concentration (200–300 mM) of monovalent cations for its activity. The purified DNaseII is also inhibited by RNA. The enzyme has an optimal activity at acid pH of 4.5. The properties of DNaseI and DNaseII are presented in Table 3.3.

Human DNaseII has been very well characterized, and its role in apoptosis has been established. The gene encoding human DNaseII has been cloned and expressed to yield a protein of 40 kDa. The human gene for DNaseII has been located on chromosome 19p13.2. The human DNaseII gene shows about 30% homology to the *Caenorhabditis elegans* gene (Krieser and Eastman, 1998).

Table 3.3 Comparison of the properties of DNaseI and DNaseII

	DNaseI	DNaseII
Substrate	DNA	DNA
pH optimum	7–8	4–5
Activation	Mg^{2+}, Mg^{2+} Ca^{2+}, Co	Na^{+} (300 mM)
Inhibitions	Citrate EDTA Actin	Mg^{2+} RNA
Product	5-Phosphate esters	3-Phosphate esters
Nature of scission	75% Single-stranded breaks 25% Double-stranded breaks	Mostly double-stranded breaks
Chemical nature	Glycoprotein of 258 amino acids with two disulfide bridges	Glycoprotein of 343 amino acids with a disulfide bridge

e. Structure-Specific Nucleases

I. NUCLEASE INVOLVED IN MATURATION OF OKAZAKI FRAGMENTS AND IN BASE EXCISION REPAIR

1. DNaseIV/FEN-1—A Family of Structure-Specific Nucleases. FEN (Flap-endonuclease) constitutes a family of structure-specific nucleases (Ceska and Sayers, 1998; Suck, 1998) that recognize and remove 5′-overhang (or flap or bifurcated) DNA structures usually produced during the process of DNA replication, repair, and recombination. These nucleases remove the branched structures endo–exonucleolitically with the aid of divalent metal cations independent of the polymerase action (Lyamichev et al., 1993). Most of these FEN-1 nucleases also have associated 5′–3′-exonuclease and RNaseH activities. These enzymes have been found to exist in all living systems from bacteria (including their bacteriophages) to arechea and to eukaryotes (including fungi, yeast, plants, and animals). Properties of these enzymes are exemplified by the 5′ domain of *E. coli* DNA pol I (Lindahl, 1994; Robins et al., 1994), mammalian DNaseIV (Lindahl et al., 1969; Harrington and Lieber, 1994a) and its homolog XPG protein (Lieber, 1997), *Schizosaccharomyces pombe* rad2 protein (Murray et al., 1994), and *Saccharomyces cerevisiae* Rad27 protein (Reagan et al., 1995). These enzymes are required for the removal of Okazaki fragments and thus have an important role in DNA replication and DNA repair (Lieber, 1997; Klungsland and Lindahl, 1997; Alleva and Doetsch, 2000). The activity of FEN nuclease is promoted by PCNA. In the absence of DNA, PCNA interacts with FEN with its interdomain connector loop (IDCL). However when PCNA encircles DNA during DNA replication, it interacts with FEN1 by its C-terminus (Gomes and Burgers, 2000; Hosfield et al., 1998). A region of conserved nine amino acids in the human FEN-1 consisting of residue Gln 337 to Lys 345 in the C terminus was found to interact with PCNA *in vitro* but not *in vivo* (Frank et al., 2001). In addition to PCNA, both Dna2 helicase (Budd and Campbell, 1997) and Werner syndrome protein (with RecQ helicase and exonuclease domains) interact with FEN-1; the Werner protein increases the cleavage activity of FEN-1 (Brosh et al., 2001). FEN1 is present in the proliferating cells and is shown to colocalize with PCNA in the proliferating cells (Kim et al., 2000).

The human FEN-1 gene is located on chromosome 11q12 and codes a protein consisting of 378 amino acids (Harrington and Lieber, 1994b; Hiraoka et al., 1995). The human FEN-1 gene is homologous to yeast YKL510 and RAD2 genes which also belong to this group of structure-specific endonuclease, that is, FEN-1(Harrington and Lieber, 1994b). The gene encoding FEN-1 has been cloned from a variety of organisms (Karanjawala et al., 2000), and the FEN-1 protein sequence structure and the amino acids involved in enzymatic cleavage of the branched structure in the substrate DNA have been determined (Shen et al., 1995). The crystal structures of a number of FEN nucleases have been solved, and the basis for their structure selectivity has been established (Lieber, 1997; Ceska and Sayers, 1998; Suck, 1998; Parikh et al., 1999). FEN-1 has a saddle-shaped 3-D structure consisting of a single domain of λ/β proteins with seven parallel β sheets surrounded by β helices creating a deep groove through which the substrate DNA is threaded in. The binding of ds

DNA is facilitated by an HTH motif that also forms a part of the conserved amino acids at the active sites. In addition to its central role in DNA replication (particularly that of the lagging strand of DNA) via maturation of the Okazaki fragments, FEN-1 plays a role in the DNA repair (i.e., long-patch base excision repair) but not in the MMR (Parikh et al., 1999) or in the maintainance of the stability of the genome. FEN-1 mutants are genetically unstable (Greene et al., 1999) and may cause cancer and/or genetic defects. FEN-1 is also involved in the maintenance of direct or CTG trinucleotide repeats, the joining of nonhomologous DNA segments, and mitotic conversion (Tishkoff et al., 1997; Freudenreich et al., 1998; Holmes and Haber, 1999). The maturation of Okazaki fragments is carried out by FEN-1 in conjunction with RNaseH1 and DNase2 protein. Further evidence suggests that in FEN-1 null mutants , the function of FEN-1 may be taken over by DNase2 protein during the maturation of Okazaki fragments (Bae and Seo, 2000).

FEN-1 nuclease also participates in the long-patch base excision repair (Klungsland and Lindahl, 1997).

2. DNase2 Nuclease. The yeast DNase2 is a 172-kDa protein with single-stranded DNA-specific endonuclease activity in the N terminus and a helicase motif in the C terminus. This protein interacts with the FEN-1 protein during the maturation of Okazaki fragments (Bae and Seo, 2000; K.-H. Lee et al., 2000; Budd and Campbell, 1995, 1997). This nuclease is essential for the processing of the Okazaki fragments *in vivo* because mutants lacking this endonuclease activity are not viable (K.-H. Lee et al., 2000).

II. NUCLEASES INVOLVED IN EXCISION REPAIR

1. XPG Protein. Humans suffering from complementation group G form of inherited xeroderma pigmentosum controlled by an autosomal recessive gene are hypersensitive to solar UV radiation. These individuals possess a defective form of a nuclease called XPG protein, which mediates the base excision repair process in conjunction with many other proteins in the repairosome. The XPG protein is a structure-specific nuclease (Constantinou et al., 1999). Both the nuclease and the gene encoding this protein has been very well characterized (Emmert et al., 2001). The gene is located on human chromosome 13q33 and span a segment of 30 kb with 15 exons (ranging from 61 bp to 1074 bp in length) and 14 introns (250 bp to 5763 bp in length). The XPG gene encodes a 133-kDa acidic protein that is specific to single-stranded DNA during excision repair. The cloned XPG (also called ERCC5) gene was found to confer UV resistance to XPG cells from XP patients or to UV-sensitive mutant rodent cells, ERCC5 (O'Donovan and Wood, 1993; O'Donovan et al., 1994; Shiomi et al., 1994; Nouspikel and Clarkson, 1994). The mutant XPG protein with Asp-77 or Glu-791 substitution was unable to cleave DNA. XPG protein exist as a dimer under physiological conditions. XPG protein could not substitute for FEN-1 protein during the long-patch base excision repair process (Klungsland and Lindahl, 1997). Mice with a defective XPG gene show degeneration of Purkinje cells; this may be the underlying neurological

Table 3.4 Properties of some of the mammalian nucleases

Enzyme	Substrate	Mode of Action	pH Optimum	Mg^{2+}	Main Reaction Product	Mol. wt.
DNaseI	ds and ss DNA	Endonucleolytic	7.1	+	5′ Oligonucleotides	31 kDa
DNaseII	ds and ss DNA	Endonucleolytic	4.1	–	3′ Oligonucleotides	38 kDa
DNaseIII	ss Duplex DNA	Exonucleolytic	8.5	+	5′ Monodinucleotides	52 kDa
DNaseIV	Duplex DNA	Exonucleolytic $3' \rightarrow 5'$	8.5	+	5′ Mononucleotides	42 kDa
DNaseV	Duplex DNA	Exonucleolytic $3' \rightarrow 5/5' \rightarrow 3$	8.8	+	5′ Mononucleotides	12 kDa
DNaseVI	ss DNA	Endonucleolytic	9.5	+	5′ Oligonucleotides	45 kDa
DNaseVII	ss and nicked and ds DNA	Exonucleolytic $3' \rightarrow 5'$	7.8	+	5′ Mononucleotides	43 kDa
DNaseVIII	5′ ss and nicked UV′d duplex DNA	Exonucleolytic $5' \rightarrow 3'$	9.5	+	5′ Oligonucleotides	31 kDa
Correxo nuclease	ss DNA nicked UV′d ds DNA	Exonucleolytic $3' \rightarrow 5'/5' \rightarrow 3'$	8.0	+	5′ Oligonucleotides	30–35 kDa
Lysosomal or spleen exonuclease	RNA or DNA with 5′-OH	Exonucleolytic $5' \rightarrow 3'$	5.5	–	3′ Mononucleotides	70 kDa

Source: Hollis and Grossman (1981), used with permission.

disorders accompanying the human XP patients (Zafeiriou et al., 2001). The transcription factor TFIIH interacts with the XPG nuclease (Araujo et al., 2001, Winkler et al., 2001). Human RPA protein also stimulate the activity of the XPG nuclease (Matsunaga et al., 1996).

The yeast excision repair RAD2 gene encodes a nuclease that is a homolog of human XPG protein (Habraken et al., 1993); similar homologs of human XPG nuclease have been found in *Drosophila* and plants. Like the human XPG protein, the yeast RAD2 protein is a single-strand-specific endonuclease that is involved in the incision step of excision repair.

f. Other Mammalian Deoxyribonucleases. Besides DNaseI and DNaseII, a number of other deoxyribonucleases have been described from mammalian cells (Hollis and Grossman, 1981). These are summarized in Table 3.4. Some of these nucleases include sugar nonspecific nucleases and damaged DNA-specific nucleases; their characteristics are described in following chapters of this book. In humans a number of DNases have been described from lymphocytes and from KB cells in tissue cultures; some of these resembled DNaseI and DNaseII in their properties (Graw et al., 1981; Zollner et al., 1974; Frankel et al., 1981). Properties of some important mammalian deoxyribonucleases are presented in Table 3.4.

g. Other Deoxyribonucleases. A large number of deoxyribonucleases have been described from different organisms. However, only a few plant nucleases have been described in any detail.

4

RESTRICTION ENDONUCLEASES

Luria and Human (1952) first described the phenomenon of host-controlled specificity in T-even phages. This was immediately confirmed by Bertani and Weigle (1953) in λ and P2 phages. In these studies it was shown that a particular bacteriophage possessed different efficiencies of infection on several closely related strains of bacteria. However, the progeny of bacteriophage which initially plated with a low efficiency was later found to plate efficiently after one generation of plating on the same bacterium. This phenomenon of host-controlled specificity was first discovered in T-even phages by Luria and Human (1952) and in λ and P2 phages by Bertani and Weigle (1953). The epigenetic nature of such host range specificity was indicated by the fact that the newly acquired high efficiency of plating on a particular bacterial host was lost by the progeny of the same bacteriophage when plated on a different hosts. The first host modification unique to T-even phages involved the glycosylation of hydroxymethyl cytosine residue (Revel and Luria, 1970). However, the modification introduced in λ phages was found to be of universal occurrence. The molecular basis of such host range specificity depended on the restriction or modification of certain DNA sequences by restriction and modification enzymes (Arber and Dussoix, 1962; Arbor and Linn, 1969). Foreign DNA molecules (lacking appropriate modification) were cleaved by the host restriction endonuclease upon entry into the bacterial cell. During this process some of the phage DNA may escape restriction by host endonucleases and instead get modified by methylation of adenine or cytosine moieties in DNA and thus acquire the ability to infect the same host efficiently in subsequent rounds of infection. In these series of events, bacterial DNAs remain protected because of previous modification of the DNA sequences by methylase. This molecular scenario for the host range specificity by Arber and Dussoix (1962) was confirmed by the discovery of bacterial restriction-modification enzymes (Messelson and Yuan, 1968; Linn and

Arber, 1968; Smith and Wilcox, 1970; Kelly and Smith, 1970; Messelson et al., 1972; Smith and Nathans, 1973; Gromkova et al., 1973; Reiser and Yuan, 1977; Gromkova and Goodgal, 1976). The restriction endonucleases widely differ in terms of subunit composition, cofactor requirements, and interaction with DNA substrate in addition to their specificity for nucleotide sequence as sites of recognition and of cleavage or modification (Smith, 1979). They are classified under three categories. These are: type I restriction endonuclease, type II restriction endonuclease, and type III restriction endonuclease. Among these three types of restriction endonuclease, type I and type III can be designated as ATP-dependent restriction endonucleases and the type II enzymes can be referred to as ATP-independent restriction endonucleases. The physiological roles of the ATP-dependent restriction endonucleases in biological restriction and modification have been genetically identified. The physiological role of the ATP-independent endonucleases remains to be elucidated. However, the discovery of this group of restriction endonucleases has revolutionized the molecular biology by facilitating the physical mapping and nucleotide sequencing of DNA segments and their amplification by molecular cloning. The properties of the three types of restriction enzymes are compared in Table 4.1.

I. OCCURRENCE, CLASSIFICATION, AND THEIR GENERAL PROPERTIES

All three classes of restriction endonucleases have been known to occur in *Escherichia coli*, *Hemophilus influenzae*, *Bacillus subtilis*, and other bacteria. *Salmonella typhimurium* SA and SB systems contain restriction endonucleases which are allelic to that of the *E. coli* B and K system (Bullas et al., 1975). In addition, *S. typhimurium* has been shown to possess an additional restriction endonuclease system called LT. Besides these, the occurrence of restriction endonuclease has been demonstrated in almost all groups of bacteria. These include *Corynebacterium diphtheriae*, *Pseudomonas arignosa*, *Staphylococcus aureus*, *Streptococcus faecalis*, and *Rhizobium leguminosarium*. No eukaryotic endonucleases have been found to possess the properties of the restriction endonucleases. However, certain site-specific recombinases such as homing endonucleases which mediate mating type switch or intron mobility in yeast possess features of restriction endonucleases; the properties of these enzymes are discussed in Chapter 7. In addition, an enzyme with some similarity in its activity to type 1 restriction endonuclease has been described to occur in *Chlamydomonas* (Burton et al., 1977).

A. Different Restriction Endonucleases and Their Properties

These deoxyendonucleases are capable of recognition of a specific oligonucleotide sequence and of making a cleavage within or beyond the recognition site in a DNA molecule. They cannot make cleavage in DNA in the absence of a recognition site

Table 4.1 Characteristics of restriction endonucleases

Features	Type I	Type II	Type III
1. Nature of enzyme containing restriction modification activities	Single multifunctional enzyme	Separate endonuclease and methylase	Single multifunctional enzyme
2. Molecular weight of the enzyme	450 kDa	11–90 kDa, usually 20–30 kDa	200 kDa
3. Protein structure	3 Different subunits	2 Distinct proteins	2 Different subunits
4. Numbers of genes involved	3	2	2
5. Location of genes	Chromosomal	Chromosomal or plasmid	Chromosomal or plasmid
6. Kinds of mutants available	r^-m rm^- (but no r^-m^-)	r^-m rm^- r^-m^-	r^-m rm^- r^-m^-
7. Requirement for restriction	Ado-met, ATP, Mg^2	Simple Mg^2	ATP, Mg^2 (adomet), stimulating but not required
8. DNA degradation after growth in ethionine medium	Yes	Yes	Yes
9. Binding of enzyme DNA complex to filter	Yes	No	No
10. Sequence of host specificity sites	TGA(N)TGCT AAC(N)GTGC	Twofold symmetry	SP^1 AGCC SP^{13} CAGCAG Hinf III CGAAT
11. Cleavage sites	Almost random 1000 bp from recognition site	Within recognition site	24–26 bp to the 3′ of recognition site
12. Enzyme turnover	No	Yes	Yes
13. DNA translocation	Yes	No	No
14. Requirement for methylation and site of methylation	Recognition site	Recognition site	Recognition site
15. Restriction versus methylation	Mutually exclusive	Separate reactions	Simultaneous

or when the latter is modified by appropriate methylation. Three types of restriction endonucleases have been described based on the properties of the endonucleases as summarized in Table 4.1. To date, over 3300 class II site-specific restriction endonculeases have been described (Roberts and Macelis, 2001) from over 230 bacterial strains; one out of every four bacterial strains have been found to possess one or more restriction endonucleases. Among 3333 restrictions endonucleases, 531 are commercially available. Unlike most enzymes, which are named based on the nature of the reaction catalyzed and the nature of substrates, the nomenclature of the restriction endonucleases in general is based on their origin— that is, the genus and species and strain of bacteria and not on the nature of substrate that they act on (Smith and Nathans, 1973). For example, the name of a particular restriction enzyme HindII refers to the fact that it has been obtained from *Hemophilus influenzae* of strain d, and the number II refers to the fact that it is the second enzyme obtained from that strain of this bacterium. Most of the site-specifc restriction endonucleases act on DNA containing a sequence of four, six, eight, or more nucleotides. These have been divided into different subgroups based on the nature of the recognition sequences (Roberts, 1982). The different subgroups include:

1. Enzymes with palindromic recognition sequence. These may consist of:
   a. Tetra- or hexanucleotide sequences with an internal AT, GC, CG, or TA palindrome such as DpnI (G*AT*C), AluI (G*AC*T), HhaI (G*CG*C), EcoRI (GA*AT*TC), or KpnI (GG*TA*CC).
   b. Pentanucleotide sequences—for example, Hha II (GANTC).
   c. Sequences of longer extension with internal (N)X sequences—for example, XmnI (GAAN*NNN*TTC).
2. Enzymes with nonpalindromic sequences—for example, Gsu I (CTCCAG) or Bth II GGATC.
3. Enzymes with unknown recognition sequences—for example, Cer I.

These enzymes may further differ in their capabilities to make a cleavage with blunt ends or with protruding 5′ or 3′ ends.

The type I enzymes include *E. coli* K and *E. coli* B systems whereas the endonucleases of *Hemophilus influenzae* and of prophages PI and P15 represent the enzymes of group III. The type I enzymes are complex multifunctional proteins capable of both cleavage and modification of a DNA molecule with a particular recognition site. The type I enzyme consists of three subunits with a molecular weight of 400 kDa or greater. The type I enzyme requires Mg^{2+}, ATP, and SAM as cofactors for cleavage reaction, and it acts as ATPase during this process. The methylation reaction requires SAM only but is stimulated by Mg^{2+} and ATP. The *E. coli* K system consists of a single complex of multifunctional enzymes in which the different subunits together catalyze both enzymatic reactions. In the *E. coli* B system, however, two distinct subunits have been shown to control the cleavage and methylation reactions. While the type I enzyme acts directly on the recognition site, the cleavage is a rather random process often between 1 and 5 kb away from the recognition site.

The type III enzymes are of intermediate complexity and consist of subunits of about 250-kDa molecular weight. This group of enzyme requires Mg^{2+} *S*-adenosyl methionine and ATP but strictly lacks ATPase activity. The methylation reaction requires SAM only. The cleavage site of the type III enzyme is roughly 10–20 nucleotides away from the recognition site.

In comparison to type I and type III enzymes, the type II enzymes are simpler with molecular weights of 80 kDa or less. Mg^{2+} alone is enough for cleavage activity, and only SAM is required for the methylation activity. The nuclease and methylase functions are controlled by distinct proteins. Also, the enzymes make cleavage or methylation within the recognition sequences. The enzyme has no ATPase activity and does not require ATP for any of its activities.

II. TYPE I RESTRICTION ENDONUCLEASES

A brief description of the purification, properties, cleavage mechanism, and genetics of this group of enzymes is presented here.

A. Purification and General Properties

The EcoK enzyme was first purified by Meselson and Yuan (1968) via a combination of methods. Both the restriction and modification activities copurify in a protein of 400-kDa molecular weight consisting of three subunits, namely, α, β, and γ. The EcoB enzyme purified in somewhat different manner has been resolved into distinct proteins containing either endonuclease or methylase activities. The EcoB enzyme consists of several forms; however, the predominant form containing both the endonuclease and methylase functions consists of three subunits. The physical properties of these enzymes are summarized in Tables 4.2 and 4.3.

B. Recognition Sequences and Nature of Substrate

The recognition sequence of EcoB consists of

1 2 3 4 5 6 7 8 9 10 11 12 13 14 15
5′ T G A* N N N N N N N N T G C T 3′
3′ A C T N′ N′ N′ N′ N′ N′ N′ N′ A* C G A 5′

Thus the recognition sites consist of three domains:

1. Trimer (TGA) domain
2. Octamer domain of variable sequences
3. Tetramer (TGCT/ACGA) domain

Table 4.2 Comparison of the properties of type I enzyme

Features	EcoK System	EcoB System
1. Enzymatic reaction catalyzed	Endonuclease methylase and ATPase	Endonuclease methylase and ATPase
2. Subunit	α 135 kDa (hsdR) β 62 kDa (hsdM) γ 52 kDa (hsdS)	α 135 kDa β 60 kDa γ 55 kDa
3. SW20	12	11–18
4. Subunit	$\alpha_2\beta_2\gamma$ (450 kDa)	$\alpha\beta\gamma$ (250 kDa) $\alpha_2 P_4 \gamma_2$ (450 kDa)
5. Relation of different enzymatic functions to different subunits	All three subunits (i.e., α, β, γ) together catalyze endonuclease, ATPase, and methylase function	Endonuclease and ATPase function require participation by all three subunits. The methylase function is controlled by β and γ subunits.
6. Recognition site	A A C $(N)_6$ G T G C T T G $(N)_6$ C A C G	T G A $(N)_8$ T G C T A C T $(N)_8$ A C G A
7. Modification product	Not precisely known but adenine residue must be involved because methylation product is invariably a 6-methyl adenine	
8. Mutants	Restrictionless (r^-m) mutant defective in hsdR is known; hsdS mutation yields mutant defective for both restriction and modification	Similar mutants known

A single base substitution mutation in either the first or third domain or insertion in second domain can interfere with the recognition of the DNA site and abolish its sensitivity to both the restriction and modification functions of the enzymes.

Once a site is recognized and the enzyme–substrate complex is formed, it may be resolved in the following three ways:

1. An enzyme complex formed with fully methylated DNA site does not proceed any further and is instead broken off as shown below:

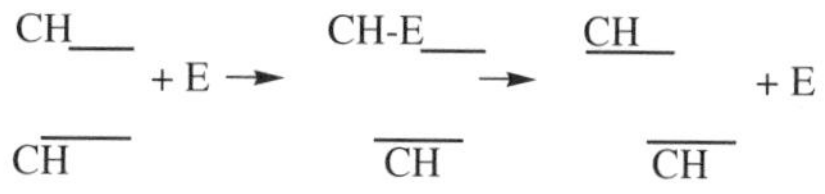

```
CH___              CH-E___         CH___
        + E →                →               + E
CH‾‾‾              ‾‾‾CH           ‾‾‾CH
```

Table 4.3 Properties of the restriction type I enzymes

System	Enzyme	Subunit	SW20	Proposed Structure	Approximate Molecular Weight	Gene	Mutation
EcoK	Endonuclease and methylase	α 135 kDa (hsdR)	12	αβγ	450 kDa	hsdR[b]	r^-m^+
		β 62 (hsdM)				hsdM	r^-m^+ mutant not known[a]
		γ 52 kDa (hsdS)[a]				hsdS[b]	
EcoB	Endonuclease	α 135 kDa	11–18	24αβγ	450 kDa		r^-m^-
		β 60 kDa		2	750 kDa		
		γ 55 kDa[a]		αβγ	250 kDa		
	Methylase	β 60 kDa	6.6	βγ	105 kDa		
			11.3	βγ	240 kDa		
			8	βγ	175 kDa		

[a] Inhibited by T3 gene 0.3.
[b] HsdM encodes a methylase function; hsdS determines the site recognition for specificity for both restriction and modification function. HsdR encodes restriction function.

2. An enzyme complex formed with a partially methylated DNA is a good substrate for methylation (but not restriction) and readily undergoes methylation in the presence of SAM.

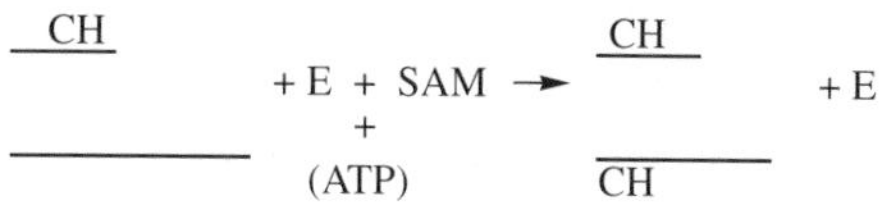

CH + E + SAM → CH + E
\+
(ATP) CH

3. An enzyme complex with unmethylated DNA is a good substrate for restriction.

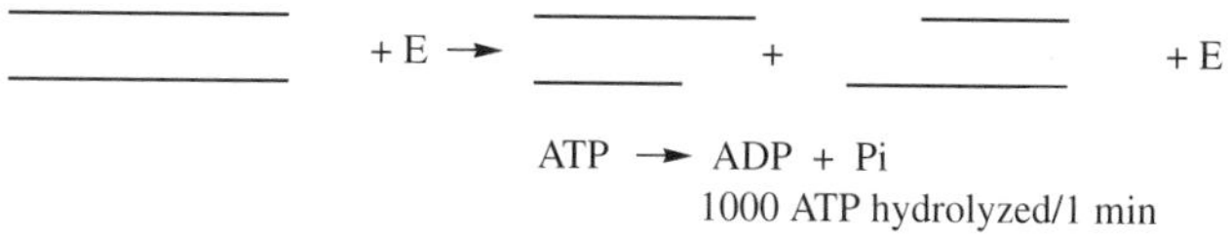

\+ E → + + E

ATP → ADP + Pi
1000 ATP hydrolyzed/1 min

ATPase function continues for hours even after endonucleolytic activity of the enzyme has stopped.

The recognition sequence of EcoK is

A A G N N N N N N G T G C
T T G N N N N N N C A C G

Both strands of EcoB and EcoK recognition sites are methylated *in vivo*. The methylation probably occurs as shown below. The product of methylation is a DNA chain containing 6-methyladenine.

EcoB	EcoK
A A C(N 6) G T G C	T G A* (N 8) T G C T
T T G (N 6) C A C G	A C T (N 8) A* C G A

In EcoB the methylation site has not yet been precisely determined. However, it is assumed that it must involve adenine residues because the methylation product is invariably 6-methyladenine.

C. Different Kinds of Type II Restriction Endonucleases

Most type II restriction endonucleases are homodimers with a molecular weight of approximately 2×30 kDa. They typically recognize a palindrome of a 4–8 base pair and then cause cleavage generating sticky or blunt ends. However, there are certain variations in their properties, and based on that they include several subgroups such as:

- Type IIS represented by FokI recognizes an asymmetric sequence and causes cleavage within it.

- Type IIE represented by NaeI recognizes two copies of recognition sequences but makes a cut within one site; the other site functions as an allosteric site only.
- Type IIF represented by NgoIV interacts with two recognition sites like NaeI but makes a cut in both sequences.
- Type IIT represented by Bpu I0I and BslI is made up of different subunits.
- Type IIB represented by BcgI or BplI is comprised of different subunits and include methylase activity and require AdoMet for the restriction activity.
- Type IIG represented by Eco571 contains both restriction and modification activities in a single polypeptide and requires AdeMet for enzymatic activity.
- Type IIM represented by DpnI causes cleavage in a methylated DNA.

D. Genetics

Extensive genetic analysis of the restriction and modification systems in *Escherichia coli* and in *Salmonella typhimurium* suggested their allelism in these two different organisms and their control by a cluster of three genes designated as hsdR, hsdS, and hsdM. The hsdR and hsdM genes control the restriction and modification functions, respectively. The hsdS gene confers the specificity for recognition site and thus controls both the restriction and modification functions and the host specificity. The restriction modification loci map left to *serB* at about 98 min in the *E. coli* map. The *S. typhimurium* SA and SB systems map at 98 min near *pyrB* locus. Transduction experiments have demonstrated the allelism of the *E. coli* K, *E. coli* B, and *S. typhimurium* SA and SB systems and have the gene order to be hsdM-hsdS-hsdR-SerB.

Only two out of the three theoretically possible mutant phenotypes have been demonstrated to occur. These include (a) restrictionless mutant (r^-m) defective in hsdR gene and its product and (b) restriction and modificationless (r^-m^-) defective in hsdS gene and its product. No mutants defective in hsdM gene leading to a deficiency in modification function alone have yet been known because such mutants will be lethal. It is possible to isolate temperature-sensitive hsdM mutants, but none has been reported so far. The presence of hsdM gene has, however, been confirmed by the stepwise derivation of rm mutants from rm bacteria. In such studies the double r^-m^- mutant (due to defective hsdR and hsdM genes) can be distinguished from single rm^- mutant (defective in hsdS gene alone) because a cross between them (i.e., hsdR$^-$ hsdM$^-$ hsdS $\times$ hsdR hsdM hsdS$^-$) restored the rm (hsdR hsdM hsdS) phenotype among the progeny. The existence of these three loci (i.e., hsdR, hsdM, and hsdS) has been further confirmed by the *in vitro* complementation of the mutant function by the purified subunits of the enzyme. Results of these studies clearly suggest that hsdR and hsdM genes are involved in controlling restriction and modification function whereas the hsdS gene is required for the identification of recognition site and host specificity. The latter view has been demonstrated by the fact that a novel recombinant host specificity designated as SQ was obtained among progeny of the cross between allelic *S. typhimurium* SA and *S. podam* SP host specificity loci (Bullas et al., 1975, 1980).

E. Cleavage Mechanism

The peculiar feature of the type I restriction endonucleases lies in the fact that the cleavage occurs at a random site almost 1 kb to 5 kb away from the site of recognition. The enzyme, while introducing a single-stranded, nick, is inactivated as a nuclease and instead is transformed into an ATPase. Thus the enzyme can turn over only as an ATPase and not as a nuclease. Two enzyme molecules are required for DNA cleavage. The DNA cleavage involves binding of the enzyme with the outflanking trinucleotide and tetranucleotide sequences of the recognition site. The enzyme remains bound to the tetranucleotide region while the DNA chain on the trinucleotide side of the recognition site is pumped randomly through the enzyme. A single-stranded nick is made randomly after translocation of a chain of an indeterminant size varying in length somewhere from 1 kb to 5 kb. ATP is not required for the formation of a recognition complex but for the initiation of the cycle of events leading to DNA cleavage. The SAM may act as an allosteric effector and thus modify the course of enzymatic action; in the absence of SAM, however, the nuclease causes the DNA cleavage.

During the process of restriction, the enzyme as well as the recognition site is inactivated, because the recognition site cannot be used for another cleavage reaction with a fresh enzyme molecule. The enzyme molecule that participated in DNA cleavage reaction is transformed into ATPase as mentioned above in this section. The enzyme remains functional as an ATPase for a long time. The ATPase function of the type I restriction endonuclease has been suggested to be an altruistic approach by bacteria to save a bacterial population at the expense of a few individuals (Bickle, 1982).

III. TYPE II RESTRICTION ENDONUCLEASES

A. Enzyme Purification and Assay

Restriction endonuclease are purified by methods first described by Smith and Wilcox (1970) for the purification of Hind II. Generally a high-speed supernatant of cell lysate is treated with streptomycin sulfate or subjected to gel filtration to remove nucleic acid and then further purified by phosphocellulose or DEAE cellulose chromatography. The enzyme so obtained is reasonably pure and free of contaminating nonspecific nucleases (Greene et al., 1978). Such enzyme preparation can be used for enzyme assay by generating a discrete banding pattern of λ or other viral or plasmid DNA analyzable on an agarose gel electrophoresis. Only in a few cases has enzyme been purified to homogeneity. Such partially purified enzymes are also free of proteases because most enzymes can remain active up to 12 h in an assay mixture.

Initially the activity of restriction endonucleases was assayed by the ability of the enzyme to degrade foreign DNA as determined by a change in the viscosity of DNA or as a loss in the biological activity of DNA during transfection experiments.

The enzyme unit is usually defined as the amount of enzyme required to obtain complete digestion of 1 μg of λDNA in 1 h at 37°C.

B. General Properties of the Enzyme

Restriction endonucleases in general are very stable. Besides Mg^{2+}, most enzymes require the proper ionic environment (6–100 mM Tris and 0–100 mM NaCl) and sulfhydryl agent (MeSOH or DTE) for enzymatic reaction at optimal level. The addition of bovine serum at 50–100 μg/ml is recommended to maintain the stability of enzymes during storage and DNA cleavage reactions. Most enzymes have an optimal activity at 37°C; however, some enzymes such as Bst E1 are heat stable and can act at a temperature of 60°C.

All type II restriction endonucleases make a cut within the recognition site (Roberts, 1982; Rosenberg et al., 1987; Pingoud and Jeltsch, 2001). Restriction endonucleases obtained from different organisms but capable of identifying and causing cleavage at a specific site within identical nucleotide sequences are called *isoschizomers* (Roberts, 1976); those causing cleavage at different sites within identical nucleotide sequences are called *neoschizomers*. Based on isochizomeric properties, over 400 restriction endonucleases have been assigned to 25 different groups (Roberts, 1987). Restriction endonucleases make either a staggered cut with protruding ends or cleavage with blunt ends. A majority of enzymes make staggered cuts with 5′ protruding ends—except certain enzymes such as Pst-1, which makes a 3′ protruding end. Most restriction endonuclease cleavage products consist of 3′-OH and 5′-PO_4 groups—except certain other enzymes such as Nca, which causes cleavage with 3′-PO_4 and 5′-OH groups. Some of the kinetic parameters showed differences depending upon the nature of the substrate used: Both EcoRI and BamHI showed different K_m values with supercoiled or linear DNA as substrates (Halford et al., 1979; Halford 1983, Hinsch et al., 1980). The kinetic parameters are, however, expected to vary with the nature and size of the substrate and number of recognition sites within the substrate. EcoRI enzyme has a much slower turnover rate than BamH1.

The only restriction endonuclease that causes cleavage in methylated DNA is DpnI (Lacks and Greenberg1975).

C. Crystal Structure of the Restriction Endonucleases

The 3-D structure of a number of restriction endonucleases have been determined (Pingoud and Jeltsch, 2001). All of the restriction enzymes analyzed have been found to possess a structural core consisting of four beta strands and an alpha helix. Howevever, they belong to two groups of enzymes with characteristic structures typical of EcoRI and EcoRV. The structural core contains the catalytic center. The EcoRI-like enzymes bind on the major groove side of the substrate DNA, causing cleavage and producing cohesive ends with 5′overhangs. The EcoRV-like enzymes bind the substrate DNA on the minor groove side, causing cleavage with blunt ends.

This structural core of the restriction endonucleases have been found to be conserved in a number of other nucleases such as λ-exonuclease, Mut H (involved in mismatch repair), Vsr endonuclease (involved in TG mismatch repair), and the TnsA subunit of the Tn7transposae. The N-terminal and the C-terminal halves of Sau3AI shows certain sequence similarity suggesting that this enzyme is a pseudodimer. The crystal structure of NaeI reveals an evolutionary relationship between restriction endonucleses and topoisomerases in view of the fact that this restriction endonuclease becomes a topoisomerase after the change of just one amino acid at the 43 position. A substitution L43K leads to conversion of NaeI with restriction endonuclease activity to topoisomerase activity. This is consistent with the finding that a 10-amino-acid sequence in N-terminal domain of NaeI shows similarity with the active site of DNA ligases and that the NaeI enzyme forms a covalent intermediate with the substrate DNA during the process of the cleavage (Huai et al., 2000). A number of other minor variations are seen in different nucleases that are of evolutionary significance (Pingoud and Jeltsch, 2001).

D. Reaction Conditions and Enzyme Specificity

Changes in ionic strength, divalent metal ions (Mg, Mn, Co, and Zn), pH, hydrophobic solvents (glycerol, dimethyl formamide, and dimethyl sulfoxide), temperature, and enzyme concentration have been shown to influence the rate and specificity of the reaction catalyzed by several restriction endonucleases. EcoRI has been shown to change its recognition specificity from the canonical sequence GAATTC to NAATTN under conditions of low ionic strength (2 mM $MgCl_2$) and high pH (8.5). The enzyme with new specificity has been designated as EcoRI* activity. Replacement of Mg^{2+} by Mn^{2+} as well as the presence of 40–50% glycerol or several other solvents can also increase the EcoRI* activity of the enzyme. EcoRI* activity was also seen to occur under low enzyme concentration of EcoRI in a standard reaction mixture. Under low enzyme concentration, EcoRI failed to make double-stranded cleavage but instead made only specific nick(s) in the same strand (Bishop, 1979).

The endonuclease Bsu has been shown to lose its cleavage specificity within the canonical sequence GGCC at higher enzyme concentration in a standard reaction mixture (Heininger et al., 1977). The changed specificity of Bsu* activity which recognized the central GC dinucleotide sequence was enhanced at high pH (8.5), low ionic concentration, and high glycerol concentration. The Bsu* activity was inhibited by NaCl. In addition, the Bsu* activity was found to act on single-stranded viral DNA. However, Hae, an isochizomer of Bsu, did not show any star enzyme activity under conditions that induced star activity in Bsu. These studies clearly indicated that the mechanisms of DNA protein interaction, recognition, and cleavage differ even among the isoschizomers (Heininger et al., 1977). Similar star enzyme activity has been reported for the enzyme BstEI in the presence of high enzyme concentration or glycerol.

Unlike the induction of star activity as shown for EcoRI, Bsu, and BstEI, BamHI shows a complete change in its activity in the presence of hydrophobic agents such as glycerol, DMSO, and ethylene glycol (George et al., 1980). In view of this fact, BamHI behaves in a rather unusual manner, although such drastic change in the behavior of an enzyme is not without precedent. Similar changes in the specificity of enzymatic reactions catalyzed by *E. coli* and *M. luteus* DNA polymerases have been shown to occur under a variety of conditions (van de Sande et al., 1972; Tamblyn and Wells, 1975; Battula and Loeb, 1974). Some of the aberrant behavior of DNA polymerases includes the incorporation of ribonucleotides instead of deoxyribonucleotides, utilization of ribonucleotide or primer template, and incorporation of wrong nucleotides.

E. Nature of Substrate

There is evidence that the restriction endonucleases use different kinds of substrates so long as they contain a particular canonical sequence. These include (a) synthetic oligonucleotides, (b) DNA with base analogs, (c) methylated DNA, (d) single-stranded DNA, and (e) DNA–RNA hybrids.

1. Synthetic Oligonucleotides. It has been shown that EcoRI, HpaI, HpaII, and MnoI can act on synthetic duplex oligonucleotides varying in length from 6 to 13 base pairs. HpaII and MnoI both recognize the same tetranucleotide sequence of CCGG even though they cut at different places. EcoRI and HpaI were found to cleave nucleotide d(pTGAATTCA) and d(GGTTAACC), respectively, containing the corresponding canonical sequences for recognition. In both cases, cleavage of octamers required that they must be base-paired.

2. DNA with Base Analogs. Many of these studies have been carried out with EcoRI and its recognition site GAATTC. Replacement by 2-amino-purine, a constituent of minor groove, had no effect on the cleavage of DNA by EcoRI. Likewise, substitution by 5-hydroxymethyl cytosine had no effect on the action of EcoRI. However, glycosylation of 5-hydroxymethyl cytosine made the DNA resistant to cleavage by EcoRI. These studies suggest that a change in major groove interfered with the action of EcoRI.

Likewise, DNA containing 5-bromouridine substitution was found somewhat resistant to cleavage by a number of restriction endonucleases (EcoRI, HaeII, HpaI, HpaII, BamHI, and HindIII). In contrast, the activity of MboI was increased fivefold on 5-bromouridine substituted DNA as substrate. Among the five EcoRI sites present in λ phage DNA, there is a tenfold difference in the rate of cleavage of the most preferred to least preferred sites; Thomas and Davis (1975) have shown that there is no adverse effect of BudR substitution on the pattern of λDNA digestion by EcoRI.

3. Methylated DNA. Interaction of restriction endonuclease with methylated DNA can provide a basis for the understanding of the biological basis for relationship between modification and restriction. Isoschizomers may act differentially on methylated DNA as substrate. HSpaII cannot cleave a methylated DNA, whereas its isoschizomer MSpI can act on DNA whether methylated or not.

4. Single-Stranded DNA. Restriction enzymes are usually considered to act on double-stranded DNA. However, it was discovered that HaeIII cleaved single-stranded φX174, M13, and f1 viral DNA. EcoRI was found to act on synthetic single-stranded octamer oligonucleotide sequences (Goodman, 1975). However, in all these cases it was shown that the restriction enzyme was active only with DNA from a duplex structure due to secondary foldback association, confirming the idea that restriction endonucleases act only on a duplex structure.

5. DNA–RNA Hybrids as Substrate. A number of restriction endonucleases have been shown to act on a heteroduplex structure of DNA–RNA (Molloy and Symons, 1980). Such cleavage required 20- to 50-fold excess of the enzyme required for cleavage for duplex DNA. HaeIII has been reported to make a cut within a DNA duplex containing ribocytidine residue.

$$5' \cdots \text{dG dG rC rC} \cdots 3'$$
$$3' \cdots \text{dC dC dG dG} \cdots 5'$$

F. Inhibition of Restriction Endonucleases

DNA cleavage by restriction endonuclease is promoted in the presence of sulfhydryl agent. Therefore an analog of sulfhydryl agents can inactivate the activity of restriction endonucleases. DNTB and *p*-mercuribenzoate have indeed been found to inhibit a number of restriction endonucleases (PvuI, HindIII, AvaI, SmaI, PstI, and SstII). The fact that methyl acetamide inhibited the activity of EcoRI suggested the involvement of lysine residue in the DNA cleavage by this restriction endonuclease (Woodhead and Malcolm, 1980). Both EcoRI and BamH1 have been shown to act as the competitive inhibitors of EcoRI cleavage of pBR322 DNA. In addition, a number of ligands that bind DNA specifically can also inhibit DNA cleavage by restriction endonucleases. Prior binding of λ or adenovirus DNA by DAPI, distamycin A, or actinomycin D preferentially inhibited the cleavage of certain specific sites (but not all sites) by EcoRI and other restriction enzymes. In addition, bacteriophages produce antirestriction protein as their defense mechanism against host-mediated restriction (Spoerel et al., 1979; Mark and Stuier, 1981). Some of the antirestriction mechanisms are listed in Table 4.4.

G. Restriction Endonuclease Genes

Almost 1% of the genome in prokaryotes encodes for restriction and modification systems that keep these organisms protected from foreign invading DNA including bacteriophages. With the completion of the genomic DNA sequences of many organisms, it has become possible to identify the genes encoding methylases (MTase) required for the modification of DNA. The genes for methylases can be identified by the presence of nucleotide sequence coding a motif of 6–12 amino acids which is typically found in the methylase proteins. Once a gene for methylase (MTase) is

Table 4.4 Antirestriction mechanism

System/Organism	Resistance to	Gene Involved	Comment
1. Mu	Types I and III	mom^+	$mom^+ + dam^+$ modified DNA
2. T_3 and T_7	Type I	0.3 gene	SAMase
3. T_5	Types I, II, and III	A gene in FTS DNA	Mutant sensitive to EcoRI because of EcoRI site in the FST (first step transfer) DNA; Wild-type FTS has no EcoRI site, but the rest of the genome has six EcoRI sites
4. λ	Type I	*ral* (early non-essential gene)	Ral and Ral^- equal plating efficiency but among progeny of primary infection; Ral^+ fully modified/Ral^- partially modified
5. P1	Type 1	*dar* (par) (non-essential gene)	*dar* gene product binds with DNA and protects from restriction

located, the gene for restriction endonucleases (Enase) could be identified by their presence close to the methylase genes. Most of the time restriction endonuclease genes are identified as ORF (open reading frames) close to the location of genes encoding methylases, even though there is not much homology in the DNA sequence of the restriction endonucleases (Enases) with a particular isoschizomeric activity.

However, the restriction endonucleases (Enase), like the methylases (MTase), show greater homology in the sequences responsible for the substrate recognition (Jeltsch et al., 1995), suggesting that both MTases and Enases evolved from a common ancestor capable of modification and restriction of DNA. In several instances as in *Neisseria gonorrhoeae*, the restriction enzyme sequences that remain unexpressed in the native bacterium could be identified via its expression after transfer into *E. coli*.

Genetic evidence for the presence of restriction endonucleases has been gathered in a number of instances. Genes for different restriction endonucleases have been located either on the plasmid DNA or on bacterial chromosome (Miller and Cohen, 1978) as summarized in Table 4.5. The gene for HhaII was the first one to be cloned (Mann et al., 1978). Since then a number of other genes for restriction endonuclease (over 60 genes) and modification enzymes (over 100 genes) have been cloned (Wilson, 1991); some of these include EcoRI, EcoRII, PstI, and Bsu. A 2.2-kb DNA segment controlling the EcoRI restriction and modification phenotypes has been cloned (Rubin et al., 1981; Greene et al., 1981), and its complete nucleotide sequence has been determined; mutagenesis experiments suggest that Glu 111 is critical for DNA cleavage activity but not for DNA binding (King et al., 1989).

Table 4.5 Genetic location of certain restriction endonucleases

Enzyme	Location
Type I	
EcoB	Chromosomal
EcoK	Chromosomal
Type II	
EcoRI	Plasmid
EcoRII	Plasmid
BsuI	Chromosomal
BsuII	Chromosomal
BsuIII	Chromosomal
BsuIV	Chromosomal
BsuV	Chromosomal
DpnI	Chromosomal
Type III	
EcoP	P1 plasmid
	P15 plasmid
Hinf	Chromosomal

Though both enzymes (endonuclease and methylase) interact with the same nucleotide sequence (GAATTC) during the process of restriction and modification of a DNA segment, there is no similarity in the amino acid sequence of these two proteins. Thus the mechanism of interaction of the two proteins with the same DNA sequence remains elusive. However, some insight is provided by the analysis of the crystal structure of the enzyme substrate complex. It is shown that EcoRI restriction endonuclease causes a kink in the DNA structure while the interaction of HhaI methyl transferase causes a flip of the base in the DNA structure and thus exposes the target sites (Becker et al., 1988; Kim et al., 1990; Klimasauskas et al., 1994; Verdine, 1994). The essential features of restriction endonucleases and methyltransferases lie in the fact that they act as dimer and monomer, respectively. The cloned EcoRI gene when introduced in yeast has been found to enter nucleus of the host cell and make changes in yeast DNA, causing the death of the cells (Barnes, 1985).

A detailed genetic analysis of five different restriction endonucleases has been carried out in *Bacillus subtilis*. Genetic segregation of loci encoding different restriction endonuclease has been utilized for the purification of a particular restriction endonuclease. A genetic approach has been also used to determine or to create and eliminate restriction sites for a particular enzyme in a bacteriophage (Murray and Murray, 1974) or other DNA molecules. Little and Mount (1982) have suggested methods to create silent mutations in a DNA sequence so that it can be converted into a restriction site for a particular restriction endonuclease. The generation of silent mutation is based on the degeneracy of genetic code.

IV. TYPE III RESTRICTION ENDONUCLEASES

Initially, type III restriction endonucleases were classified as type II restriction endonucleases. However, it soon became obvious that the substrate requirement for these enzymes was much different than for those belonging to type II enzymes (Kauc and Piekarowicz, 1978). These restriction endonucleases were therefore included in a new group designated as type III.

There are three type III restriction enzymes reported to this date. These are EcoP1, EcoP15, and HinfIII. EcoP1 is coded by *E. coli* prophage P1; EcoP15 is coded by a plasmid resident in *E. coli* 15. The HinfIII is coded by a chromosomal gene of *Hemophilus influenzae* Rf. The genes encoding EcoP1 and EcoP15 are allelic (Arber and Wausters-Willems, 1970).

V. EVOLUTIONARY SIGNIFICANCE AND BIOLOGICAL ROLE

Yuan (1981) has suggested an evolutionary scheme for the three different types of restriction endonuclease based on their characteristics. The type II restriction endonucleases represent the simplest form among them because the restriction and modification phenotypes are each controlled by a distinct gene. The enzyme recognizes and cleaves the same sequence of nucleotides except for a few enzymes, such as HpaI, MboII, and HgaI, which represent a subgroup between the type I and type II restriction endonucleases. This subgroup recognizes a specific nucleotide sequence but makes a cleavage in a sequence ordinarily 8–9 base pairs away from the site of recognition. All type II enzymes have a minimal requirement for a divalent cation (Mg^{2+}) for enzymatic reaction. The type II enzymes do not require ATP, nor is any ATP hydrolase activity associated with the type II restriction endonucleases.

This group of enzymes is also coded by two genes: one for the restriction and the other for the modification polypeptides. Almost all type II restriction endonucleases act as dimers during DNA sequence recognition and cleavage. However, unlike most restriction systems, in FokI the recognition and cleavage functions are determined by two distinct domains of the same polypeptide (Li and Chandrasegaran, 1993; Kim and Chandrasegaran, 1994). The enzymes recognize and methylate a specific site, but cleavage is made within a sequence of 5–6 base pairs which lies 24–27 bases away from the site of recognition. It is possible to alter the distance between recognition and cleavage sites (Li and Chandrasegaran, 1993) or to create restriction endonuclease with a specific recognition site different from the recognition site of the native enzyme (Kim and Chandrasegaran, 1994). In this evolutionary scheme, next comes the restriction endonuclease type III. The DNA cleavage requires ATP, whereas ATP can only stimulate methylation. The restriction and methylation reactions seem to compete with each other and are seemingly modulated by the level of adenosylmethionine.

Type I restriction enzymes seem to be somewhat more complex. Their system is controlled by three genes, one each for specific recognition, restriction, and methylation. However, unlike the type III enzyme, DNA cleavage and methylation reactions are mutually exclusive. The system is also known to engage in DNA translocation, which is related to its ATP hydrolase activity. The DNA translocation, however, is brought to a halt by the DNA cleavage event. Yuan (1981) has further suggested a particular configuration of the recognition site (called cage model) in which the internal nucleotide sequence may be matched with the translocated DNA to identify the cleavage site that is ordinarily several thousand base pairs away from the site of recognition and/or methylation.

The role of type I restriction endonucleases and some of the type II restriction endonucleases such as EcoRI and BsuR in controlling the host range specificity is very well elucidated. These enzymes limit the phage infection, transduction, and bacterial conjugation besides the DNA uptake by sphaeroplasts. The role of the other type II and type III enzymes in site-specific recombination has been invoked (Roberts, 1976). It is further suggested that type I enzymes may play a role in generalized recombination besides their involvement in determining the host specificity (Endlich and Linn, 1981).

VI. APPLICATION OF RESTRICTION NUCLEASES

Restriction enzymes are used to generate the restriction map of a DNA segment by idenytifying the site(s) of digestion; these sites are used as markers; loss or gain of restriction site due to a mutation in the DNA fragment containing a gene or within a gene could lead to polymorphism in the length of the DNA fragments of the wild-type and mutant DNA generated after digestion with a particular restriction enzyme. Such restriction fragment length polymorphism (RPLP) is used as a marker in genetic analysis and in identificaion and cloning of genes causing human diseases. Hemoglobin genes in normal human and those suffering from sickle cell anemia were the first example of showing RFLP and paved the way for identification and cloning of several genes controlling human diseases such as the gene controlling Huntington's disease in humans, which was the first gene to be identified and cloned with respect to RFLP. Use of RFLP was instrumental in mapping of human chromosomes. The presence of a restriction site in the wild type and its absence in the mutant DNA fragment can be ascertained readily after digestion with a particular restriction enzyme and seperation of fragments of various lengths upon eletrophoresis in a gel. The location of a gene to a particular chromosomal fragment is readily established by hybridization of the clond gene used as a probe on a Southern blot (Southern, 1975).

Restriction enzyme is used to generate the map on the cDNA cloning module or that of a PCR fragment in order to decide which fragment to subclone for DNA sequencing or for cloning into protein expression vector. The fact that restriction enzymes can generate staggered ends in a DNA fragment was crucial in developing the recombinant DNA technology, which was a major factor in the molecular cloning of genes and in the identification of genes causing diseases in humans and is of commercial importance, leading to gene therapy and designer drugs and finally to genomics. The restriction enzymes were utilized for the undertaking of human genome project and other genome projects; the completion of the genome projects has led to a new chapter in understanding the evolutionary history of all organisms—particularly that of humans, which could be used for shaping our future on this planet!

VII. GENERAL TIPS FOR BEGINNERS OR THE FIRST-TIME USERS OF RESTRICTION ENZYMES

To be used as a substrate for a restriction enzyme, DNA must be free of any contaminating phenol and/or alcohol used during the extraction of DNA. Proper ionic concentraion is very critical; in addition, certain enzymes are more tolerant to higher salt concentration or to temperature than other enzymes; therefore buffers and the assay conditions should be as recommended by the commercial supplier of the enzymes. Usually a small amount of bovine serum albumin, provided by the commercial supplier, is added to overcome the problem associated from the impurities in DNA and to maintain the stability of the low concentration of the restriction enzyme in the reaction mixture. Ordinarily, 1 μg of DNA is used in a reaction mixture of 20 μl containing the restriction enzyme at a temperature 37°C in a water bath for couple of hours. Incubation of DNA with restriction enzyme for a longer period would not influence the results (i.e., the size of the DNA fragments) unless the restriction enzyme is contaminated with nonspecific DNases. If the restriction enzyme is contaminated with DNases, then a smear will appear upon electrophoresis of the DNA after incubation with restriction enzymes instead of the fragments of particular lengths generated by the pure restriction enzyme. As mentioned previously in this chapter, restriction enzyme digestion carried under improper conditions may yield fragments of sizes other than expected to result under proper conditions; such activity of a restriction enzyme is called star (*) activity.

All restriction enzymes are quite expensive and heat labile. They should be stored at low temperature as suggested by the commercial supplier; only a small sample should be dispensed with micropipette and kept on ice. Residual amount of enzyme should be kept frozen to be used later. A micropipette tip used for dispensing the enzyme into the tube containing the reaction mixture should not be used again to a dispense the same enzyme or any other enzyme or reaction mixture into the same or another tube in order to avoid costly contamination. Finally, one must not forget to add the restriction enzyme into a reaction mixture containing the DNA because without enzyme the DNA would remain intact and no expected restriction digestion pattern would be seen upon gel electrophoresis! Use of these tips will prevent loss of time, loss of money, and confusion in the interpretation of restriction digestion patterns to investigators and other collaborators working on a research project.

5

DAMAGE-SPECIFIC NUCLEASES

Nucleases that act on damaged DNA play a central role in DNA repair. At times they have been called an *excision* or *incision* nuclease depending on whether the enzyme completely removes the damaged DNA region or just makes the initial incision in the damaged DNA region, leaving its removal to other exonucleases.

I. CLASSIFICATION AND ASSAY

A recent classification of the enzyme is based on the nature of phosphodiester linkages hydrolyzed (Friedberg, 1985). Two groups of enzymes have been identified based on such criteria.

A. AP Endonucleases

This group includes enzymes that specifically act on phosphodiester linkages containing AP (apurinic or apyrimidinic) sites. This group can be further classified into:

1. AP endonucleases associated with DNA glycosylase activity. These include:
   a. T_4UV endonuclease
   b. *Micrococcus luteus* enzyme
   c. *E. coli* endonuclease III
2. AP endonucleases associated with other nuclease activity
   a. *E. coli* exonuclease III AP endonuclease activity
   b. AP endonuclease associated with exonuclease III-like enzymes from other organisms.

3. AP endonucleases
   a. *E. coli* endonuclease IV and VII
   b. *M. luteus* AP endonuclease
   c. Eukaryotic AP endonucleases characterized from *Neurospora*, yeast, and mammals including humans

B. Enzymes that Directly Attack Phosphodiester Linkages in the Damaged DNA Region

1. *E. coli* UV endonucleases
2. Endonuclease V of *E. coli*
3. Endonucleases specific for photoalkylated purines in DNA
4. Eukaryotic UV endonucleases

C. Assay

The enzyme can be assayed by measuring the radioactivity released into a TCA-soluble fraction after incubation of the radioactively labeled DNA containing DNA-damaged regions as substrate. UV endonuclease activity can be measured as the amount of pyrimidine dimers released following incubation of UV-irradiated DNA with an enzyme preparation. The AP endonuclease activity is usually measured by the release of radioactivity from DNA containing AP sites. The UVRABC nuclease activity is measured as the release of oligonucleotides by treatment of UV-damaged DNA with appropriate enzyme preparation (Sancar and Rupp, 1983). The enzyme activity is also measured as single-stranded breaks in DNA. The different methods for the assay of the enzymatic methods have been described in detail (Ciarrocchi and Linn, 1988; Hanawalt et al., 1978).

II. PROPERTIES OF TWO GROUPS OF ENZYMES FROM DIFFERENT ORGANISMS

A. AP Endonucleases

This group of enzymes acts on apurinic or apyrimidinic sites created by the loss of a base that may occur due to spontaneous hydrolysis of the *N*-glycosidic bond after a chemical modification or due to the activity of DNA glycosylases. Different kinds of AP endonucleases based on the position of the phosphodiester cleavage made by the enzyme with respect to AP site as depicted in Figure 5.1 have been described (Lloyd and Linn, 1993). The different AP endonucleases can also be distinguished based on the nature of the product of their action: class I, III, and IV enzymes produce a 3 end product that cannot be acted upon by the DNA polymerase I of *E. coli* (Mosbaugh and Linn, 1980; Warner et al., 1980a,b), whereas the class II

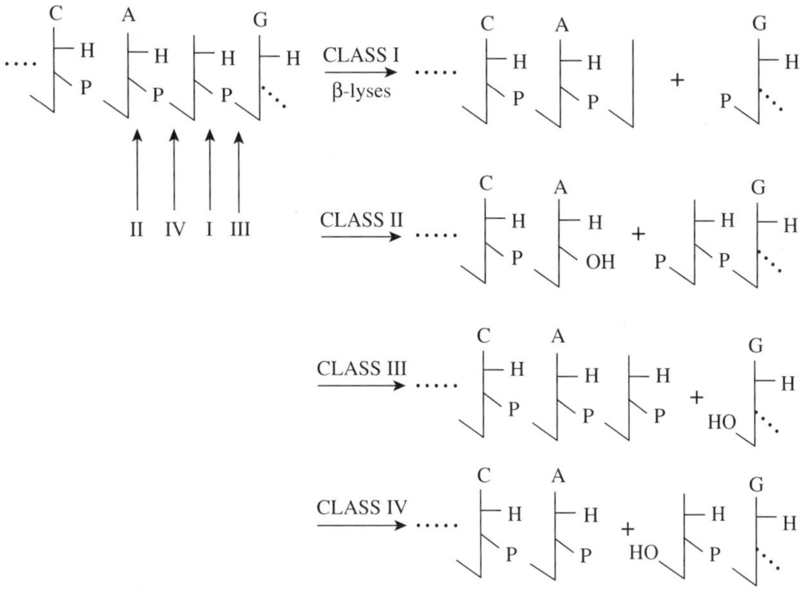

Figure 5.1. Action of different classes of AP endonucleases. [From Lloyd and Linn (1993), copyright Cold Spring Harbor Laboratory Press, used with permission (Courtesy of Professor Stuart Linn).]

enzymes produces a 3 end that serves as a good substrate for the DNA polymerase I. It is believed that these AP endonucleases may act in unison to eliminate the baseless site from the DNA chain which is subsequently repaired.

A number of AP-endonuclease activities have been shown to be associated with other enzyme activities; the latter may include related DNA glycosylase activity or exonuclease activity or unrelated redox activity (Babiychuk et al., 1994).

1. AP Endonucleases Associated with DNA Glycosylase Activity

a. T_4UV Endonuclease V. This is the only known DNA-damage-specific enzyme encoded by a virus. The enzyme is specific for UV-induced pyrimidine dimers (Simon et al., 1975). This gene has been cloned and expressed in *E. coli* (Lloyd and Hanawalt, 1981). The native enzyme is a monomer of 18-kDa molecular weight encoded by the viral gene V. The enzyme has an associated DNA glycosylase activity that is responsible for the creation of an AP site by the hydrolysis of a glycosidic bond of the thymine in a dimer and its respective sugar moiety (Radany and Friedberg, 1980). A single domain in the same polypeptide carries two distinct catalytic activities such as AP-endonuclease and DNA-glycosylase activities (Nakabeppu and Sekiguchi, 1981; Warner et al., 1980a,b; Morikawa et al., 1992). The T_4UV endonuclease is a class II AP endonuclease.

Table 5.1 Properties of certain DNA damage-specific nucleases

Protein	Gene	Site Attacked	Molecular weight	pH Optimum	Metal Cofactor	Remarks
Bacteria or phase						
E. Coli					Mg^{2+}	ATP-dependent
uvrA	uvrA		104 kDa		Mg^{2+}	
uvrB	uvrB	UV damage	84 kDa		Mg^{2+}	
uvrC	uvrC		61 kDa		Mg^{2+}	
EndoIII	nth	AP, x-rays, acid damage	27 kDa	7	None	
mutY	mutY	AP site	39.1 kDa	—	—	
EndoIV	nof	Uracil AP site OsO4, x-rays	33 kDa	8–8.5	None	
EndoV			20 kDa	9.25	Mg^{2+}	
ExoIII, (EndoII, Endo VI)	Xth	AP site	32 kDa	8.5	Mg^{2+}	Preference for ssDNA
Endo VII			60 kDa	7.0	None	0
			30 kDa		Mg^{2+}, Mn^{2+}	
T4 endo	gene V	UV damage	18 kDa	None		Complement uvrA and human mutant
M. luteus		UV damage	11 kDa	None		Acid pI 4.8 stimulated by Mg
M. luteus		UV damage	11 kDa	None		Alkaline pI 8.8
Drosophila						
AP endo I		AP sites	66 kDa	None		
AP endo II			63 kDa			
Mammalian		UV	30 kDa	None		Rat liver 15–20 kDa
Calf thymus						
Lymphoblast						
Rat liver						
Calf thymus		AP site	32 kDa	None		Stimulated by Mg^{2+}
Human lymphoblast						
Human placenta						

2. M. luteus Enzyme. This was first described by Strauss (1962) in *Micrococcus luteus* as the enzyme which attacks UV-irradiated DNA. Later the enzyme was shown to include two enzyme activities, both active on UV-induced DNA-damaged sites (Riazzudin and Grossman, 1977). It is believed that these two activities correspond to the two AP-endonuclease activities described in *Micrococcus luteus* (Tomlin et al., 1976). This enzyme is a class I AP endonuclease. The properties of the two kinds of *M. luteus* enzyme are summarized in Table 5.1.

3. E. coli Endonuclease III. This enzyme was first described by Radman (1976) and later investigated by Linn and others (Gates and Linn, 1977a,b; Demple and Linn, 1980). The enzyme seems to remove a photoproduct other than thymine dimer from the UV-irradiated DNA; the photoproducts are 5,6-dihydrothymine or 5,6-dihydroxydihydrothymine. The enzyme is an AP endonuclease of the class I type. The glycosylase activity associated with the enzyme has not yet been physically separated from the AP-endonuclease activity. The properties of the enzymes are summarized in Table 5.1. In addition to endonuclease III, mutY encodes a 39.1-kDa protein that possesses both AP-endonuclease and DNA-glycosylase activity. However, the mutY gene product is involved in the correction of A-G mismatch to AT or CG (Tsai-Wu et al., 1992).

B. AP Endonuclease Associated with Other Enzyme Activities

1. E. coli Exonuclease III AP-Endonuclease Activity. An AP-endonuclease activity initially identified as endonuclease II (Friedberg and Goldthwait, 1969; Hadi and Goldthwait, 1971) or as endonuclease VI (Verely and Paquette, 1972; Verely and Rassart, 1975) was found to be physically inseparable from exonuclease III of *E. coli* (Milcarek and Weiss, 1972; Yajko and Weiss, 1975; Weiss, 1976). This was demonstrated with the use of mutant defective in gene *Xth* which controlled the structure of the enzyme exonuclease III of *E. coli*. All *Xth* mutants were shown to be simultaneously defective for both the exonuclease III and AP-endonuclease functions. The fact that the exonuclease III-associated AP endonuclease is the same as the endonuclease II or the endonuclease VI has led to the suggestion of omission of the latter term (i.e., endonuclease II and endonuclease VI). The properties of exonuclease III are summarized in Table 5.1. A similar enzyme has been described from *Hemophilus influenzae* (Clements et al., 1978). The properties of this enzyme are also summarized in Table 5.1.

C. AP Endonucleases

This group of enzymes shows only AP-endonuclease activity without any other enzymatic activity associated with it. The properties of this class of enzymes are described below and are also summarized in Table 5.1.

1. E. coli AP Endonucleases. *E. coli* possesses at least two enzymes that are exclusively AP endonucleases. They are endonuclease IV and endonuclease VII.

These two enzyme activities are present in the *Xth* mutants (lacking exonuclease III). The endonuclease IV is identified as the activity of Xth mutant capable of removing AP sites from a duplex DNA, whereas the endonuclease VII is identified as the enzyme capable of degrading single-stranded DNA containing depyrimidinated sites. The properties of these enzymes are summarized in Table 5.1.

2. Fungal Apurinic Endonuclease. In *Neurospora crassa*, two forms of apurinic endonuclease have been found to exist. These two forms can be described as follows: One form does not bind to the DEAE cellulose column, and the second form binds to the DEAE cellulose column and eluted with a higher concentration of KCl. Activity of both forms of AP endonucleases is simulated by Mg^{2+} and inhibited by EDTA (Mishra, 1986; Mishra, unpublished data). At present it is not determined if these two forms of *Neurospora* AP endonucleases resolved by DEAE cellulose chromatography indeed correspond to the two forms of AP endonucleases (i.e., class I and class II) found in *E. coli* or in human cells (Linn, 1982b) or if *mms* mutants of *Neurospora* (Delange and Mishra, 1981, 1982) are defective for this enzyme. A number of yeast endonucleases acting on apurinic DNA have been characterized (Armel and Wallace, 1978). The gene (APN1) for major yeast AP endonuclease has been identified, cloned, and sequenced (Popoff et al., 1990).

3. Drosophila AP Endonucleases. Two distinct AP endonucleases have been extensively purified from *Drosophila* embryo. The *Drosophila* AP endonuclease I has a molecular weight of 66 kDa; this enzyme flows through phosphocellulose column. The AP endonuclease II has a molecular weight of 63 kDa and is eluted later from the phosphocellulose column. The distinction of AP endonucleases is based on the presence of 3′ OH in the nick introduced in AP DNA by *Drosophila* enzyme. The 3′ nucleotide is recognized by its ability to support DNA synthesis by DNA polymerase. The *Drosophila* AP endonuclease I creates a nick with 3′ OH in the DNA, whereas the AP endonuclease II causes a nick without 3′OH. The *Drosophila* AP endonucleases I and II as well as the *E. coli* endonuclease IV have been found to cross-react with antibody prepared against human AP endonuclease. Thus these enzymes of DNA repair are immunologically related.

4. Human AP Endonucleases. Human fibroblast cells have been found to contain two forms of AP endonucleases. The class I AP endonuclease was not retained by the phosphocellulose during chromatography. This endonuclease was found to be missing from the extracts of xeroderma pigmentosum cells of the complementation group D (Kuhnlein et al., 1976, 1978). The human class I AP endonuclease is quite unstable and not yet purified. The class II AP endonuclease is obtained as the one that is retained by phosphocellulose during chromatography and then eluted by increase in the molarity of KPO_4 buffer. This form of enzyme has been obtained from different cell lines including human placenta (Linsley et al., 1977) and HeLa cells (Kane and Linn, 1981). An exonuclease involved in DNA repair has also been characterized from human placenta (Doniger and Grossman, 1976).

5. Plant AP Endonuclease. A plant AP endonuclease has recently been described from *Arabidopsis*; the cDNA encoding this enzyme has been cloned (Babiychuk et al., 1994). This plant AP endonuclease, like human and other AP endonucleases, was found to possess additional functions.

D. Direct-Acting Enzymes

This group of enzymes acts directly on the damaged DNA without any baseless site. Such enzymes most likely recognize the distortion in the DNA structure caused by specific damages.

1. E. coli UV Endonuclease

a. UVR ABC Excision Nuclease. This enzyme acts on DNA containing modified nucleotide (Sancar and Sancar, 1988; Jones and Young, 1988). The enzyme consists of three subunits—uvrA, uvrB, and uvrC—which act together as an excision nuclease and cleave phosphodiester linkages to the 5′ and 3′ ends of the damaged nucleotides (Yeung et al., 1983; Sancar and Rupp, 1983; Sancar et al., 1981a,b, 1985). The physical properties of the uvr ABC proteins are summarized in Table 5.1. The enzyme activity is dependent on ATP and requires both DNA polI and helicase (uvrD protein) for turnover (Carson et al., 1985). The uvrA, uvrB, and uvrC proteins are encoded by uvrA, uvrB, and uvrC genes of *E. coli*. These genes are under the control of recA and LexA genes of the SOS system. The uvrABC system has been shown to participate in other DNA repair in addition to its repair of UV damage (Chambers et al., 1985).

b. E. coli Endonuclease V. Besides the uvrABC system, *E. coli* possesses endonuclease V, which preferentially acts on DNA damaged by OsO_4, methyl benzanthralene, and other agents. Similar enzymes are described in *Micrococcus luteus* and in *Drosophila* (Sander et al., 1991).

2. Human Excision Nuclease. Comparable to *E. coli* uvrABC excision nuclease, a human exonuclease has been identified (Huang et al., 1992). This human enzyme removes thymine dimer in an oligomer that is 29 nucleotides in length, unlike the *E. coli* enzyme that yields a 13-mer containing the thymine dimer. The human exonuclease removes the thymine dimer by making an incision in the 22nd phosphodiester bond 3′ to the thymine yielding a 29-mer containing the photodimer. This enzyme may play a role in removal of photodimer *in vivo* because human cells are known to remove thymine dimer as an oligomer of 20–30 nucleotides.

6

TOPOISOMERASES

I. CHOREOGRAPHY AND TOPOLOGY OF DNA

The genetic material of an organism does not exist only in the form of a simple DNA duplex structure as originally proposed (Watson and Crick, 1953a,b). Instead, several choreographic and topological isomers of DNA are now known (Pohl and Jovin, 1972; Saenger, 1983); some of these forms may even exist in living organisms (Wang et al., 1979). The different choreographic forms include the distinct helical families of DNA molecules such as the right-handed B-DNA, the left-handed Z-DNA, and many other forms. Among these different forms of DNA, the right-handed B form exists *in vivo*; there is some evidence that the left-handed Z form may be assumed by certain nucleotide sequences, under physiological conditions (Haniford and Pulleyblank, 1983). In addition, each helical form may include duplexes that are severally folded and condensed into complex higher-order structures of different topology. Such a higher order of DNA must exist simply for the reason of the accommodation of an extensive DNA molecule in a relatively smaller space. For example, *E. coli* DNA, which is slightly over 1 mm in length, must be folded in order to be accommodated in a cell of much smaller dimension. Besides the spatial considerations, there is physical evidence for the existence of different structural polymorphic forms inside a cell. Vinograd et al. (1965) first identified the existence of the different topological forms of polyoma virus DNA in the mammalian cells. Since then the different topological isomers of the prokaryotic and eukaryotic chromosomes have been demonstrated (Stonington and Pettijohn, 1971; Worcel and Burgi, 1972; Cook and Brazel, 1975; Benyajati and Worcel, 1976; Piñon and Salts, 1977). At present, no enzymes that can cause the interconversion of the different choreographic forms of DNA or act on a particular choreographic form of DNA are known. For example no nuclease, except for certain chemzymes, is known which specifically acts on the Z-DNA alone. Certain artificial nucleases (chemzymes) have now been described which act specifically on B-DNA or on Z-DNA. However, a class of enzymes called *topoisomerases* has been shown to cause the interconversion of the different topological forms of DNA. Topology of DNA plays a great role in the DNA replication and recombination and gene expression (Wasserman and Cozzarelli, 1986; Snapk, 1986; Wasserman et al., 1985; Weintraub et al., 1986; Yamamoto and Droffner, 1985).

Topoisomerases are enzymes that catalyze the concerted breakage and reunion of the phosphodiester bond(s) in the DNA duplex (Liu and Wang, 1979; Wang et al., 1979; Wang, 1981). A topoisomerase is equivalent to the combined activity of an endonuclease that introduces nicks or breaks in the DNA duplex and that of a ligase which reseals them. Topoisomerases either can relax a supercoiled DNA or can introduce superhelicity in a relaxed DNA duplex and thus play an important role in the interconversion of the different topological forms of DNA which are now known to exist in a cell. Topoisomerases can be divided into two groups: Topoisomerase I makes a transient single-stranded break in DNA; Topoisomerase II makes a transient double-stranded break. The ω protein of *E. coli* and an activity from mouse embryo cells were the first topoisomerases identified as capable of relaxing a negatively supercoiled DNA (Wang, 1971; Champoux and Dulbecco, 1972). Since then a number of enzymes or proteins with topoisomerase activity have been described from a variety of organisms including prokaryotes and eukaryotes and their different cell types (Vosberg, 1985). Enzyme with topoisomerase activity has been described from rat liver mitochondria (Fairfield et al., 1979). A number of recombinases (site-specific endonucleases) have also been found to possess topoisomerase activity; these include λ Int protein, φ X174 gene A protein, and fd gene 2 protein (Kikuchi and Nash, 1979; Ikeda et al., 1976; Meyer and Geider, 1979). Likewise, a certain transcription factor shows gyrase activity (Kmiec and Worcel, 1985; Kmiec et al., 1986). The topoisomerase activity of other recombinases such as HO endonuclease of yeast and V(D)J protein of mammalian cells remains to be investigated.

II. ENZYME ASSAY

A linear DNA duplex is ordinarily visualized as a right-handed helix with a pitch of 10 bp per helical turn. A circular DNA with about one helical turn for every 10 bp is considered to be relaxed. A circular DNA with a helix pitch of less or more than 10 bp causes the folding of duplex on its own axis and leads to supercoiling. The intertwining of the two strands of DNA leads to the *twist*, whereas the folding of the DNA helix upon its axis is designated as *writhe*. The supercoiling of DNA can be expressed as linking number α which is represented as $N/10.4$, where N is the total number of base pairs in a DNA molecule. DNA molecules with a mean linking number smaller than $N/10.4$ are called negatively supercoiled or underwound, whereas DNA with a larger linking number is positively supercoiled or overwound. Enzymes of both topoisomerase class I and class II affect the superhelicity of DNA, which can be determined by any of the following methods.

A. Electron Microscopy

The difference among the highly supercoiled DNA, the completely relaxed DNA, and those with different degrees of supercoiling can be readily distinguished upon visualization by electron microscopy (Germond et al., 1975).

B. Sedimentation Methods

Vinograd et al. (1965) first demonstrated the different topological forms of polyoma DNA by sedimentation analysis. The different state of the supercoiling of DNA can be somewhat more accurately determined by the sedimentation velocity measurements of DNA treated with intercalating dyes such as ethidium bromide. The sedimentation values for the negatively supercoiled DNA change inversely with increasing amounts of intercalation with ethidium. However, as more and more ethidium is intercalated, DNA is supercoiled in the opposite direction and the degree of positively supercoiled DNA is directly related to the amount of intercalation by ethidium.

C. Agarose Gel Electrophoresis

This method can resolve the DNA molecules differing by a linking number of 1 based on the compactness of the DNA molecules. A highly supercoiled DNA molecule, being the most densely compact, moves very rapidly through the gel, whereas a fully relaxed DNA molecule has the slowest mobility. This method is commonly used to determine the activity of different topoisomerases (Keller, 1975a,b; Trash et al., 1984).

The superhelicity of a DNA molecule can also be determined by its susceptibility to S1 nuclease. DNA molecules with different degrees of supercoiling may act as substrate for transcription or for binding by a repressor molecule. Thus the degree of hyperhelicity of DNA molecules may as well be determined by their ability to act as substrate in these assays (Kornberg, 1980).

III. PROPERTIES OF ENZYMES FROM DIFFERENT GROUPS OF ORGANISMS

A. Prokaryotic Topoisomerases

1. Prokaryotic Topoisomerase I. *Escherichia coli* ω protein is the most well-characterized DNA topoisomerase I among prokaryotes. This enzyme can be readily purified to homogeneity by routine enzyme purification procedures (Depew et al., 1978). The native protein is a 105-kDa monomer. The enzyme requires Mg^{2+} for its activity. The enzyme activity is inhibited by single-stranded DNA (Wang, 1971) and by the antibiotics neomycin sulfate and actinomycin D.

The *E. coli* topoisomerase I catalyzes four distinct reactions. These are: (i) the relaxation of negatively supercoiled DNA (Wang, 1971); positively supercoiled DNA are not relaxed by this enzyme; (ii) knot formation in single-stranded DNA (Liu et al., 1976); (iii) intertwining of the complementary single-stranded DNA (Kirkgaard and Wang, 1978); and (iv) catanation and decatanation of circular DNA duplex (Marini et al., 1980; Tse and Wang, 1980).

Other bacterial topoisomerases I that have been well-characterized include those from *Micrococcus luteus* (molecular weight 120 kDa), *Salmonella typhimurium*, and *Agrobacterium tumefaciens* (molecular weight 100 kDa). The *Salmonella typhimurium* enzyme is encoded by the *supX* gene (Overbye and Margolin, 1981), which is located between *trp* and *cysB* loci on *Salmonella* genetic map. This position of the *supX* gene corresponds to that of the *top* gene encoding *E. coli* topoisomerase I (Margolin et al., 1985).

2. Prokaryotic Topoisomerase II. Among this group of enzymes the most important enzyme is the DNA gyrase, which catalyzes the conversion of a relaxed DNA into negatively supercoiled form. To date, such gyrase-like activity has been associated only with the bacterial topoisomerase and never with the eukaryotic enzyme. The gyrases from *E. coli* and from *Micrococcus luteus* are the most well-characterized enzymes. Gyrases have been described from other bacterial sources such as *Bacillus subtilis* (Sugino and Bott, 1980) and *Pseudomonas aeruginosa* (Miller and Scurlock, 1983).

3. Properties of Gyrase. The *E. coli* gyrases consist of the *gyrA* and *gyrB* proteins of molecular weight 105 kDa and 95 kDa, respectively (Sugino et al., 1977; Mizuuchi et al., 1980, 1982a,b; Higgins et al., 1978). These two protein subunits and their active complex have been purified to homogeneity (Mizuuchi et al., 1982a,b; Sugino et al., 1977; Higgins et al., 1978). These two subunits are the product of the *gyrA* and *gyrB* genes located at 48 and 82 min, respectively, on the *E. coli* genetic map (Bachman and Low, 1980). The *E. coli* gyrase makes site-specific cleavages in DNA (Morrison and Cozzarelli, 1979). Mutation in these genes also conferred resistance to *E. coli* against nalidixic acid and coumermycin (Hane and Wood, 1969; Ryan, 1976), which are inhibitors of DNA replication (Gross et al., 1965; Gross and Cook, 1975; Staudenbauer, 1975, 1976; Ryan, 1976). Nalidixic acid and oxolinic acid specifically inhibit the *gyrA* subunit (Gellert et al., 1977; Sugino et al., 1977), whereas novobiocin, coumermycin and chlorobiocin inhibit the *gyrB* subunit (Gellert et al., 1976a,b; Fairweather et al., 1980). The *M. luteus* gyrase resembles *E. coli* enzyme and the two subunits have an apparent molecular weight of 115 kDa and 95 kDa, respectively. The enzyme requires Mg^{2+} and ATP for the supercoiling of DNA, which is the principal reaction catalyzed by this enzyme. The oxolinic acid and nalidixic acid act by blocking the reactions involving breakage and rejoining of DNA, whereas the novobiocin, coumermycin, and chlorobiocin act by blocking ATP requiring steps in the enzymatic reaction.

The native enzymes make a noncovalent but stable complex with the substrate DNA. Like topoisomerase I, the gyrase is linked to DNA by a phosphotyrosine bond. Both subunits of the enzyme are required for such binding with the substrate. The enzyme acts processively during the catalysis of supercoiling. Enzyme binding protects a DNA sequence of 140 bp from digestion by *Staphylococcus* nuclease (Gellert et al., 1980; Liu and Wang, 1987). This result suggests that the gyrase is wrapped by a DNA segment in a manner very much analogous to that of the wrapping of histone molecules by DNA duplex in a nucleosome (Kirchhausen et al., 1985; Liu and Wang, 1987). Although supercoiling is the main activity of this enzyme, gyrase can catalyze relaxation of the negatively supercoiled DNA in the absence of ATP (Sugino et al., 1977; Gellert et al., 1977). The enzyme cannot, however, relax the positively supercoiled DNA. The oxolinic acid (but not novobiocin or coumermycin) can inhibit the relaxation activity of the enzyme. This finding suggests that relaxation indeed requires subunit A for cleavage and rejoining of DNA. *E. coli* topoisomerase I can act on oligonucleotides (Tse-Dinh et al., 1983).

4. Other Activities of Gyrase. The enzyme is capable of the catanation, decatanation, and resolution of other topological knots (Kreuzer and Cozzarelli, 1980; Mizuuchi et al., 1980; Liu et al., 1980). These activities of the enzyme are concomitant with the associated ATPase function of the gyrB subunit of the enzyme. The hydrolysis of ATP is, however, inhibited by novobiocin and coumermycin. These two drugs act as a competitive inhibitor of ATP binding to gyrB subunit of the enzyme (Sugino et al., 1978; Staudenbauer and Orr, 1981). *E. coli* gyrase acts in a site-specific manner.

5. Prokaryotic Topoisomerase III. An *E. coli* topoisomerase that can relax negatively supercoiled DNA has been described (Srivenugopal et al., 1984). This enzyme consists of a single polypeptide with a molecular weight of 74 kDa and possesses properties similar to those of topoisomerase type I. However, it significantly differs in property from both type I and type II topoisomerases (Srivenugopal et al., 1984). The enzyme is the type I topoisomerase present in the topA mutant of *E. coli*. The properties of *E. coli* topoisomerases are compared in Table 6.1.

B. Eukaryotic Topoisomerases

1. Eukaryotic Topoisomerase I. Topoisomerases from a variety of eukaryotic organisms have been obtained. The topoisomerase I can be readily isolated from chromatin preparations after treatment with high salt to dissociate the enzyme and then purified following routine methods available for enzyme purification (Champoux and McConaughy, 1976; Prell and Vosberg, 1980). Enzymes characterized from a wide variety of organisms showed a wide range of molecular weights varying from 66 kDa to 135 kDa. There is roughly 1 topoisomerase per 10–15 nucleosomes (Vosberg, 1985). This value may correspond to 1 topoisomerase per gene. Topoisomerase I requires monovalent cations (Na) for enzymatic activity. Mg^{2+} may stimulate the enzymatic reaction but cannot substitute for monovalent

Table 6.1 Comparison of the properties of *E. coli* topoisomerases

	Topoisomerase I	Topoisomerase II
1. Function	Relaxes DNA	Supercoils DNA
2. Nature of cleavage	Single-stranded break	Double-stranded break
3. Consequences of enzyme inhibition or mutation	Increase in negative superhelicity, increases the linking number of helix by a step of 1	Decrease in negative superhelicity, decreases the linking number by a step of 2
4. Gene	Top	gyrA (= nalA), gyrB
5. Requirement for ATP	No	Yes
6. Associated ATPase activity	No	Yes
7. Effect of coumermycin and novobiocin	None	Inhibits of ATPase activity
8. Effect of oxolinic acid nalidixic acid	None	Inhibits breakage and rejoining of DNA

cations. The enzyme has a broad pH optimum of 6.5–8. The enzyme is inhibited by *N*-ethylmaleimide, actinomycin D, and heparin.

The eukaryotic topoisomerase I can catalyze the following reaction:

1. Relaxation of both the negatively and positively supercoiled DNA (Champoux and Dulbecco, 1972)
2. Intertwining of complementary single-stranded DNA circles (Champoux, 1977a,b)
3. Catenation (Brown and Cozzarelli, 1981; Krasnow and Cozzarelli, 1982; Baldi et al., 1988)
4. Intermolecular chain transfer (Been and Champoux, 1981; Halligan et al., 1982)

2. Eukaryotic Topoisomerase II. This enzyme has been obtained from a variety of eukaryotic organisms. The HeLa cell enzyme has been purified to homogeneity (Miller et al., 1981). The molecular weight of the native enzyme is 300 kDa; the enzyme consists of two identical subunits. The enzyme requires ATP for relaxation of supercoils and for decatenation and unknotting of DNA. Thus the enzyme has a DNA-dependent ATPase activity. The enzyme is inhibited by nalidixic acid, oxolinic acid, and novobiocin. This eukaryotic enzyme resembles T4 topoisomerases in its catalytic properties. The human topoisomerase II cDNA has been identified, cloned, sequenced, and localized to human chromosome 17q 21–22 (Tsai-Pflugfelder et al., 1988). A comparison of the properties of prokaryotic and eukaryotic topoisomerases is presented in Table 6.2.

Table 6.2 Comparison of the properties of prokaryotic and eukaryotic topoisomerases

Prokaryotic	Eukaryotic
A. Type I topoisomerase function	
(a) Can relax only negative supercoils.	(a) Can relax both positive and negative supercoils.
(b) Enzyme binds to 5 end of the nick.	(b) Enzyme binds to the 3 end of the nick.
(c) Enzyme does not cause complete relaxation of DNA.	(c) Cause complete relaxation and cause nicking and closing of fully relaxed DNA.
(d) No intermolecular strand transfer catalyzed by the enzyme.	(d) Enzyme catalyzes intermolecular strand transfer.
B. Type II topoisomerase function	
(a) ATP-dependent catalysis of negative supercoiling.	(a) No gyrase function.
(b) Decatanate or catenate DNA.	(b) Decatanate or catenate DNA.
(c) —	(c) Knot and unknot DNA molecules.
(d) Inhibition by nalidixic or oxolinic acid. These drugs stabilize the topoisomerase–DNA complex, causing single- or double-stranded breaks, and make a covalent linkage of the drug to the 5-phosphoryl end of DNA through a tyrosyl phosphate linkage.	(d) Inhibition by adriamycin, mAMSA, ellipticines, and epipodophyllotoxins. These drugs stabilize the topoisomerase–DNA complex, and the drug is covalently linked to the 5-phosphoryl end of DNA through a tyrosyl phosphate linkage.

C. Mitochondrial Topoisomerase

A topoisomerase with distinct properties of being inhibited by ethidium bromide and by another trypanocidal drug Berenil was first described from rat liver mitochondria (Fairfield et al., 1979). The presence of mitochondrial enzymes was further indicated by the fact that the mitochondrial DNA was found to exist in relaxed form when the rat liver was treated with nalidixic acid, oxolinic acid, novobiocin, and coumermycin (Castora and Simpson, 1979). The mitochondrial topoisomerase has been extensively characterized from the mitochondria of *Xenopus laevis* oocytes (Brun et al., 1981). The purified enzyme has a molecular weight of 65–70 kDa. The mitochondrial enzyme, however, resembles the nuclear topoisomerase I in catalytic properties (Vosberg, 1985), and its possible origin from the nuclear enzyme cannot be ruled out. However, this remains to be established by immunological evidence. In addition to the mitochondrial topoisomerase, a topoisomerase from spinach leaf chloroplast has been described.

D. Viral Topoisomerases

These enzymes are of two kinds: first, those that are encoded by the viral genes and, second, those that are of host origin but associated with the virus. Among bacteriophages T4, φX174, λ, and vaccinia virus, each encodes for a topoisomerase of its own, whereas host enzyme has been found to be associated with SV40, polyoma, and Rous sarcoma viruses. Of these viral topoisomerases, T4 enzyme is the most well-characterized. The λ Int protein, φX174A protein, and fd gene-2 protein, which are essentially recombinase with topoisomerase activity, are described later in the subsequent chapter on recombinases. The T4 topoisomerase is a multimeric enzyme that consists of three subunits with molecular weights of 64 kDa, 57 kDa, and 16 kDa, encoded by the viral genes 39, 52, and 60, respectively (Liu et al., 1979). The viral enzyme is resistant to novobiocin or coumermycin. McCarthy (1979) has suggested that *E. coli* gyrase can substitute for the viral enzyme because the wild-type *E. coli*, but not the mutants, can support the growth of T4 phage. The T4 enzyme catalyzes the following reactions: (a) relaxation of (positively or negatively) supercoiled DNA; (b) catenation and decatenation, and (c) knotting and unknotting of DNA molecules. The enzyme requires ATP and changes the linking number in a step or two. These properties suggest that this viral enzyme is similar to eukaryotic topoisomerase II. The vaccinia virus topoisomerase is activated by trypsin treatment (Reddy and Bauer, 1989). The gene for the vaccinia virus topoisomerase has been cloned, and this viral enzyme has been found to cause a site-specific recombination in *E. coli* (Shuman, 1991). T4 virus topoisomerase interacts with several other proteins for its action (Stetler et al., 1979).

IV. GENETICS AND BIOLOGICAL ROLE

The physiological role of an enzyme is usually obtained by genetic analysis of appropriate mutants. DNA topoisomerase mutants of both prokaryote and eukaryote have been characterized. Effect of inhibitors on topoisomerase has also been utilized to understand the role of these enzymes on DNA synthesis and cell division (Fairweather et al., 1980).

A. Prokaryotic Topoisomerase Mutants

1. Topoisomerase I. The gene encoding topoisomerase I of *E. coli* has been identified (Trucksis and Depew, 1981)). Mutants of *E. coli* topoisomerase I were isolated by "brute force method" by examining over 800 temperature-sensitive mutants. A number of mutants deficient in topoisomerase I activity were identified, and the gene encoding topoisomerase I activity has been designated as topA. The topA gene was mapped at 28 min of *E. coli* chromosome and linked to cysB gene. It has also been shown that *supX* mutant of *Salmonella*, which mapped near the cysB gene in both *Salmonella* and *E. coli*, is indeed allelic to the topA gene. The *supX* gene of *Salmonella* was originally identified by its ability to suppress promoter mutation in leucine and other genes. The topA gene has been cloned and sequenced (Goto and Wang, 1984, 1985). The fact that mutants carrying deletion of the *topA* gene remain viable supports the idea that topoisomerase I is not essential for the survival of the organism. It is possible that certain compensatory mutations may occur in the topA mutant to ensure the viability of such mutants (Goto and Wang, 1984, 1985). Furthermore, a new enzyme designated as topoisomerase III has been shown to be present in the *E. coli* topA mutant. However, the topA mutants were altered in transcription and in transposition. The topA mutant showed enhanced induction of certain catabolite-sensitive enzymes. The transposition of Tn-5 (but not that of Tn-3) was decreased in topA mutant. *Escherichia coli* mutants defective in topoisomerase I are also defective in transcription and transposition (Sternglanz et al., 1981).

2. Topoisomerase II. *Escherichia coli* topoisomerase II or gyrase mutants have been identified by their resistance to nalidixic and/or oxolinic acids. Genes gyrA and gyrB have been identified as the structural gene of the two polypeptides that contribute the gyrase enzyme. It is known that the level of expression of a number of genes (including catabolically regulated ones) and rRNA genes is decreased by the inhibition of gyrase (Smith et al., 1978; Sanzey, 1979; Yang et al., 1979).

B. Eukaryotic Topoisomerase Mutants

Topoisomerase mutants of yeast have been well-characterized. The characterization of mutants suggests the role of these enzymes in different aspects of chromosome duplication, transcription, and recombination. The yeast topoisomerase I and topoisomerase II genes have been cloned, and mutants have been generated and characterized (Goto and Wang, 1982, 1984, 1985).

1. Topoisomerase I Mutants of Yeast. The yeast topoisomerase I mutants were identified by brute force method after screening of 280 temperature-sensitive mutants. The yeast topoisomerase I (topI) gene is tightly linked to the centromere of chromosome XV (Thrash et al., 1984) and is allelic to the previously described Mak1 gene; the latter is required for the maintenance of an RNA killer virus in yeast. The yeast topoisomerase I gene has been cloned and sequenced (Thrash et al., 1985). The topI mutant of yeast is able to grow normally but unable to maintain the killer virus. The possible role of the topoisomerase in the maintenance of the killer virus was discussed earlier (Thrash et al., 1984).

2. Topoisomerase II Mutants of Yeast. A temperature-sensitive topoisomerase-deficient mutant was identified after screening of 150 temperature-sensitive mutants by the method of brute force. The gene coding the topoisomerase II has been designated as *top2*. This gene has been now mapped to chromosome XIV near the met-4 gene. The temperature-resistant revertant of the top2 mutant was found to regain normal enzyme activity and wild-type phenotype of the cell. The top2 mutants have been shown to be defective in DNA replication (DiNardo et al., 1984). The yeast top2 gene has been cloned (Goto and Wang, 1984, 1985), and its role in segregation of daughter chromosomes during mitosis has been invoked (Holm et al., 1985). The mutants of top1 and top2 genes encoding topoisomerase I and II of the fission yeast have also been identified and characterized (Uemura and Yanigida, 1984). The type I mutants were viable, but type II mutants were defective in segregation of daughter chromosomes during mitosis. The yeast mutants defective in topoisomerase II have been complemented with *Drosophila* gene encoding this enzyme (Wycoff and Hsieh, 1988), and thus the *Drosophila* topoisomerase gene II has been cloned and analyzed; also the homology between their cleavage sites has been determined (Sander and Hsieh, 1985).

3. Topoisomerase Mutants of Higher Eukaryotes. Even though topoisomerases from higher eukaryotes were the first ones to be described, no mutants of these enzymes were identified until recently (Andoh et al., 1987), mostly because of the lack of a mutant library or of known inhibitors of the enzyme. However, the tsA159 locus of mouse L cells has been reported to encode a novobiocin-binding polypeptide required for DNA topoisomerase II activity (Colwill and Sheinin, 1983). Now a number of antitumor drugs have been identified as inhibitors of topoisomerases in higher organism. Some of these drugs that can modify the mammalian topoisomerase I and II activities are camptothecin, adriamycin, and acridine derivative 4(9-acridinylamino) methane sulfonic *m*-anisidide (m-AMSA) (Tewey et al., 1983; Nelson et al., 1984; Bodley and Liu, 1988). These drugs have been used to generate the topoisomerase mutants of mammalian cells (Andoh et al., 1987). An antitrypanosomal and antifalarial drug suramin has been shown to be an inhibitor of mammalian topoisomerase II (Bojanowski et al., 1992). Polyamines seem to influence the activities of topoisomerase either directly or indirectly (Zwelling et al., 1985).

Human topoisomerase mutants resistant to a number of antitumor drugs (such as anticyclines, epipodophyllotoxins, and amino acridines) possess altered topoisomerase II or reduced activity of this enzyme. Molecular analysis of topoisomerase II from a human leukemic cell line shows a GA base change in the topoisomerase II gene. This base change is located in the consensus sequence required for the ATP interaction of the topoisomerase II (Bugg et al., 1991). Two forms of human topoisomerase II have been identified and characterized. These include a 174-kDa protein and another 180-kDa protein. The cDNA for both proteins have been isolated and characterized. These two forms of topoisomerase II differ in their drug sensitivity, cleavage site, thermostability, and inhibition by (A + T)-rich oligonucleotides. The 174-kDa topoisomerase is more sensitive to teniposide and merbarone than the 180-kDa form of the enzyme (Drake et al., 1989). In *Drosophila*, topoisomerase II has been shown to be localized in nuclear matrix (Berrios et al., 1985) and makes cleavage in the 5-untranscribed region of the gene but not within the coding region of a gene. The micrococcal nuclease-hypersensitive sites correspond to the Topo II cleavage site.

7

RECOMBINASES

I. GENERAL DESCRIPTION AND CLASSIFICATION

Genetic recombination involves breakage and reunion of DNA segments. Nucleases involved in recombination of nucleic acid molecules are grouped here as recombinases. Some of these enzymes mimic the effect of topoisomerases because they can cleave and rejoin DNA phosphodiester bonds. However, unlike topoisomerases, most recombinases are nucleases devoid of ligation activity. Recombinases catalyze the breakage (and rejoining) of DNA molecules with a long stretch of homology (i.e., general or homologous recombination), or with a short stretch of homology (i.e., site-specific recombination), or with homology of only a few base pairs (illegitimate or transpositional recombination). However, all recombinations (including those involved in general recombination) are initiated from specific sites. Site-specific recombination involving defined sites occurs independently of the general homology between the stretches of DNA sequences undergoing recombination. During general recombination, nucleases are involved in the different steps leading to the formation and resolution of Holliday structures and correction of mismatch in the heteroduplex. Recombinases play an important role within the living systems by facilitating gene rearrangements, including integration, excision, inversion, transposition, resolution, and conversion of specific DNA sequences. The different recombinase activities may be classified as follows.

A. General Recombinase

Nucleases involved in general recombination are as follows:

1. *Initiase*: Nuclease involved in the initial steps leading to the formation of a Holliday structure. It is proposed here that these enzymes may be called initiation-nickase or initiase.

2. *X-Solvase*: Enzymes involved in the resolution of a Holliday structure; these enzymes have been named X-solvase (Jensch et al., 1989).
3. *Correctase*: Mismatch repair enzymes involved in the removal of mismatch in a heteroduplex. These enzymes have been named correctase because they are responsible for the correction of mismatch in the heteroduplex. Alternatively, these enzymes may be called convertase because the correction of mismatch may lead to conversion of genetic loci.

B. Site-Specific Recombinase

1. Prokaryotic. Nucleases involved in site-specific recombination are as follows:

(i) Integrase and other related proteins
(ii) Invertase
(iii) Resolvase

2. Eukaryotic

(i) Yeast FLP protein
(ii) Yeast Y-Z endonuclease
(iii) Mammalian V(D)J endonuclease
(iv) Intron encoded endonuclease or homing endonuclease
(v) Viral integrase

C. Transpositional Recombinase

Nucleases are involved in transpositional or illegitimate recombination. Such recombination was called *illegitimate recombination* because it was thought that it did not involve DNA with homology; this was mainly because of the absence of methodology to detect homology of a few base pairs. Now it is known that even the so-called illegitimate recombination involves a small stretch of homology. Transposases are best described in the following groups:

(i) Prokaryotic transposases
(ii) Eukaryotic transposases

D. RNA Recombinase

Nucleases involved in the recombination of viral and other RNA molecules.

II. PROPERTIES OF DIFFERENT RECOMBINASES

A. General Recombinase

These enzymes are involved in the formation of a Holliday structure, its resolution, and finally the correction of heteroduplex in the recombinant DNA molecules leading to gene conversion. Unlike the site-specific recombination, the general recombination involves the homology of DNA sequences over a greater stretch of DNA segments. However, even the general recombination is initiated from specific sites. These sites have been identified in a variety of prokaryotes and eukaryotes and have been designated as Chi elements. The *Escherichia coli* Chi element consists of an eight-nucleotide sequence (5′GCTGGTGG3′). The prokaryotic enzyme ExoV, the product of recBCD genes, acts by causing cleavage in the phosphodiester linkage near the Chi element in *E. coli*. The three groups of enzymes involved in the steps leading to the formation and resolution of a Holliday structure and correction of mismatch within it are described below.

1. Initiase. These nucleases are involved in initiation of the formation of Holliday structure.

The different models of recombination require a single-stranded (Holliday, 1964; Meselson and Radding, 1975) or a doubled-stranded break (Szostak et al., 1983) to initiate the process of recombination. No specific enzyme has been identified which makes such a nick or cleavage in DNA required for the initiation of recombination process. Based on the genetic studies of rec mutants in *E. coli* (Clark, 1973; Smith, 1988) it seems very likely that ExoV of *E. coli* has such attributes because it can unwind the DNA and make a single-stranded nick in the vicinity of the Chi element found in the bacterial chromosome; however, no such role of these enzymes has yet been demonstrated in *in vitro* assays. The properties of this *E. coli* enzyme (i.e., ExoV) encoded by the recBCD gene have been discussed earlier in Chapter 3. It is further possible that ExoI and ExoVIII may have a role in the initial steps of recombination (Smith, 1988). Some of these nucleases are listed in Table 7.1.

Table 7.1 Nucleases Involved in general recombination

Enzyme	Gene and/or Protein Subunit	Reference
I. Initiases: Nucleases Involved in the Steps Leading to the Formation of Holliday Intermediate		
E. coli		
ExoV	recBCD	
	recB 140 kDa	Smith (1988)
	recC 130 kDa	
	recD 58 kDa	
ExoI	sbcB	
ExoVIII	recE structural gene	
	$sbcA^+$ repressor gene	
	recF	
ssDNA exo	recJ	
H. influenzae		
ExoV		

Table 7.1 *(continued)*

Enzyme	Gene and/or Protein Subunit	Reference
II. X-Solvases(s): Enzymes Involved in Resolution of Holliday Structure		
T4 endonuclease VII[a]	43 kDa (gel filtration)	Jensch and Kemper (1986)
	36 kDa (sedimentation)	Mizuuchi et al. (1982a)
	18 kDa (SDS page) (gene 49)	Barth et al. (1988)
		de Massy and Weisberg (1987); Kemper and Brown (1976)
		Connolly and West (1990)
T7 endonuclease I	40 kDa (gene 3)	Barth et al. (1988)
		Dickie et al. (1987)
		Lee and Sadowski (1981)
E. coli resolvase	20 kDa (ruvC gene)	Connolly et al. (1991)
Yeast		
Endox-1	>200 kDa	
Endox-2	41 kDa	Symington and Kolodner (1985)
		Evans and Kolodner (1987, 1988)
Endox-3	43-kDa (gel filtration) dimer	Jensch et al. (1989)
	18-kDa (SDS gel) monomer	
CCE-1[b]	41 kDa	
Mammalian cells		
Calf thymus	75 kDa	Elborough and West (1990)
HeLa cells	—	Waldman and Liskay (1988)
Human placenta	—	Jeyaseelan and Shanmugam (1988)
III. Correctase: Correction of Mismatch in Holliday Structure		
E. coli	a. Dam instructed system	Pukkila et al. (1983)
	mutH 35.4 kDa	Su and Modrich (1986)
	mutL 67.7 kDa	Mankowich et al. (1989)
	mutS 97 kDa	Lu et al. (1983, 1988)
S. tymphimurium	b. mutY (or micA) hex system	Tsai-Wu et al. (1992)
Streptococcus pneumoniae	Hex A 95 kDa	Claverys et al. (1983)
	Hex B 86 kDa	Lacks et al. (1982)
	PMS-1 103 kDa	Priebe et al. (1988)
Yeast		Muster-Nassal and Kolodner (1986)
		Bishop et al. (1989)
		Varlet et al. (1990)
Drosophila melanogaster	—	Holmes et al. (1990)
KC cells	—	Varlet et al. (1990)
Xenopus		
HeLa cells	—	Holmes et al. (1990)

[a] Immunological relatedness is shown by the fact that the resolution of Holliday structure is inhibited by either of these enzymes in the presence of an anti-Endo VII antiserum (Jensch et al., 1989).

[b] CCE-1-encoded endonuclease is identical to endox-2 and is allelic to the mgt-1 gene also located on chromosome XI close to the centromere. This enzyme plays a role in the maintenance of the mtDNA because CCE-1 mutants yield petite colonies at a very high frequency.

2. X-Solvase. These enzymes are involved in the resolution of Holliday structure. A number of such nucleases have been identified by their ability to resolve DNA structures like Holliday intermediates in *in vitro* assays. At least three different kinds of DNA resembling Holliday structure have been used to identify and characterize the X-solvases. These artificial Holliday structures include (a) DNA molecules in the form of a figure eight, (b) plasmids containing palindromic structures capable of forming cruciform structures at predetermined sites near to a site for a restriction enzyme, and (c) DNA duplexes formed among four oligonucleotides with complementary base sequences that lead to formation of a four-way branched duplex DNA. These DNA substrates equivalent to Holliday junction and the effect of resolvase on such substrates are shown in Figures 7.1, 7.2, and 7.3.

The bacteriophage G-4 DNA occurring in the form of a figure eight, when acted upon by an X-solvase, has been shown to yield monomer and dimers. The plasmid with cruciform structure, when digested with a restriction enzyme following action by X-solvase, was found to yield DNA fragments of predicted sizes. The four-way branched DNA molecule (see Fig. 7.1c) generated by the hybridization of oligonucleotides was resolved by X-solvase into linear DNA duplex structures and

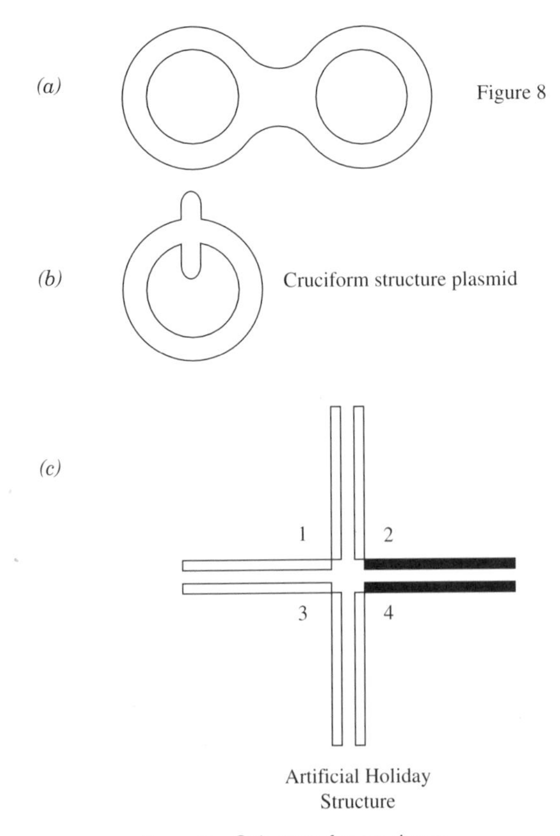

Figure 7.1. Substrate for resolvase.

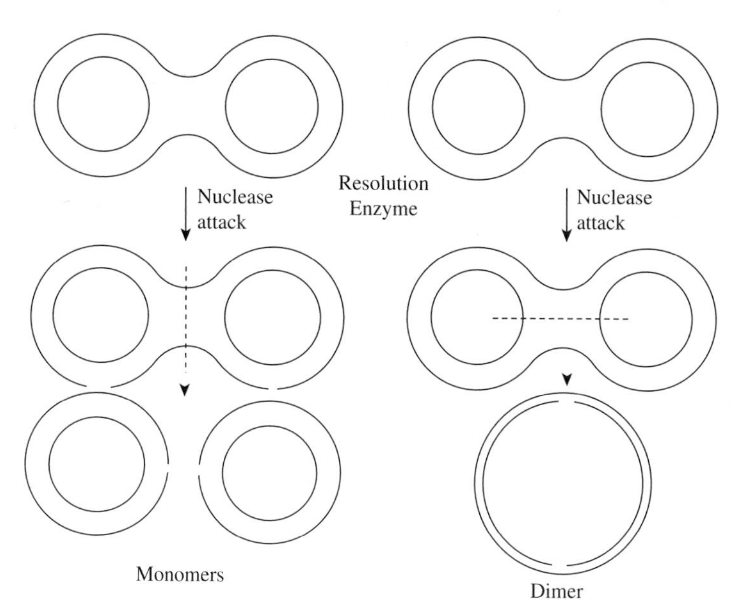

Figure 7.2. Resolution of figure eight by resolvase. [From Symington and Kolodner (1985), used with permission.]

visualized by electron microscopy. A number of such enzymes from bacteriophage to human cells have been identified, and their properties are summarized in Table 7.1. A brief description of X-solvase from different sources is presented below.

a. Bacteriophage. Endonuclease VII encoded by gene 49 of phage T4 has been very well characterized. The bacteriophage T4 gene 49 encoding endonuclease VII has been cloned and overexposed in bacteria (Kosak and Kemper, 1990).

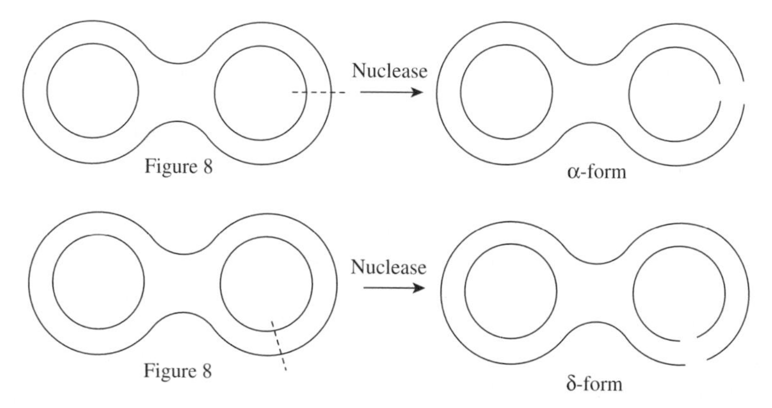

Figure 7.3. Degradation of the figure eight by nucleases other than resolvase. [From Symington and Kolodner (1985), used with permission.]

This enzyme can act on any kind of Holliday structure and resolve it. However, it seems that the *in vivo* role of this enzyme is primarily in the resolution of the branched DNA molecule formed during the replication of the bacteriophage (Broker and Lehman, 1971). It has been shown that branched T4 DNA accumulates in *E. coli* when infected by T4 carrying a mutation in gene 49. These DNA molecules could, however, be resolved into genome-length sizes by purified T4 endonuclease or by yeast X-solvase-3 in *in vitro* assay.

The antibody prepared against the T4 endonuclease-7 has been shown to inhibit the X-solvase activities of the T4 (endo-7) as well as yeast (X-solvase-3) enzymes. Similarities in the molecular structure, enzymatic action, and immunological cross-reactivity suggest that T4 and yeast enzymes are related. It is suggested that T4 enzyme might have been transposed to yeast cells. It would be of further interest to examine the nucleotide and amino acid homologies of the bacteriophage and yeast X-solvase and genes encoding them. The endonuclease I of bacteriophage T7 has been shown to possess X-solvase activity in *in vitro* assays. This enzyme is encoded by gene 3. The relationship of T7 enzyme to other X-solvase has not yet been determined.

b. Escherichia coli. Earlier it was thought that exonuclease V of *E. coli* may have a role in the resolution of Holliday structure. However, the result of *in vitro* assay negates such a role for this *E. coli* enzyme. Thus the role of exonuclease V may be limited to initial steps of recombination leading to the formation of Holliday intermediates.

Another *E. coli* protein, a product of ruvC gene, has been shown to possess the ability to resolve Holliday structures in *in vitro* assay. This gene has been cloned, and the size of the protein encoded by ruvC gene has been determined upon agarose gel electrophoresis to be 20 kDa as predicted from the nucleotide sequence analysis. It is suggested that the *E. coli* ruvC gene product acts in concert with ruvA and ruvB gene products analogous to uvr ABC proteins. It is possible that ruvA and ruvB proteins may enable the *E. coli* X-solvase ruvC gene products to recognize the Holliday structures.

c. Yeast. Three different yeast enzymes X-solvase-1, X-solvase-2, and X-solvase-3 have been identified to possess X-solvase activities in *in vitro* assays. Of these, X-solvase-3 has been characterized in detail and found to show immunological relation to the bacteriophage T4 enzyme as discussed above. Based on the properties of this yeast enzyme such as (a) the ability to resolve branched T4 replication intermediate structures, (b) the fact that yeast enzyme is not UV-inducible, and (c) the fact that its expression is not related to meiosis in yeast, it is suggested that this yeast enzyme-like T4 enzyme may be involved in the maintenance of the chromosome structures during cell division. Thus it is implied that other yeast enzymes such as X-solvase-1 and X-solvase-2 may have a role per se in the resolution of Holliday structures formed during the meiotic recombination. It is known that DNA structure and nucleotide sequence have effects on the resolution of Holliday structure (Evans and Kolodner, 1988).

d. Mammalian Cells. The enzyme X-solvase has been identified from at least three different mammalian sources. The calf thymus enzyme, a 75- kDa protein, has been shown to resolve Holliday structures in *in vitro* assays. This enzyme resembles T4 enzymes in its X-solvase activities but is not inhibited by the antibody prepared against the T4 enzyme. The X-solvase has also been identified from HeLa cells as well as from human placenta. These enzymes have not yet been purified to any extent, and details about their molecular size and structure are lacking.

3. Correctase. These enzymes are responsible for correction of mismatch in a heteroduplex DNA and are described below. The correctase enzyme is usually assayed by its ability to convert a mutant into wild-type gene or by its ability to restore a restriction enzyme cleavage site in the region containing base pair mismatch; the latter is corrected by the activity of the correctase enzyme (Bishop et al., 1989; Muster-Nassal and Kolodner, 1986; Varlet et al., 1990) as shown in Figure 7.4.

a. Prokaryotic Correctase. Among bacteria, the enzymes involved in mismatch correction have been very well characterized from *E. coli* and *Streptococcus aureus*. The properties of these enzymes are summarized in Table 7.1.

E. coli Mismatch Repair Enzyme. The mutH gene product has been identified as mismatch repair enzyme in *E. coli*. This enzyme has been purified to homogeneity and has a molecular weight of 25.4 kDa as based on analysis in SDS gel electrophoresis. The mutH gene has been cloned and the entire nucleotide sequence has been determined. The molecular weight of the mutH protein is in agreement with the amino acid deduced from the nucleotide sequence of the cloned gene. The promoter region of the cloned gene has been characterized. MutH protein is a strand-specific endonuclease that cleaves the strand containing the unmethylated GATC sequences (Grafstrom and Hoess, 1983, 1987; Welsh et al., 1987). The MutH gene product has been characterized in an *in vitro* transcription–translation system as well as after expression in maxicells (Grafstrom and Hoess, 1983). The mismatch repair by MutH endonuclease has been based on its ability to restore a site for restriction endonuclease in a known plasmid. Analysis of the restriction pattern of the plasmid before and after incubation with the MutH gene product can reveal its correctase activity (Welsh et al., 1987). The *mutH* gene product depends on the product of the *dam* gene of *E. coli* for its activity.

In addition to dam-dependent MutH, *E. coli* possesses the mutY system for the mismatch correction. The MutY gene product, a 39.1-kDa protein, possesses both *N*-glycosylase and AP-endonuclease activities and acts preferentially on A-C and A-G mispairs (Tsai-Wu et al., 1992).

b. Eukaryotic Correctase. The presence of this enzyme in the eukaryotic cell free extracts has been identified by its ability to correct a mismatch at a predetermined position in a plasmid such that the correction of mismatch generates a new restriction site in the plasmid DNA molecule which can then be identified by the comparison of the restriction digestion pattern of the plasmid molecules before

A. Assay for correctase enzyme

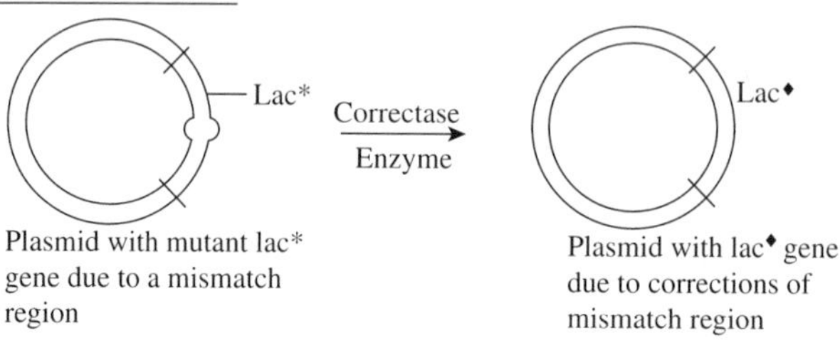

The plasmid containing Lac* or lac$^{\blacklozenge}$ gene can be identified after transfection into Lac* *E. coli* by their ability to produce white or blue colonies respectively on *X*-gal medium.

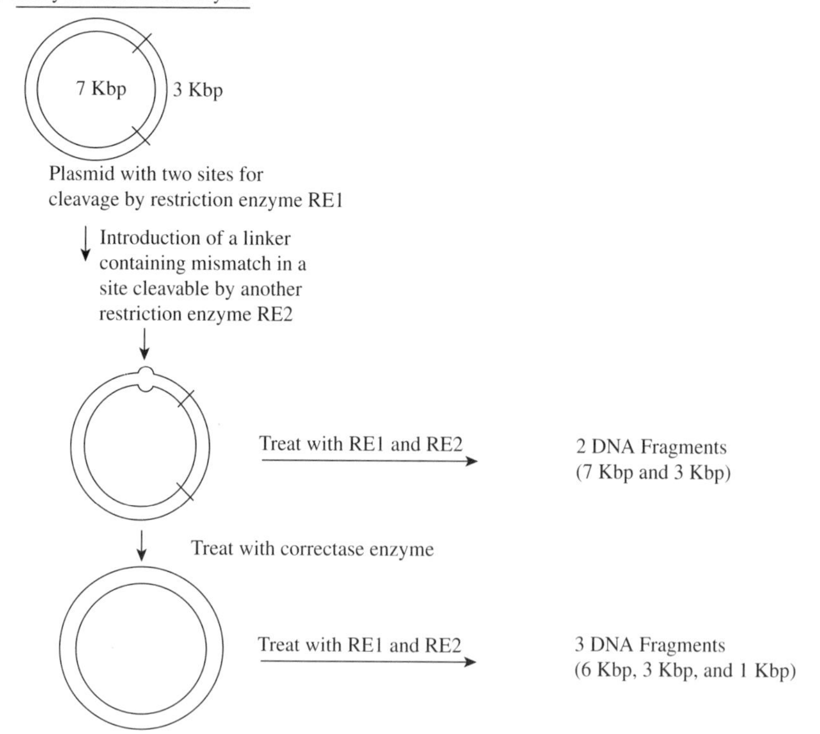

Figure 7.4. Methods for the assay of correctase enzyme.

and after incubation with correctase. The strategy for developing assay for correctase is depicted in Figure 7.4. Later the mismatch correction can be confirmed by the nucleotide sequence analysis. Such correctase activities *in vitro* have been described from the yeast cells (Muster-Nassal and Kolodner, 1986; Bishop et al., 1987) as well as from oocytes of *Xenopus* (Varlet et al., 1990). Also, the presence of heteroduplex in the meiotic product of yeast has been demonstrated (Lichten et al., 1990). Plasmid DNA containing a heteroduplex region has been found to undergo mismatch repair in yeast when such plasmids are introduced into yeast cells via transfection (Bishop and Kolodner, 1986). The *in vivo* activity of correctase has been assayed by introducing plasmids containing known mismatch of base pairs in a reporter gene such as the β-galactosidase gene. Correction of mismatch in a reporter gene can be identified by the ability of yeast transformants with blue colonies on plates containing X-gal. A particular yeast mutant PMS-1 was found to yield mixed colonies (i.e., blue and white) at a much higher frequency as compared to the wild-type cells when transformed with a plasmid containing Lac reporter gene with a mismatch of base pairs (Bishop et al., 1989). These findings suggested that the PMS-1 gene may control or encode a correctase enzyme in yeast. Genetic studies indicate the role of other genes in controlling the mismatch repair process in yeast. It is not known whether these genes encode or control the structures of correctase or proteins that influence the activity of correctase. However, it has now been shown that PMS-1 mutants are deficient in mismatch repair enzyme proteins that enable the correctase to recognize the mismatch in a heteroduplex DNA. For mismatch correction in yeast, in addition to PMS-1, another gene product, MLH-I, is required (Prolla et al., 1994). The yeast PMS-1 and MLH-I proteins are similar to *E. coli* mut-L protein in amino acid sequence and in its function or to HexB of *S. pneumoniae* (Kramer et al., 1989; Prudhomme et al., 1989; Prolla et al., 1994). Likewise, it has been shown that the mutS gene of *E. coli* or of *S. tymphimurium* is similar to HexA of *S. pneumoniae* (Balganesh and Lacks, 1985; Haber et al., 1988; Kramer et al., 1989). Thus mutL, HexB, and PMS-1 are similar; also, HexA is similar to MutS. The human genes in MLHI and hMSH2, located on human chromosome 3 and 2 respectively, are equivalent to *E. coli* mut L and mut S involved in the repair of base-pair mismatch by recognizing the mismatch region and by orchestrating the involvement of the appropriate DNA repair nucleases and other enzymes (Service, 1994). It has been further shown by *in vitro* assays that a certain mismatch such as T/C is corrected more efficiently by the yeast enzyme. In addition, a number of other genes such as Mut A, Mut B, Mut D, and Mut T participate in mismatch repair (Michaels et al., 1990; Schaaper, 1988). A similar correctase has been described from *Xenopus* oocyte cells by the ability of the cell-free extract to generate a new restriction site in the region of mismatch as a result of mismatch correction by the enzyme in *in vitro* assay. However, the information regarding the molecular size or structure of yeast or *Xenopus* enzymes is lacking.

B. Site-Specific Recombinase

The site-specific recombination involves a number of events; these include pairing of two recombination sites, four-strand cleavage, and rejoining of the DNA strands. At least in some cases this complete set of events is mediated by a single protein (Reed, 1981a,b; Abremski and Hoess, 1984, Broach et al., 1982; Meyer-Leon et al., 1987; Silverman and Simon, 1983; Vetter et al., 1983). Thus, the site-specific recombination provides an excellent opportunity to study the interaction between a protein and DNA segment (Nash, 1981; Craig and Nash, 1983; Craigie and Mizzuchi, 1986; Huess et al., 1987; Kitts and Nash, 1987; Nash and Robertson, 1981; Nash et al., 1987). Most site-specific recombinases such as λ Int protein, P-1 cre protein, Tn3 and γδ resolvases, and Hin, Cin, Pin, and Gin invertase and yeast FLP protein recognize a specific site in target DNA (Kamp and Kahman, 1981; Sternberg and Hoess, 1983; Hoess and Abremski, 1984, 1985; Hoess and Landy, 1978; Hoess et al., 1980). The prokaryotic site-specific recombinases have been classified into two main groups. These are the Int family and Hin family (Argos et al., 1986; Jayaram, 1988). The first family consists of 7–8 Int recombinase proteins consisting of 336–440 amino acids and includes λ phage and related phage (80, P22, P2, 186, P4, and P1) (Ljungquist and Bertani, 1983; Leong et al., 1986). In contrast to the Hin family of recombinases, the Int family proteins are divergent (Argos et al., 1986). However, a 40-amino-acid stretch at the C-terminus bears great homology, and remarkably three amino acids such as His 396, Arg 399, and Tyr 433 are exactly the same in all proteins of the Int family. The yeast 2u circle plasmid recombinase Flp also belongs to the Int family because of its similarities of protein including the homology of the stretch of 40 amino acids at the C-terminus and the presence of the conserved amino acids such as His 396, Arg 399, and Tyr 433. The recombinases belonging to the Int family are very site-specific and cannot complement each other *in vivo* or *in vitro*. Unlike the Int family, the Hin family, which consists of invertases such as *Salmonella* Hin, phage Mu Gin, phage P1 Cin, and *E. coli* Pin proteins, is highly homologous. These proteins can readily complement for each other *in vivo* and *in vitro*. The Hin family also includes Tn3- and γδ-encoded resolvases. Recently a number of site-specific recombinases have been identified and characterized from eukaryotes. In mammals, recombinase plays a great role in the organization of the antibody gene, the T-cell receptor gene (Tonegawa, 1983), and in the development of neurons (Chun et al., 1991). In addition, the site-specific recombinase gene, along with the DNA sequence that its product recognizes, has been introduced in a variety of cell lines including bacteria, animals, and plants to cause site-specific recombination and expression of a reporter gene (Sauer and Henderson, 1988). A general description and properties of the recombinases are presented below. Certain characteristics of the site-specific recombinases are summarized in Table 7.2.

Integrases represents a diverse group of tyrosine recombinases. A comparison of 81 such proteins via alignment of their amino acid sequences suggest the conservation of many more amino acids in key positions beside the tyrosine. These include Arg, His-X-X-Arg, and Tyr in the C-terminal half. Mutations in any of these amino acids render the integrase enzyme defective in recombination (Esposito and Scocca, 1995).

C. Prokaryotic Site-Specific Recombinase

1. Integrase

λ Int Recombinase. The λ Int recombinase native protein is a monomer with a molecular weight of 40 kDa. The Int protein in combination with an integration host factor (IHF), a 20-kDa protein encoded by *E. coli*, catalyzes the integration of the λ phage into *E. coli* chromosome by site-specific recombination between the λ (att-p) and bacterial (att-b) attachment sites (Campbell, 1962). The λ Int protein, the gene encoding this protein, and the sites involved in recombination are fully characterized at the molecular level (Hsu and Landy, 1984; Landy, 1989). The λ Int gene and the phage recombination site (att-p) as well as the bacterial recombination site (att-b) have been cloned and the role of different nucleotides in recombination has been demonstrated by *in vitro* mutagenesis. The intermediates in the site-specific recombination have been identified and characterized by the use of suicide substrates (i.e., altered form of att sites) that fail to release the enzyme integrase (Nunes-Düby et al., 1987).

Int proteins are synthesized during the lysogenic cycle but not during the lytic cycle. Excision (or reversal of integration) is catalyzed by Int protein in combination with phage Xis protein and host-encoded IHF factors. The Xis protein appears to be a polypeptide of 8 kDa. The properties of Xis protein remain to be elucidated. The IHF and Xis proteins have no detectable endonuclease and topoisomerase activities. The Int protein has been shown to possess topoisomerase activity. The int and xis genes are sequenced and found to overlap a region of nucleotide segment (Davis, 1989).

A mutant Int-h protein has been described which does not require *E. coli* IHF protein for its function, suggesting that the IHF acts only as an accessory protein (Lange-Gustafson and Nash, 1984; Miller et al., 1980). It is further shown that Int-h protein can recognize mutant attachment sites; this property of the mutant Int-h suggests that Int-h is a proficient integrase endowed with the ability to recognize an att site not recognized by the wild-type Int protein (Miller et al., 1980). Integrase (Int) and excisionase (Xis) proteins have been described from a number of bacteriophages as listed in Table 7.2.

a. P1 Recombinase. In bacteriophage P1, a gene encodes for a recombinase called Cre protein which efficiently carries out recombination between two p lox sites. The Cre protein is a monomer with a molecular weight of 35 kDa (Abremski and Hoess, 1984). The Cre protein, unlike the λ Int (Kikuchi and Nash, 1978) or $\gamma\delta$ resolvase, can act on a linear, supercoiled, or nicked DNA equally well. The recombination mediated by Cre protein causes the circularization of P1 DNA and the breakdown of dimer P1 (Austin et al., 1981; Sternberg et al., 1986). Both the Cre gene as well the p lox site have been cloned, and an *in vitro* recombination of p lox sites by Cre protein has been described (Abremski and Hoess, 1984). Unlike λ int protein, Cre protein does not show any classical topoisomerase activity.

2. Invertase. In certain instances, the site-specific recombination involves the inversion of a DNA segment, and the recombinase involved in such transposition by inversion of a specific DNA segment is called *DNA invertase* or *DIN* (for DNA inversion) *recombinase*. The DNA segment undergoing inversion is called an *inverton*. At least four invertase systems have been very well described (Kutsu Kake and Iino, 1980; Craig, 1985; Glasgow et al., 1989). These include P1 *Cin*

Table 7.2 Characteristics of site-specific recombinase

	Size, Molecular Weight, and Amino Acids (AA)		Inversion Region
	Prokaryotic Enzymes		
Inv Family			
Cin	21 kDa	186 AA	4200 bp
Gin		193 AA	3000 bp
Hin	19 kDa	190 AA	1000 bp
Pin		174 AA	1800 bp
Tn-3		185 AA	
τδ			
Int family			
λ Int (xis)		356 AA	
P2 Int		(72 AA)	
186 Int		—	
P22 Int (xis)		—	
Q80 Int (xis)		388 AA	
P4 Int		(116 AA)	
PI cre		338–416 AA	
		(65 AA)	
		343 AA	
	Eukaryotic Enzymes		
Yeast FLP		423 AA	
HO endonuclease	65 kDa	586 AA	
V(D)J recombinase			
	Intron Encoded Endonucleases (Homing Enzymes)		
Yeast ω protein		234 AA	
Yeast al_4 protein		258 AA	
Bacteriophage T4 td intron nuclease		245 AA	

(Iida et al., 1982), Mu *Gin* (Kamp et al., 1978), *Salmonella typhimurium Hin*, and *Escherichia coli Pin* invertase (van de Putte et al., 1984). Expression or nonexpression of the specific gene controlled by site-specific inversion leads to phase variation in *Salmonella typhimurium* or determines host specificity of phage Mu. Except for Hin recombinase of *Salmonella*, none of the invertase have been purified or their physicochemical properties characterized to any extent. However, the structural genes for *Cin*, *Gin*, *Hin*, and *Pin* (also the resolvase trpR gene) possess a considerable extent of DNA homology (Zieg and Simon, 1980; Plasterk et al., 1984; Reed et al., 1982; Hiestand-Nauer and Iida, 1983; Halford et al., 1979; Rogowsky and Schmidt, 1984). These genes (except for trpR) complement each other *in vivo*.

a. Cin Invertase. This enzyme is involved in the inversion of the C region of bacteriophage P1. The Cin gene has been cloned, and its sequence as well as the sequence of the substrate DNA have been determined (Hiestand-Nauer and Iida, 1983; Iida, 1984; Huber et al., 1985).

b. Gin Invertase. Mu G region 3000-bp inversion determines the host range specificity among *E. coli* K-12, *E. coli* C, and *Citrobacter freundi* (Kamp et al., 1978; Giphart-Glasser et al., 1982; Van de Putte et al., 1984; Kahmann et al., 1985; Kanaar et al., 1988). The G(+) phages can grow on *E. coli* 12 by controlling the expression of s and u genes, whereas G(−) phages controls the expression of another set of proteins such as s′ and u′. This inversion of the "G" region is catalyzed by the Gin invertase.

c. Hin Invertase. Hin invertase causes the inversion of a 1000-bp DNA segment adjacent to H2 operon. The invertible segment encodes the Hin protein that controls the expression of H2 operon in *Salmonella*. The Hin protein consists of 190 AA. It has both trans and cis activity function (Simon et al., 1980). The crystal structure of Hin invertase has been reported (Feng et al., 1994), which provides insight into the mechanism of the action of this enzyme. Hin invertase interacts with the recombination sites as well as with recombination enhancer sequence in conjunction with a DNA bending protein, FIS, to facilitate the site-specific inversion (Johnson et al., 1988). A complex of these proteins with the DNA sequences is called Hin invertasome (Heichman and Johnson, 1990).

d. Pin Invertase. Pin invertase catalyzes the inversion of a 1800-bp region adjacent to it on the *E. coli* chromosome. The gene encoding Pin invertase is located between *pur*B and *fab*D on the *E. coli* map. The Pin gene and the P region are parts of the e14 genetic element that is a cryptic UV excisable prophage.

The presence of Pin invertase was first identified by its ability to complement to the Gin^- Mu phage grown in the *E. coli* K-12 strain (van de Putte et al., 1984). Cin, Gin, Pin, and Hin systems have been found to complement each other (Iida, 1984; Kahmann et al., 1985; van de Putte et al., 1984). The possible role of P inversion mediated by *Pin* invertase in *E. coli* has not yet been determined.

e. Min Invertase and Shufflon-Recombinase. Min invertase causes the inversion of DNA segments in the phage P1-related plasmid of p15B of *E. coli* $15T^-$ (Sandmeier et al., 1990). This enzyme is a member of the invertase family but causes site-specific recombination among six sites as opposed to two sites involved in inversion of DNA by other invertase of the Din family. The Min invertase has been shown to complement with Cin invertase of the phage P1. In addition to the Min system, the R64 plasmid of *E. coli* contains an invertase that is also involved in multiple DNA rearrangement (Komano, 1999). This enzyme is called *Shufflon-recombinase*. However, unlike the Din enzyme, the Shufflon-recombinase shows similarities to the integrase family of site-specific recombinase (Kubo et al., 1988).

3. Resolvase. A number of transposons such as $\gamma\delta$ (a 5700-bp element related to Tn3 transposon; Heffron et al., 1979), Tn3 (Reed, 1981a), Tn21, and Tn17-21 (Halford et al., 1979; Rogowsky and Schmidt, 1984) possess two genes, TrpA and TrpR. The first gene (TrpA) encodes for a 120-kDa transposase that is responsible for the formation of a cointegrate, whereas the second gene (TrpR) encodes a 21-kDa resolvase protein that catalyzes the resolution of cointegrates as depicted in Figure 7.3. The resolvase acts at a specific site called Res, located cis to the trpR cistron (Kitts et al., 1982).

The resolvase protein (Tn3 and Tn21) has 185 amino acids and requires Mg^{2+} for its action (Halford et al., 1979). TrpR genes of different transposons (Tn3, Tn21, and Tn17-21) have been cloned in expression vectors, and resolvase proteins have been extensively purified and their action in resolution of cointegrates has been demonstrated *in vitro* (Reed, 1981a; Halford et al., 1979; Rogowsky and Schmidt, 1984). The resolvases consist of a short 40-AA C-terminus domain involved in DNA binding and a large N-terminal domain with catalytic function. Resolvase is a site-specific recombinase that acts at the Res site. Likewise, the $\gamma\delta$ resolvase consists of 185 amino acids. The 145-AA N-terminal and 40-AA C-terminal fragments can be obtained after mild proteolysis; the N-terminal fragments possess the enzymatic activity, whereas the C-terminal AA possess the DNA binding specificity. This view is supported by the study of recombination-deficient and DNA-binding-proficient mutants. The analysis of crystal of $\gamma\delta$ resolvase shows N-terminal and C-terminal amino acid sequences joined by a hinge. The resolvase forms a linkage with DNA-phosphoserine.

D. Eukaryotic Site-Specific Recombinase

The site-specific recombinases are involved in the rearrangement of chromosomal sequences leading to development of immune system. This idea was first put forth by Tonegawa (1983). A number of site-specific eukaryotic recombinases have been described in somewhat greater detail. These include FLP protein, HO protein, V(D)J protein, intron-encoded homing endonucleases, and other recombinases.

1. Eukaryotic Site-Specific Recombinase

a. Yeast FLP Protein. This protein is responsible for the flip-flop recombination of the 2 μCi in yeast (Broach et al., 1982; Andrews et al., 1985; Govind and Jayaram, 1987; Senecoff et al., 1985; Gates and Cox, 1988). The FLP gene has been cloned (Vetter et al., 1983; Cox, 1983; Sadowski, 1986; Meyer-Leon et al., 1987), and the enzyme has been purified to homogeneity and characterized with respect to certain physicochemical properties including its amino acid sequence and its recognition and cleavage of recombination sites. The yeast FLP recombinase belongs to the integrase family of proteins and possesses the conservative amino acid residues (i.e., His 396, Arg 399, and Tyr 433) of the Int family (Jayaram, 1988). Mutants with altered amino acids in these positions are defective in the cleavage of recombination sites but not in the recognition and binding to specific sites

(Jayaram et al., 1988; Prasad et al., 1987). The FLP protein is covalently linked to DNA via a phosphotyrosyl linkage (Gronostajski and Sadowski, 1985). The role of the flanking region of FLP gene in site-specific recombination has been established (Jayaram, 1985; Andrews et al., 1985, 1987).

b. Yeast Endonuclease SceI. This yeast endonuclease makes site-specific cuts with 5-PO_4 and 3-OH ends that can be rejoined by the ligase. The enzyme has been purified to near homogeneity and found to consist of two dissimilar subunits of 70 kDa and 50 kDa. These subunits do not show endonucleolytic activity when alone. The gene for the 70-kDa subunit has been cloned and its product has been found to be identical to the yeast heatshock proteins HSP-70. This discovery adds a new dimension to the control of the action of this enzyme. The enzyme is considered to form recombination initiation sites in yeast. In view of this possibility, the fact that this enzyme may act in response to certain environmental factors is of significance. The characteristics of this enzyme have been described in detail (Watabe et al., 1983; Morishima et al., 1990; Monteilhet et al., 1990). Sce-I can be used as a rare cutter (restriction) endonuclease (Thiery et al., 1991).

c. Yeast HO Gene Product or Y-Z Endonuclease or EndoSceII. This protein catalyzes the interconversion of the yeast's two mating types α and a. Proximal as well as distal to *mat* locus, additional mating types designated as *HMLα* and *HMLa* exist on the third chromosome of yeast. The conversion of one mating type into another may occur due to recombination between *mat* locus and the *HMLα* or *HMLa*. Such interconversion is very frequent, almost during every cell division, in the homothallic yeast strain carrying the HO gene. In the heterothallic strain lacking the HO gene, such interconversion of mating types occurs at a very low frequency. It has now been shown that the HO gene encodes for a site-specific endonuclease. The HO gene has been cloned and its product is a protein containing 586 AA with a molecular weight of 65 kDa. The purified HO endonuclease has no detectable topoisomerase activity. Its endonucleolytic activity is not influenced by the presence of dNTPs, NTPs, or adenosylmethionine (Kostriken et al., 1983). It requires Mg^{2+} for its activity. The presence of Mn leads to a relaxation in the site specificity of the enzyme. The recognition site for the HO gene product has been characterized *in vitro* using mutant obtained after site-directed mutagenesis (Nickoloff et al., 1990). The HO endonuclease is specific for the cleavage of Y-Z junction in the *mat* locus. At least 16 base pairs are involved in the recognition of site for cleavage (Kostriken et al., 1983). Mutations in the HO gene or in the recognition site lead to reduced frequency of mating type switch. HO endonuclease is produced only during the early S phase of haploid yeast cells. It has been suggested that HO endonuclease is like class II restriction endonuclease and is most likely suitable for involvement in mitotic recombination leading to mating type switch. The finding that *E. coli* cells carrying the plasmid-borne yeast HO gene are unable to survive suggests that HO endonuclease can recognize certain sequences in *E. coli* genome and cleave them, leading to the death of the bacterial cells.

d. Mammalian V(D)J-Endonuclease. This is a site-specific endonuclease responsible for the assembly of functional immunoglobulin and T-cell receptor genes in mammals (Tonegawa, 1983). Enzymes for site-specific cleavage of the J segment in the immunoglobin genes have been described (Kondo et al., 1984). Such enzyme activity has been described from extracts of chicken bursa of Fabricius and from mouse liver and HeLa cell extracts (Kondo et al., 1984). A similar endonuclease activity has been described by Desidederio and Baltimore (1984). The physical properties of the V(D)J-endonuclease have not yet been available. However, the enzyme has a specificity for cleavage of sites rich in (G)n/(C)n sequences (Kondo et al., 1984). It is suggested that J-endonuclease may cleave immunoglobin gene specifically in conjunction with some other protein, possibly a putative recombinase (Kondo et al., 1984; Yancopoulous et al., 1986; Lieber et al., 1988; Blackwell and Alt, 1989). Two genes (*RAG-1* and *RAG-2*) have been identified (Oettinger et al., 1990) whose products may act as site-specific recombinase during V(D)J recombination in murine, human, and chicken cells (Chun et al., 1991; Schatz et al., 1989; Oettinger et al., 1990). V(D)J recombinase is active only in the precursors of B and T lymphocytes, not in the mature B cells and T cells producing immunoglobulin and T-cell receptor molecules. Interleukin (IL-7), which is in excess in embryonic thymus, has been found to promote the sustained expression of *RAG-1* and *RAG-2* genes involved in the rearrangement of genes that encode the antigen receptors of lymphocytes (Muegge et al., 1993).

The *RAG-1* gene product has been shown to possess homology to C-terminus half amino acids of yeast topoisomerase I and also with yeast HPR1 gene product, which also acts as topoisomerase and affects recombination (Schatz et al., 1990; Silver et al., 1993). The *RAG-1* gene product is not only specific with regard to cell producing antibody but has been found in cells of murine nervous systems, suggesting the role of this site-specific recombination in the neuronal and other development (Chun et al., 1991; Robinson and Kindt, 1987; Schuler et al., 1986). In addition to RAG genes, the SCID gene product seems to be involved in V(d)J recombination as well as in double-stranded break repair pathways (Hendrickson et al., 1991; Moiser et al., 1988). Mice homozygous for scid mutation are hypersensitive to X-ray and bleomycin, which cause a double-stranded break in DNA and possess large deletion in immunoglobulin and T-cell receptor genes.

2. "Homing" Nuclease (Intron-Coded Nuclease). Certain introns encode an endonuclease that facilitates the transfer of the intron from an intron-containing gene to an intronless gene in genetic crosses (Dujon, 1989). Such site-specific endonucleases have been called *homing endonuclease* because the transfer of intron is always targeted to a particular site or home (Perlman and Butow, 1989). Such mobile introns are responsible for the phenomenon of gene conversion of an intron-minus gene to an intron-plus gene, first discovered in yeast (Dujon, 1989). It has been possible to construct an artificial mobile element containing, the EcoRI restriction endonuclease gene (Eddy and Gold, 1994). A number of introns encoding such homing endonucleases have been identified and well investigated both in bacteriophage T4 and in yeast. Some of these are described below.

a. Bacteriophage (T4) Intron-Encoded Homing Enzymes Td and SunY. The bacteriophage T4 thymidylate synthetase gene (Td) and SunY gene of unknown function contain introns that encode endonuclease responsible for their mobility. The ORF (open reading frame) for these T4 enzymes has been identified and characterized after the *in vitro* transcription and translation of the ORF of these introns. The T4 Td ORF is 735 bp in length and encodes an endonuclease that has a molecular weight of 28 kDa and contains 245 amino acids (West et al., 1986, 1989; Quirk et al., 1989). The SunY ORF is 774 nucleotides in length and codes an endonuclease that is rich in basic amino acid (such as lysine) like the T4 Td homing enzyme. The expression of the Td and SunY ORFs and the mobility of these introns are controlled by the T4 late promoter and gp55 protein (Quirk et al., 1989). The T4 homing enzymes identify a 21-bp nucleotide sequence during the transfer of intron. Mutation in this region abolishes such intron transfer.

b. Yeast Homing Endonuclease. The most well-characterized yeast homing endonucleases are the yeast omega (ω) protein and other two nucleases all encoded by introns of different mitochondrial genes. Yeast ω (omega) protein is an endonuclease encoded by the 1.1-kb intron of the mitochondrial gene encoding 21S rRNA. The ω ORF encodes a protein of 235 AA which has been identified as a 26-kDa protein upon cloning and expression in maxicells of *E. coli*. Another yeast intron called 4d of the CoxI mitochondrial gene also encodes a homing endonuclease. The third yeast mitochondrial intron al4 encodes a homing endonuclease containing 258 AA. These yeast enzymes recognize an 18-bp nucleotide sequence where it makes a 4-bp staggered cleavage with 3-OH hangover. The yeast homing endonucleases resemble HO and other endonucleases in their recognition of the homing sites. These two intron-encoded endonucleases in yeast share a short peptide containing a LAGI-DADG (i.e., leucine, alanine, glycine, isoleucine, proline, adenine, proline, glycine) sequence that is found in a number of other group I introns but not in the T4 introns. Homologous introns are inserted at the same positions in different fungi (Lang, 1984).

3. Viral Integrase. A number of DNA and RNA viruses are integrated into eukaryotic chromosomes. The DNA viruses can be integrated directly, whereas the RNA viruses and retroposons are integrated after reverse transcription into their DNA copies. In any event, the process of integration requires recombinase function that is mediated by integrase encoded by the viral *Pol* gene. A number of integrases have been characterized from different sources. Recently an integrase encoded by the human immune deficiency virus (HIV) which may cause the AIDS disease in humans has been described. The HIV integrase has been identified and characterized as a 32-kDa protein after cloning of the HIV integrase gene. The HIV Int gene product was shown to carry out integration *in vitro* (Bushman et al., 1990). It has been further shown that the HIV integrase can catalyze the reversal of integration leading to splicing of DNA substrate analogous to the action of RNA splicing enzyme. Viral integrase including HIV integrase can catalyze two nucleolytic reactions: first the cleavage of the viral DNA and second the cleavage of the host DNA in site-specific manner. In addition, the integrase catalyzes the DNA joining reaction like most recombinases (Katz et al., 1990; Katzman et al., 1991). The viral integrase must require several host proteins for its action which have not yet been identified.

E. Transpositional Recombinase

Transposable elements or transposons, first identified in corn (McClintock, 1951; Döring and Starlinger, 1984, 1986), have been found to exist throughout different living systems (Kleckner, 1981; Boeke and Corces, 1989). They can bring about several kinds of genetic rearrangements via the action of nucleases or transposases encoded by them. Different kinds of transposable elements have been identified. They are known as insertion sequences (IS), transposon (Tn), and transposable genetic elements (TGE).

A transposase gene of an insertion sequence (IS) or of an inverton or of a transposable genetic element usually encodes for a protein that can bring about concerted breakage and reunion of phosphodiester bonds within specific DNA sites leading to insertion (integration), excision (deletion), or inversion of specific DNA segments. These events are exemplified by the transposase activity of IS-2 sequence or $\gamma\delta$ of F plasmid. In addition to the inversion mediated by the invertase determining the flagellar phase variation in *Salmonella* (Simon et al., 1980), the Mu transposon that controls the host range variation in phage Mu causes inversion (Shapiro, 1979; van de Putte et al., 1984; Berg and Howe, 1989; Mizuuchi, 1992). Transposases do not necessarily need any DNA homology during the genetic recombination; hence, the latter is called *illegitimate recombination*. Transposases may be encoded by a gene or by the ORF of an intron in an organism. These are classified as described below.

1. Prokaryotic transposases
   Class I insertion sequence (IS) transposases
   Class II transposasable element transposases
   Class III bacteriophage transposases
2. Eukaryotic transposases
   Ac-element and spm transposases in corn
   P-element and other transposases in *Drosophila*
3. Retro transposable elements and retroposases
   Human retrotransposases

A brief description of certain important transposases is given below. An extensive description of different transposases is presented elsewhere (Berg and Howe, 1989). Certain transposases and nucleases involved in transposition (including retroposition) are summarized in Table 7.3.

1. Prokaryotic Transposases. Prokaryotic transposases have been grouped into three classes based on the properties of the transposons that encode them.

Table 7.3 Transposases and nucleases involved in transposition and retroposition

	Prokaryotic
IS-4 transposases	47- to 52-kDa proteins that require host protein Hu and IHF
IS-10 transposases (Tn10)	42-kDa (402 AA) protein that also requires host protein Hu and IHF (Tn50 transposon contains IS-10 as inverted terminal repeat; IS-10 on the left is inactivated by accumulation of mutation, and IS-10 on the right is functional and provides the transposases for the mobility of Tn10)
IS-50 transposases (Tn50)	Also requires host dnaA protein
Tn7 transposases	Five proteins—two involved in target selection
Tn3 transposases resolvases	185 AA (120 kDa) and (21 kDa) proteins; resolvases also involved in transposition
Mu transposases	gpA gene product 75 kDa (663 AA), gpB gene product; 33-kDa protein helps in integrative transposition (Mizuuchi and Craigie, 1986)
	Eukaryotic
Maize Ac transposases	807 AA (Gierl et al., 1989)
Drosophila P element transposases	87 kDa
	Retroposases
Retroviral integrase	100 S particles, 32–46 kDa (Katz et al., 1990)
ty Integrase	58 kDA Eichinger and Boeke (1988)
HIV integrase	32 kDa

The Class I transposases belong to the bacterial insertion sequences that contain only one gene encoding the transposase. The class II transposases belong to transposing elements designated as Tn series which encode for proteins other than the transposases. The Class III transposases are encoded by bacteriophages that facilitate their mobility.

a. Class I Transposases. Certain bacterial insertion sequences (IS) encode a transposase required for their mobility. The class I transposases are highly basic proteins that function only in cis positions—that is, act on DNA segments that encode them. The complementation of a mutant IS by wild-type IS *in trans* is not possible. Some of the class I transposases have been cloned and identified by an *in vitro* transcription and translation system and by expression in minicells after cloning in *E. coli* (Trinks et al., 1981). The IS-4-encoded transposase was found to show a molecular weight of 47 kDa, which is somewhat less than its molecular weight (54 kDa) as predicted from the nucleotide sequences. Likewise, the molecular weight of the transposase encoded by IS-10 was determined to be 42 kDa, which is again somewhat less than the 46 kDa predicted by the nucleotide sequence of the gene (Marisato and Kleckner, 1984; Mahillion et al., 1985). The IS transposases show variation in the number of amino acids in the peptides: 270 AA (ISH1), 442 AA (IS4), 402 AA (IS10), 461 AA (IS50), and 478 AA (IS231). However, these disparities can be due to the nature of proteins. The IS-10 transposases possess a tyrosine residue in their C-terminus. The C-terminus is involved in recognition of target DNA during transposition. Two classes of mutants of IS-10 have been identified and characterized. The first group of mutants, called $ExC^{+}Int^{-}$, are proficient in excision but defective in integration, whereas the second group of mutants, called ATS, are altered in target site specificity (Haniford et al., 1989; Bender and Kleckner, 1992). These two groups of mutants significantly differ in transposition frequency. The action of IS-10 transposases is controlled by an antisense RNA.

b. Class II Transposases. Both transposases and resolvases have been very well characterized from several Tn transposons such as Tn3, Tn7-21, Tn21, and γδ (Stark et al., 1989; Grindley, 1983; Grindley and Reed, 1985; Wasserman and Cozzarelli, 1985). The transposase is encoded by the tpnA gene. The tpnA gene product leads to the formation of cointegrate. Tn3 resolvase selects the recombination sites by a tracking device (Benjamin et al., 1985).

c. Class III Transposases. The transposase encoded by bacteriophage Mu has been well described. This is a 70-kDa protein encoded by the MuA gene. The transposase has been characterized after cloning in *E. coli* and expression in minicells. The purified protein can catalyze the transposition *in vitro* and complement *in trans* (Mizuuchi, 1983). MuA protein occurs as a tetramer. The amino acids Asp 269 and Glu 392 are crucial for the transposase activity (i.e., cleavage and rejoining of DNA) but not for tetramer formation of the MuA protein. MuA protein possesses the D-D35-E motif possessed by several viral integrases (Baker and Luo, 1994). The transposition by the MuA gene product is further assisted by the MuB gene product, which is a 33-kDa protein. In addition, the transposition is facilitated by a host protein, HU. Mizuuchi and Craigie (1986) have reviewed the role of MuA transposases and other proteins in the transposition of Mu phage.

2. Eukaryotic Transposase. Transposition of a DNA segment was first described in corn by Barbara McClintock (1951). She described the presence and transposition of certain DNA segments designated as Ac, Ds, and Spm. She also showed that mobility of Ds is dependent on the presence of Ac. It has now been shown that both Ac and Spm encode transposases required for the transpositional recombination or mobility of these genetic elements in corn. It has further been shown that Ds is a mutant of Ac and that Ds has lost the ability to encode for its transposase; therefore its mobility is dependent on the presence of Ac which provides the transposase required for the mobility of transposons. The properties of some of these corn transposases as revealed by the techniques of molecular genetics have been reviewed (Gierl et al., 1989; Döring and Starlinger, 1986; Federoff, 1989) and are summarized below.

a. Ac Transposases. The Ac element has an ORF that is transcribed into a 3.5-kb mRNA. The latter is translated into a protein consisting of 807 AA. The tranposase identifies an 11-bp terminal inverted repeat sequence where the enzyme makes the incision. The Ds element of corn is a mutant of an Ac element which has a deletion in the region encoding the Ac transposase. Therefore, the Ds element requires the presence of the Ac element to provide the transposase during its movement from one location to another in the corn genome. In addition to the Ds element, several other mutants of the Ac element with deletion or mutation in the transposase gene are known.

b. Spm Transposase. The evidence for the presence of spm transposase is based on the genetic analysis of this element in corn (Gierl et al., 1989). The Spm has two ORFs. The tpnA gene product encoded by ORF1 is involved in DNA binding. The tnpB gene product is involved in transposition. This is transcribed in part both from ORF1 and ORF2 and yields a 6-kb mRNA. The tnpB transposase has not yet been biochemically characterized due to the low expression of this protein.

c. Drosophila Transposase. The P element of *Drosophila melanogaster* is responsible for the phenomena of hybrid dysgenesis. The P element contains four exons (ORF0, ORF1, ORF2, and ORF3) interrupted by three introns. In the germline cell, all three introns are spliced out to yield an mRNA that is translated into a 87-kDa protein. This protein has the transposase function. In the somatic cell, however, the third intron is not spliced out from the transcript. This transcript containing a stop translational signal in the third intron yields a 66-kDa protein. This protein in the somatic cell acts as a repressor of the transcription of the transposase gene in the P element (O'Hare and Rubin, 1983; Rio et al., 1986; Rio, 1990). This repressor activity of the P element gene product has been demonstrated *in vivo* as well. The differential splicing of the P element transcript is the basis for the tissue-specific effect of P elements. How the differential splicing is controlled and what roles are played by the splicing nucleases remain to be elucidated. *Drosophila* P element has been used for insertion of designed DNA segment into its genome.

3. Retrotransposable Elements and Retrotransposases. These are the most abundant form of transposable elements. They can be classified into two groups: One contains long terminal repeats (LTR) and genes encoding proteins similar to retrovirus. The other group, also called retroposon, does not contain LTR and encodes proteins that differ in amino acid sequences of the retroviral proteins. Like retroviruses, these transposons carry a gene that encodes integrase. Such an integrase facilitates the integration of the DNA copy of the retroviruses (such as HIV or AIDS virus) or of retroposons (such as copia in *Drosophila* or Ty element in yeast) into host chromosomes. The yeast ty-1 retroposon integrase is encoded by the tnpB (ORF-2) gene, which encodes a 58-kDa protein (Boeke et al., 1985). This gene is located upstream of the reverse transcriptase gene. In addition to integrase, the other nuclease involved in the process of retroposition is RNaseH. This enzyme (RNaseH) has also been characterized from several retroelements, including the yeast ty-1 element (Boeke and Corces, 1989). Retroposons (non-LTR-retrotransposable elements called R1 and R2) have been found in 28SrRNA genes of a large number of insects. The insertion of these elements in ribosomal genes of insects may be facilitated by a site-specific integrase encoded by them (Jakubczak et al., 1991).

Ribozymes may be involved in retrotransposition of copia-like elements by providing tRNA fragments as primer via cleavage of tRNA by RNaseP (Kikuchi et al., 1990); RNaseP ribozymes are described in Chapter 9.

a. Human Retrotransposases. All human transposases identified so far are retroposons. No transposase enzyme has been identified from humans, but its existence is implied by the fact that a transposable element has been isolated and characterized from human individuals with hemophilia A (Dombroski et al., 1991, Mathias et al., 1991). This human transposable element has been found to encode reverse transcriptase, which is presumably involved in the retrotransposition of the human transposon. In addition, a human genome possesses LINES and SINES repeat sequences (Fanning and Singer, 1988; Korenberg and Rykowski, 1988; Hutchinson et al., 1989; Deininger, 1989; Chen and Manuelidis, 1989), which are also transposable genetic elements.

F. Control of Recombinases

Nothing is known about the control of the recombinases involved in homologous and site-specific recombination. In site-specific recombination, the recombinases also interact with several host factors (IHF) as well as with Xis protein to form into a complex structure called *intasome*, which facilitates the site-specific integration and expulsion of DNA segments.

Transposases provide an interesting group of recombinases so far as the control of their activity is concerned. Many of them require host factors (IHF and HU proteins). IS transposases are unique in control of their action by an antisense RNA (called RNA out, transcribed from a strong promoter) (Simon and Kleckner, 1988). The pairing of RNA out transcript with the transposase transcript (called RNA in) leads to the blocking of the region that binds with ribosome. Most transposons encode for a repressor of its transposase. A most interesting situation is presented by the P-element transposase of *Drosophila* where N-terminus amino acid sequences of the repressor and transposase proteins are exactly the same (Rio, 1990). *Drosophila* P element transposase has been shown to repress transcription *in vitro* as well. The unique control mechanism of tranposase might have played a greater role during the process of evolution of organisms, via gene arrangement as discussed in Chapter 14.

G. RNA Recombinase

In addition to recombinases that recombine DNA segments, there must exist recombinases that recombine RNA molecules. However, little is known about the RNA recombinase that can recombine RNA segments even though the recombination of RNA is an established fact (Lai, 1992; White and Morris, 1994). Results of genetic studies provide evidence for the recombination of RNA in RNA virus or of retroelements during the process of reverse transcription. However, no RNA recombinase has yet been characterized which catalyzes these recombination reactions. Recently, certain maturases have been shown to possess RNA recombinase activity. In *Neurospora*, a tRNA synthetase, a product of a nuclear gene cyt18, known to function as RNA maturase by facilitating the self-splicing of certain mitochondrial introns, has now been known to function as a RNA recombinase because it can reinsert the intron RNA into outflanking exon transcripts (Mohr and Lambowitz, 1991). Thus the identification of RNA recombinases and their characterization remain to be investigated. It is possible to develop *in vitro* assays to identify and characterize these RNA recombinase. Their characterization will certainly enhance our understanding of the process.

Furthermore, the enzymes involved in RNA splicing during maturation of pre-mRNA into mRNA in eukaryotes may be considered as RNA recombinase. This type of RNA recombinase capable of alternate splicing or trans splicing between two different pre mRNA plays a significant role during the development of multicellular organisms (Sharp, 1987; Leff et al., 1986; Breitbart et al., 1987). In view of this possibility the identification and characterization of RNA recombinase (including RNA splicing enzymes) are very important.

8

SUGAR-NONSPECIFIC NUCLEASES

I. GENERAL DESCRIPTION, CLASSIFICATION, AND METHODS OF ASSAY

Sugar-nonspecific nucleases can utilize both DNA and RNA as substrates. The classical examples of the enzymes are the snake venom and spleen phosphodiesterases, micrococcal nuclease, S_1 nuclease, and *Neurospora* nucleases. These enzymes have been very well investigated. Micrococcal nuclease is the most well-studied enzyme both biochemically and physically. The sugar-nonspecific nucleases are classified as exonucleases and endonucleases producing 3′ or 5′ phosphomonoesters. The most common examples of sugar-nonspecific nucleases are:

1. Exonuclease
   a. Producing 3′ phosphomonoesters (e.g., spleen phosphodiesterase)
   b. Producing 5′ phosphomonoesters (e.g., snake venom phosphodiesterases)
2. Endonuclease
   a. Producing 3′ phosphomonoesters (e.g., micrococcal nuclease)
   b. Producing 5′ phosphomonoesters (e.g., *Aspergilus* S_1 nuclease)

The properties of these nucleases are summarized in Table 8.1.

A number of enzymes belonging to this group are specific for single-stranded DNA or RNA as substrates. These enzymes are metalloproteins with endonucleolytic activity producing 5′ phosphomonoesters. The properties of these enzymes are summarized in Table 8.1. The majority of the enzymes belonging to this group have been used as a tool in different scientific investigations or commercial preparations. This aspect of the enzyme is discussed separately in Chapter 13.

Table 8.1 Properties of the sugar-nonspecific nucleases

Enzyme	Molecular Weight	pH Optimum	Mode of Attack	Nature of Product
1. Snake venom phosphodiesterase		pH 9	Exonuclease	5′ Mononucleotides
2. Spleen phosphodiesterase		pH 4.8	Exonuclease	3′ Mononucleotides
3. S_1 nuclease *Aspergillus oryzea*	32 kDa	pH 4.5	Endonuclease	5′ Phosphomonoester
4. Micrococcal nuclease *Staphylococcus aureus*	11 kDa	pH 9	Endonuclease	3′ Phosphonucleotides

The enzyme activity is usually determined by measuring the optical density of the solution after TCA precipitation of the undigested substrate (DNA/RNA) (Linn and Lehmann, 1965). An increase in optical density at 260 nm is directly proportional to the enzyme activity. The enzyme activity is also determined by the mobility of the nucleic acid molecules upon electrophoresis in agarose gel after treatment with enzymes such as Bal 31.

II. PROPERTIES OF ENZYMES FROM DIFFERENT GROUPS OF ORGANISMS

A. Microbial Nucleases

1. Neurospora crassa Endonuclease. Among the different single-strand-specific endonucleases, *Neurospora* enzyme was the first one to be discovered (Linn and Lehman, 1965). The enzyme has been extensively purified from both the conidia and the mycelium of *Neurospora crassa* and has a molecular weight of 55 kDa (Linn, 1967). However, Fraser and his co-workers (Fraser et al., 1976, 1980; Fraser, 1979; Ramotar et al., 1987) have recently shown that the enzyme most probably occurs as a prenuclease protein of molecular weight 88 kDa. This protein, when acted upon by endogenous protease or by added trypsin, yields an active enzyme with both endo- and exonucleolytic functions. Further proteolysis of this endo–exonuclease yields the single-strand-specific endonuclease with a molecular weight of 55 kDa, similar to the form of the enzyme first described by Linn and Lehmann (1965). The *Neurospora* enzyme hydrolyzes DNA and RNA with equal efficiency and can effectively attack single-stranded regions in a double-stranded DNA. The latter ability of the enzyme was successfully utilized during the first isolation of a gene (Shapiro et al., 1969). The properties of *Neurospora*

Table 8.2 Properties of the single-strand-specific nucleases

Enzyme and Source	Molecular Weight	pH Optimum	Cofactor	Inhibitors	Comments and Reference
1. Bal31 nuclease *Alteromonas espejiana*	73–83 kDa	pH 7 ssDNA 8.8 dsDNA 8	Ca, Mg		Slow and fast forms of enzyme; active in presence of SDS, urea (Gray et al., 1975, 1981)
2. Neurospora crassa	55 kDa	pH 7.5–8.5	Ca, Mg, Fe	EDTA phosphate	90-kDa precursor yields a 75-kDa exonuclease or 65-kDa endo–exonuclease latter on further digestion yield a 55-kDa protein with endonucleolytic activity. Mycelial enzyme 65-kDa protein dependent on Ca with endo activity.
3. S_1 nuclease *Aspergillus oryzae*	32 kDa	pH 4.5–5	Zn	EDTA phosphate nucleotides	Heat-stable (Ando, 1966; Vogt, 1973)
4. R nuclease *Penicillus citrinum*	42–50 kDa	pH 5–6	Zn	EDTA	Heat-stable (Fujimoto et al., 1974)
5. *Saccharomyces* nuclease	—	pH 7.6	Mg	—	Lee et al. (1968)
6. *Physarum polysephalum*	32 kDa	pH 8	Zn	EDTA, PO	Like S nuclease except for pH optimum (Waterborg and Kuyper, 1979)
7. DNaseI *Ustilago maydis*	42 kDa	—	Zn, Co	EDTA	Recombination-deficient mutants lack DNaseI (Holloman and Holliday, 1973; Holloman, 1973)
8. Mung bean nuclease I	39-kDa glycoprotein	pH 5	Zn	EDTA	Sung and Laskowski (1962), Krocker et al. (1976)
9. Wheat seedling nuclease	43 kDa	pH 4.5	Zn		Exonuclease for RNA. Endonuclease for DNA properties similar to mung bean nuclease (Krocker et al., 1975; Wani and Hadi, 1979)
10. Sheep Kidney nuclease (SK)	52–53 kDa	pH 7.6	Mg	EDTA	Nucleolytic product includes all 16 deoxynucleotides (Kasai and Grunberg-Munago, 1967; Watanabe and Kasai, 1978)

enzyme are summarized in Tables 8.1 and 8.2. Genetic mutants *uvs*-2, *uvs*-3, *uvs*-6, and *nuh*-4 are deficient in the release of ss specific nuclease.

Mutants of *Neurospora crassa* deficient in sugar-nonspecific nuclease have been described (Ishikawa et al., 1969; Forsthoefel and Mishra, 1983). In *Neurospora* a mutation in either of several *nuc*-1 to *nuc*-7 loci can cause deficiency in sugar nonspecific nucleases. However, it is not known if any of these genes are the structural gene for the sugar-nonspecific nucleases (Forsthoefel and Mishra, 1983; Käfer and Fraser, 1979). There is some indication that *nuc*-2 may be a structural gene because a ts *nuc*-2 mutant was found to possess temperature-sensitive sugar-nonspecific nuclease (Ishikawa et al., 1969). Similar mutants deficient in sugar-nonspecific nucleases of *Ustilago maydis* have been described. These mutants seem to be defective in DNA repair or in genetic recombination (Mishra and Forsthoefel, 1983; Holloman, 1973; Holloman and Holliday, 1973).

2. S_1 Nuclease. This enzyme was first described by Ando (1966) from *Aspergillus oryzae*. This enzyme has been purified to homogeneity (Vogt, 1973) and has a molecular weight of 32 kDa. Like other single-strand-specific nucleases, this enzyme is a metalloprotein and acts on both DNA and RNA; however, it is five times more active on DNA than on RNA. The enzyme acts on a partially denatured region caused by superhelicity of DNA or caused by mismatch due to mutation in a duplex DNA (Beard et al., 1973; Kato et al., 1972; Shenk et al., 1975; Rushizky et al., 1975) or on a non-hydrogen-bonded loop of tRNA. These properties of the enzyme have been utilized in the so-called S_1 mapping of a DNA region. Other properties of the enzyme are summarized in Tables 8.1 and 8.2. The enzyme has been used to map DNA and RNA (Lycan and Danna, 1984).

3. Yeast Nucleases. Two sugar-nonspecific enzymes have been described from yeast. These include a nuclear enzyme encoded by the *Rad*-52 gene (Chow and Resnick, 1988) and a mitochondrial enzyme encoded by the *Nuc*-1 gene. Both *Rad*-52 and *Nuc*-1 are nuclear genes located on chromosomes XIII and X of yeast genome. Surprisingly, both are immunologically related to *Neurospora* endo–exonuclease. It is of interest that in yeast the two distinct genes encode these enzymes. However, in *Neurospora* the nuclear and mitochondrial enzymes are derived from the same preprotein. The role of both *Neurospora* and yeast enzymes in DNA repair and recombination is indicated by the fact that mutations in genes encoding these proteins are defective in the DNA metabolism. The yeast *Rad*-52 enzyme is a 72-kDa protein that is present in logarithmically growing cells but not in stationary cells; the *Rad*-52 mutant cells are deficient in this enzyme and possessed only 10% of the wild-type level. The yeast *Rad*-52 endo–exonuclease has a pH optimum of 7.5 and requires Mg^{2+} or Mn^{2+} for its activity (Chow and Resnick, 1988). The fact that yeast *Rad*-52 mutants are defective in meiotic synapsis, recombination, and DNA repair suggests the role of this enzyme in all these processes (Klein, 1988; Alani et al., 1990).

The yeast *Nuc*-1 enzyme is a 37-kDa protein. This enzyme represents 50% of all cellular activity and occurs bound to mitochondrial inner membrane; the gene for this enzyme has been cloned, and the entire nucleotide sequence and its deduced amino acid sequence have been determined (Dake et al., 1988). The yeast *Nuc*-1 enzyme showed close homology to C-terminus amino acids of *E. coli Rec C* enzyme; this yeast enzyme also showed immunological relatedness to *Neurospora* endo–exonuclease (Fraser et al., 1990). Future studies of these enzymes could reveal the relationship of these enzymes and their role in DNA repair and recombination.

4. Micrococcal Nuclease. This enzyme is the most well-characterized among all sugar-nonspecific nucleases. This enzyme was first described from *Staphylococcus pyogenes* (Cunningham et al., 1956). It acts on both DNA and RNA with almost equal efficiency yielding 3′ nucleotides. The enzyme is an endo–exonuclease because it first makes endonucleolytic cleavages and then acts exonucleolytically. The enzyme has a molecular weight of 11 kDa but occurs as a dimer with a molecular weight of 20 kDa (Anfinsen et al., 1971). The entire amino acid sequence and three-dimensional structure of the enzyme is known. The enzyme contains 149 amino acids and has a single tryptophan residue but lacks any sulfhydryl groups or disulfide bond. Calcium, essential for the activity of the enzyme, may be required for the stabilization of the tertiary structure of the enzyme. The enzyme isolated from *S. fogii* has properties identical to those of the enzymes obtained from *S. aureus*. The gene for the *fogii* enzyme has been cloned, and mutants of enzymes have been generated. The micrococcal nuclease has been used initially for nearest-neighborhood analysis of bases in a DNA molecule and subsequently for the analysis of the nucleosomal organization of the chromatin. The properties of the enzyme are summarized in Tables 8.1 and 8.2.

Benzonase. Benzonase is the trade name for *Serratia* nuclease (marketed by Merck), which has found extensive use in pharmaceutical and biotechnological applications. This endonuclease cleaves both double-stranded and single-stranded DNA and RNA without any sequence specificity (Benedik and Strych, 1998). This enzyme retains catalytic activity under adverse conditions such as presence of reducing agents, chaotropic agents, and 4 M urea. This enzyme is routinely used for the removal of nucleic acids from cell extracts. The activity of this enzyme is 15 times faster than that of DNaseI and is a dimer under physiological conditions. The gene for this enzyme has been cloned, and its catalytic site established by *in vitro* mutagenesis studies. This enzyme possesses homology to yeast NUC1. The crystal structure of this enzyme is known (Miller et al., 1994). This enzyme is characteristic of *Serratia marcescens*, which distinguishes it from other Enterobactericeae (Eisenstein, 1995).

5. Bal-31 Nucleases. Fast and slow forms of this extracellular enzyme have been purified from the culture medium of *Alteromonas espejiana* (Gray et al., 1975, 1981). The molecular weights of these two forms of enzyme are 109 kDa (slow) and 85 kDa (fast). Both enzymes have an isoelectric pH of 4.2. They are active at neutral pH and are remarkably resistant to inactivation by a number of denaturing agents or conditions such as 5% sodium dodecyl sulfate, 6 M urea, and electrolytes or upon extended storage at cold temperature. Both forms of the enzyme attack linear duplex DNA at termini acting at both strands in a manner to yield shorter duplexes, which can be easily ligated. This property of the enzyme makes it an unusual tool for experiments in molecular biology. This ability of the enzyme has been exploited for rapid mapping and for the production of deletion in cloned DNA segments. The fast form of the enzyme possesses a novel exonucleolytic action on linear duplex RNA. Bal-31 nuclease is also active endonucleolytically on the junction between B and Z regions in a DNA molecule. The single-stranded DNA is degraded by the enzyme to yield 5 mononucleotides. The slow and fast forms of the enzyme are equally active on a linear single-stranded DNA but differentially active on a linear double-stranded DNA; the fast form is 27 times more active than the slow form. Both forms of enzymes are active at neutral pH; this is unlike the *Neurospora*, *Aspergillus* S1, or mung bean nucleases.

B. Animal Nucleases

Two enzymes belonging to this group have been very well characterized; they are the mammalian spleen exonuclease and the snake venom phosphodiesterases. The properties of these enzymes are described below as well as in Table 8.1.

1. Spleen Exonucleases. This enzyme, also called a spleen phosphodiesterase, is an exonuclease that attacks sequentially an DNA and RNA from the 5-OH end, producing 3-monophosphate esters. The enzyme completely lacks endonuclease activity. The enzyme has been extensively purified as a 46S particle. The enzyme is heat-labile, and a 50% inactivation is obtained at 50°C for 20 min. Its reaction mechanism has been extensively investigated and is similar to the mechanism of action of the snake venom enzyme. The native calf thymus DNA is resistant to digestion by this enzyme. Also poly C is resistant to the enzyme, but poly A, poly U, and poly I are readily digested. The glucosylated DNA of T_4 virus is also resistant to digestion by this enzyme. The acetylation of tRNAs completely renders them resistant to digestion by spleen exonuclease but not by snake venom phosphodiesterase. The enzyme has been extensively used in the nucleotide sequence analysis of both RNA and DNA.

2. Snake Venom Exonuclease. This enzyme, also called a snake venom phosphodiesterase, has been very well described (Laskowski, 1957). The enzyme attacks both DNA and RNA and produces 5-monophosphate esters. It also attacks nucleic acid molecules containing the arabinose moiety. The enzyme also attacks native DNA molecules, but glucosylated DNAs are resistant to the enzyme. The chemical and physical properties of the enzyme are not well characterized because of the scant availability of the enzyme. The snake venom enzyme has been extensively used as a tool in analysis of nucleic acid. The enzyme has been extensively used to determine the 5 or 3 end of the nucleic acid molecule (Laskowski, 1967).

C. Plant Nucleases

1. Mung Bean Endonuclease. Among the plant nucleases, this is the most well-characterized enzyme. The enzyme was first described by Sung and Laskowski (1962) from *Phaseolus ouircus* or mung bean. The enzyme has been purified to homogeneity. The enzyme has a molecular weight of 39 kDa and is a glycoprotein. The enzyme catalyzes the hydrolysis of DNA and RNA with equal efficiency. The enzyme is a metalloprotein and readily inactivated by EDTA. The enzyme is an endonuclease as established by the facts that the high viscosity of DNA solution drastically disappears as soon as the enzyme is added to the solution and that the products of degradation are usually up to heptanucleotide (Privat de Garilhe, 1964). However, the enzyme must have some exonucleolytic activity since there is abundance of mononucleotides among the products of enzymatic action (Privat de Garilhe, 1964). Other properties of the enzyme are summarized in Table 8.1. A novel property of the enzyme includes its ability to recognize nucleotide sequences in the beginning and in the end of a gene as shown by the fact that the cleavage of SV40 genomic DNA produces DNA fragments corresponding to the number of viral genes (McCutchen et al., 1984). Two other plant nucleases that are involved in programmed cell death of barley endosperm during germination or trans differentiation of zinnia mesophyll cells into tracheary elements have been very well characterized (Aoyagi et al., 1998). These genes encoding a S1-type DNases have been cloned; the barley cDNA called BEN1 was found to encode for a protein consisting of a 288-amino-acid residue and a putative signal sequence of 23 amino acids. The amount of BEN1 mRNA was found to increase in response to gibberellic acid. The zinnia cDNA called ZEN1 encodes a protein of 303 residues with a putative signal sequence of 25 amino acids. The amount of ZEN1 mRNA increases during the transdifferentiation of the mesophyll into trachea.

2. Other Plant Nucleases. Many fruit plants possess nucleases that produce 5′-GMP, which provide the tasty flavor to the fruits. Such enzymes are used in industry to produce large amounts of tasty substance 5′-GMP as discussed in Chapter 14. Two such acid nucleases (Le1) have been described from *Lentinus edodes* (Kobaayashi et al., 2000). The gene encoding Le1 nuclease has been cloned and was found to encode a glycoprotein containing 290 amino acids and 2–6 residues of hexosamine and neutral sugar. Nuclease Le1 has a signal sequence of 20 amino acids and possesses extensive sequence homology to fungal nucleases P1 and S1.

D. Parasitic Protozoan Nuclease

A group of protozoans such as trypanosomes, trypanozomatids, and leischmania live as parasites in humans and domestic animals and are important because of the diseases that they cause worldwide. These organisms requires purines for their growth but cannot synthesize purines by themselves. They possess surface anchored 3′-NT/NU (nucleotidase/nuclease) to acquire purine by salvage of host nucleic acids. This enzyme is highly inducible by purine starvation. These parasitic protozoan nucleases have been very well characterized from a number of species, and the gene encoding a 40-kDa protein with bifunctional (i.e., nucleotidase and nuclease) activities has been cloned and characterized (Debrabant et al., 1995; Yamage et al., 2000). These protozoan nucleases show extensive homology to fungal S1 and P1 nucleases. The characterization of these nucleases from parasitic protozoans may provide clues to control diseases caused by them.

9

NONPROTEIN NUCLEASES

Molecules other than protein nuclease can possess the nucleic acid strand scission activity. These include ribozymes and chemzymes. Ribonucleic acid molecules that possess the catalytic properties of a protein enzyme have been designated as ribozymes (Cech and Bass, 1986). Likewise, chemical reagents that mimic the catalytic functions of an enzyme have been designated as *chemzymes* (Corey and Reichard, 1989; Waldrop, 1989).

I. RIBOZYMES

Like protein enzymes, ribozymes require native structures to specify the catalytic activity, show high substrate specificity, and form substrate enzyme complex intermediates and demonstrate Michaelis–Menton kinetics during the process of catalysis (Cech and Bass, 1986; Pyle et al., 1990). The properties of these nonprotein enzymes and their role as nucleases are discussed in this chapter. Ribozyme nuclease activity is possessed mostly by eukaryotic intron transcript. Introns with nuclease activity have been found to occur in eukaryotes; the only exception is the T-even series of bacteriophages (Belfort et al., 1986; Bell-Pederson et al., 1989, 1991; Shub et al., 1988; Chu et al., 1984–1988, 1990). However, the bacteriophage intron facilitates splicing by encoding protein nucleases. Properties of different ribozyme nucleases are summarized in Table 9.1.

A. RNaseP

This is among the first ribozymes described and one of the best characterized ribozymes. This is an endonuclease and is required for the processing of all precursor tRNA transcripts to yield their mature 5 termini (Reich et al., 1988). This ribozyme has been characterized from a number of prokaryotic and eukaryotic organisms. Their properties are summarized in Table 9.1. Both the RNA and protein components of this ribozyme have been identified. *In vitro,* the demonstrated RNase activity is entirely due to the RNA component, and the ribozyme acts like a restriction endonuclease (Zuag et al., 1986). The structure of the ribozyme has been very well defined (Latham and Cech, 1989). The *in vitro* protein component may facilitate the enhancement of the ribozyme activity. Genes encoding both the RNA and protein components of this *E. coli* and *Salmonella* ribozyme have been cloned (Baer and Altman, 1985). In *in vitro* reconstitution experiments, it has been possible to reassemble RNaseP enzyme activity using RNA and protein components from different organisms. It has been shown that the presence of 3-terminal CCA nucleotide residues is required for the processing of a tRNA molecule by the M_1RNA of the RNase P enzyme.

The *E. coli* rnpA and rnpB genes encode the protein and RNA components, respectively, of the *E. coli* RNase P enzyme. The rnpB genes from different members of the enterobacteria have been cloned by their ability to complement *E. coli* mutants temperature-sensitive for RNaseP activity. The RNA components of *E. coli* (M_1RNA) and *B. subtilis* (PRNA) enzymes differ significantly in their nucleotide sequences, suggesting the role of protein in recognition of RNA components during the process of catalysis (Cech and Bass, 1986).

In yeast the structure of the anticodon–intron structure of tRNA has been shown to influence the cleavage activity of the RNP (Willis et al., 1986). Studies with RNase P of fission yeast show that the protein component of the enzyme possesses the substrate binding activity while the RNA component has the endonucleoytic activity (Krupp et al., 1986; Nichols et al., 1988). These studies definitely show that yeast RNase P protein plays a greater role than the corresponding components in *E. coli* in enzyme activity.

The RNase P has been purified from a variety of organisms upon DEAE Sepharose column chematography. The *E. coli* enzyme, however, has been found to elute at a higher salt level than the eukaryotic enzyme. The HeLa enzymes purified in this manner can process tRNA precursor molecules form a variety of organisms. The HeLa enzyme is nonspecific to the sources of tRNA precursor molecules. However, the *E. coli* enzyme shows some sensitivity to the origin of the tRNA precursor molecules because it has been shown to process HeLa tRNA precursor molecules at only 10% of the rate of the cleavage of the *E. coli* tRNA precursor molecules. Furthermore, HeLa RNase P is more labile than the *E. coli* enzyme.

1. Protein Component of RNaseP. The *E. coli* RNaseP protein (C-5) has a molecular weight of 13.8 kDa, and a similar protein component from *Bacillus subtilis* has a molecular weight of 14.1 kDa. These values are derived from the nucleotide sequences of the *E. coli* rnpA gene and similar *B. subtilis* gene encoding these proteins. The protein component by itself has no nuclease activity; however, it increases the ribozyme activity of the RNA by over 10-fold.

The role of RNaseP in bacterial tRNA processing has been very well established. Initially, it was thought that both the RNA and protein components are required for

Table 9.1 Properties of different ribozyme nucleases

I. RNaseP		
A. RNA component		Encoded by rnpB gene, *E. coli* and *S. typhimurium* show changes in six nucleotide position. *B. subtilis* RNA is completely different from the *E. coli* RNA enzyme (Baer and Altman, 1985).
E. coli M_1RNA	375 nt	
B. subtilis M_1RNA	400 nt	
S. typhimurium M_1RNA	375 nt	
HeLa cells	400 nt	(Gold et al., 1988, 1989; Gold and Altman, 1986)
B. Protein component		
E. Coli C5 protein	11.3 kDa	Encoded by rnpA gene
B. subtilis C5 protein	14 kDa	
Yeast	—	(Koski et al., 1976).
II. Group I intron		**Mechanisms of splicing**
A. Fungal mitochondrial pre-mRNAs pre-tRNAs pre-rRNAs		Intramolecular self-cleavage is achieved by a two-step transesterification process (Zinn and Butow, 1985).
B. *Tetrahymena* or *Physarium* nuclear pre-rRNA		In some cases, RNA catalysis is facilitated by participation of proteins (Lazowska et al., 1980; Gariga and Lambowitz, 1986; Muscarella and Vogt, 1989).
C. Bacteriophage (T4, T2, and T6) pre- mRNAs		
III. Group II intron RNA		
A. Mitochondrial and chloroplast group II mRNAs		Self-splicing leads to formation of lariat structure. The group II intron is characterized by a conserved structure.
IV. Splicosomal SnRNAs		
A. U2U4 and U6		Splicing of nuclear pre-mRNAs mediated by small nuclear RNAs in conjunction with several proteins; lariat formation.
V. Hammerhead RNA ribozyme	Virusoid	
A. Cis-acting (self- cleavage)		
Naturally occurring	359 nt	(Satellite RNA of ring spot virus STRSV)
hammerhead RNA	274 nt	(Buzayan et al., 1986) (ASBV viroid)
In vitro-synthesized hammerhead RNA	55 nt	(Froster and Symons, 1987; Froster et al., 1988; Uhlenbeck, 1987)

Table 9.1 *(continued)*

B. Trans-acting (cleavage of substrate RNA)	
Catalytic RNA	13 nt
Substrate RNA	13 nt
SnRNA 4	Nuclear
SnRNA 6	Pre-mRNA
VI. Engineered ribozyme with endonuclease activity	
A. T4SUN Y introns reengineered to contain 184 nt	
B. RNasep reengineered to contain 263 nt instead of 359–417 nt	
C. *Tetrahymena* intron missing both 5′ and 3′ splice sites	

the RNase activity of the enzyme RNaseP. However, it has now been established that the nuclease activity is primarily due to a ribonucleic acid component and not due to the protein component.

The RNase activity of this RNA component may, however, be increased by the protein component. It has been shown that the RNA moiety of RNaseP contains the active site for the nuclease activity (Guerrier-Takada et al., 1983; Guerier-Takada and Altman, 1984), and under high Mg^{2+} concentration the RNA component alone can perform the catalysis of RNA maturation *in vitro*. The presence of RNA component in eukaryotic RNaseP suggests the universality of catalysis by RNA in nature (Akaboshi et al., 1980; Kline et al., 1981). Thus, the role of an RNA molecule in self-splicing and RNA maturation clearly establishes its nuclease activity. However, not all RNA molecules possess RNase activity. In the majority of instances, RNA molecules have been known to act as a passive transmitter of genetic information. In certain other instances, RNA molecules have been shown to possess a structural role as is true in the case of ribosomes. However, in some other cases, RNA molecules may control the RNase activity of a protein; such a role of RNA has been documented in the case of RNaseMRP. Chang and Clayton (1987) have described a site-specific ribonuclease involved in primer RNA metabolism in mammalian mitochondria. This ribonuclease has been designated as RNaseMRP (i.e., ribonuclease involved in mitochondrial RNA processing). A 125-ribonucleotide RNA species was required for the activity of this enzyme. This RNA coded by the nuclear gene is transported to mitochondria where it interacts with the protein component coded by mitochondrial DNA to yield the functional RNaseMRP. Properties of the RNA and protein components of RNaseP are presented in Table 9.1.

B. Introns as Ribozymes

Until recently, proteins were the only macromolecules considered to be endowed with the enzymatic property. However, this notion has now changed with the discovery of the role of ribonucleic acids in the process of RNA processing and splicing (Cech and Bass, 1986; Sharp, 1985; Burke, 1988; Dujon, 1989; Hickey et al., 1989). RNA splicing leads to the removal of intervening sequences or introns. During these processes the ribonucleic acid molecules have been shown to possess endonuclease activity (Peebles et al., 1986; Van derveen et al., 1986; Zuag and Cech, 1985; Epstein and Gall, 1987). Besides this enzymatic activity, certain ribonucleic acids have been shown to function as complementary factors essential for the RNase activity of certain proteins (Chang and Clayton, 1987). Much of the evidence for the RNase activity of ribonucleic acid molecules comes from the study of the process of tRNA maturation and the splicing of RNA in different living systems. The first evidence for the RNase activity of an intron came from the self splicing of rRNA of *Tetrahymena* (Cech, 1983). Cech et al. (1981) showed that a 414-nucleotide-long intron was self-spliced, forming the mature ribosomal RNA from the precursor rRNA molecule. This process of self-splicing occurred without any external energy or participation by any protein molecule, and the intron was released in the form of a circular RNA molecule. The only external factor required for the RNase activity of the intron was a guanosine molecule. Such RNase activity of introns has been shown to occur in a variety of organisms besides *Tetrahymena* (Peebles et al., 1986; Van derveen et al., 1986; Chu et al., 1985). The RNase activity of introns has been very well investigated in the self-splicing of RNA precursors both from eukaryotic mitochondria and nucleus and from viral intron (Chu et al., 1985). Thus the RNA precursor molecules can be self-spliced as a result of the RNase activity of intron, and the intron may be released in the form of either a circle or a lariat during this process (Peebles et al., 1986; Reed and Maniatis, 1985 Van derveen et al., 1986).

Based on the structure of the intron molecules after the process of splicing, the introns have been designated as intron I (circular form) and intron II (lariate form). The group I introns have been shown to possess a series of conserved sequences that may be required for the proper folding of the RNA precursor molecules during the process of splicing. The group II introns have been shown to possess a different set of conserved sequences. The characteristics of the different groups of introns have been reviewed elsewhere (Cech, 1990; Michel et al., 1989; Perlman and Butow, 1989; Celander and Cech, 1991; Been et al., 1987). In addition, the self-splicing of nuclear RNA intron may involve participation by several proteins and some small nuclear RNA molecules known as U1, U2, U4, U5, and U6 snRNAs. The splicing of nuclear pre-RNA intron also results in the formation of the RNA-spliced product and the lariat form of the intron. During the splicing of nuclear pre-mRNA intron it is possible that either the intron itself or the small RNA (such as U1, U2, U4, U5, and U6) may possess the RNase activity that controls the process of self-splicing. The RN activity of RNA molecules in intron group I, intron group II, and nuclear RNA self-splicing has been discussed (Cech, 1986; Cech and Bass, 1986; Sharp, 1985). Properties of different intron ribozymes are presented in Table 9.1.

1. Group I Intron Ribozymes. These are characterized by the presence of an invariant uridine (U) residue at the 3′ end of the preceding exon, a G residue at the 3′ end of the intron, and an internal guide sequence (IGS). In addition, the introns contain four highly conserved sequences designated as P, Q, R, and S. These adjacent conserved sequences form base pairs (i.e., P forms a base pair with Q, whereas R forms a base pair with S) to yield structures that are required in splicing. In addition to these major four conserved sequences, there may be extra sequences located prior to P and adjacent to R between R and S sequences that may be conserved. Mutations in any of these conserved paired regions as well as in the internal guide sequence (IGS) can decrease the efficiency of splicing. However, no loss in splicing was observed when mutations were caused in nucleotide sequences peripheral to these conserved sequences (Been et al., 1987; Cech, 1988, 1990). A number of group I introns have been well-characterized, and their properties are summarized in Table 9.1. These include *Tetrahymena* nuclear ribosomal-RNA, fungal mitochondrial introns, bacteriophage genes, and plant chloroplast genes. The nucleotide sequences of the conserved regions in a set of 66 group I introns have been compared, and a three-dimensional model of the active center of the ribozyme has been presented (Cech, 1988).

2. Mechanism of Catalysis by Group I Intron Ribozyme. The self-splicing of group I intron RNA after transcription *in vitro* by purified *E. coli* RNA polymerase using *Tetrahymena* ribosomal DNA as template has been demonstrated. The

Table 9.2 Different chemzymes with DNA cleaving activity

I. Small molecules
- A. H_2O_2
- B. Piperdine
- C. EDTA–Fe
- D. Pb

II. DNA intercalators
- A. Metal intercalators
  1. Phenathroline
     Phen–Cu, Phen–Rh, or Phen–Co complex
  2. Methidium propyl–iron complex
- B. Antibiotic intercalators
  1. Bleomycin–glycopeptide antibiotics
     require Fe and oxygen for DNA scission
  2. Daunamycin–anthralin intercalator between GC in CGTACG sequence
  3. Adriamycin–anthralin intercalator between GC in CGTACG sequence
  4. Aclacinomycin–anthralin intercalator between GC in CGTACG sequence
  5. Epperamicin–enedyne core
  6. Calicheamicin–anthralin intercalator between GC in CGTACG sequence
  7. Neocarzinostatin–anthralin intercalator between GC in CGTACG sequence
  8. Dynemicin–anthracycline and enedyne cores
  9. Kedarcidin

III. Oligopeptides
- A. lys–trp–lys

self-splicing involves two transesterification reactions involving the splicing of the 5′ end first and then splicing at the 3′ end of the intron, causing the removal of an intron as a circular product and also causing the joining of the exons preceding and following the intron. During the splicing process a guanosine residue is added to the 5′ end of the intron. The enzyme activity required Mg^{2+}. This requirement for Mg^{2+} can be met by Mn^{2+} but not by Ca^{2+}, Zn^{2+}, Co^{2+}, and Pb^{2+} (Cech and Bass, 1986).

3. Assay of Ribozyme Activity of Intron RNA or Other RNA. First, the activity of group I intron RNA has been accessed by the incorporation of radioactive G residue into product of self-catalysis activity. The ribozyme activity of some of the bacteriophage T4 introns was detected in this way.

Second, the ribozyme self-cleavage activity can be determined by the inclusion of the ribozyme into a reporter gene such as LacZ gene carried in a chimeric plasmid. Insertion of ribozyme into the LacZ gene inactivates its β-galactosidase activity. However, under condition of self-cleavage of ribozyme *in vivo* or *in vitro*, the ribozyme RNA within the reporter LacZ gene can be spliced out, leaving the reporter gene in frame to code for an active β-galactosidase. The prescence of active and inactive β-galactosidase can be demonstrated by the presence of blue and white colonies of bacteria harboring the chimeric plasmids with reporter gene on X-gal medium.

C. Group II Intron Ribozymes

The nucleotide sequences of over 70 group II introns from plant and fungal mitochondria and plant chloroplasts have been examined for the conservation of primary sequences, secondary structures, and three dimensional structures (Michel et al., 1989; Perlman and Butow, 1989).

Group II introns possess sequences that are organized into six radiating helices or spokes; these spokes define the intron substructures. The first domain possesses complex substructures including five exon binding sites (EBS), whereas the fourth domain is highly variable and the fifth domain is highly conserved. In addition, the group II intron, like the nuclear pre-mRNA genes, possess dinucleotides at the 5′ and 3′ ends of the intron. The *in vitro* self-splicing ribozyme activity of a number of group II introns has been demonstrated (Michel et al., 1989).

It has been further shown that the six radiating spokes of the group II self-splicing intron ribozyme are not necessarily contained within the same intron; instead, they may be derived from a stretch of nucleotide sequences derived from different RNA segments, which, with appropriate base pairing, may be organized to yield the conserved secondary structure of group II intron in the form of six radiating helices with ribozyme activity (Sharp, 1991). Such prototype group II intron RNA effective in splicing of *Chlamydomonas* chloroplast psa A exons 1 and 2 have been assembled in conjunction with a 430 nt RNA of another gene (tscA) (Sharp, 1991). A mutation in the psaA gene adversely affects the splicing of the psaA gene.

D. Splicosomal snRNA Ribozyme

A number of small nuclear RNAs (snRNAs) in conjunction with a large number of proteins forms an organelle called a *splicosome*, which is involved in the splicing of nuclear pre-mRNAs. The ribozyme activity of snRNAs in splicosomes is also based on the assumption that snRNAs can form structures with radiating spokes like the group II intron RNA with ribozyme activity (Sharp, 1991). Among the different snRNAs, the role of the two domains of U6 snRNA has been demonstrated in *in vitro* splicing of the nuclear pre-mRNA (Fabrizio and Abelson, 1990). Single-base-substitution mutants of snRNA U6 are defective in cleavage steps involving splicing of the 5′ or 3′ sites (Cotten et al., 1988; Fabrizio and Abelson, 1990).

The ultimate proof that snRNAs are involved in splicing of nuclear pre-mRNA would require the demonstration of the involvement of snRNAs alone in this process. However, this may not be possible because snRNAs have evolved to work in conjunction with several proteins that constitute the splicosome (Guthrie, 1991). The protein facilitation of the ribozyme activity is not limited to snRNAs but has been seen as a requirement for the proper functioning of several group I and group II intron ribozymes.

1. Proteins that Facilitate the Ribozyme Activity of RNA Nucleases. Several RNA molecules such as RNaseP, group I intron RNA, group II intron RNA, and of course the spicosomal snRNAs require participation by one or more proteins for their RNase activity *in vivo* as well as *in vitro*. The K_m of the RNase activity of RNaseP is the same with or without the protein components. However, the main fact that this ribozyme has evolved as a two-component complex implies the unidentified role of the protein components in the RNase activity of the M_1RNA. Furthermore, the fact that RNA components of several RNase P (such as *E. coli* and *B. subtilis*) differ entirely in their nucleotide sequence but are still capable of interacting with protein components of a heterologous source points to the fact that protein components may be an integral part of the *in vivo* ribozyme activity of the RNA components of RNaseP.

E. Maturase

A number of proteins either encoded by the ORF of an intron (Davies et al., 1982; Macreadie et al., 1985; Gariga and Lambowitz, 1986; Colleaux, 1986; Colleaux et al., 1988; Delahodde, 1989) or encoded by the nuclear genes (Collins and Lambowitz, 1985; Banroques et al., 1987; Karsen et al., 1987; Weiss-Brummer et al., 1982) have been shown to mediate the self-cleavage activity of the group I and group II introns. Such proteins have been designated as *maturases*. Ordinarily a maturase encoded by an intron usually facilitates the ribozyme activity of that intron RNA (Carignani et al., 1983; Wenzlau et al., 1989). However, sometimes a maturase encoded by ORF of one intron may facilitate the self-cleavage activity of a second intron RNA. In yeast the maturase encoded by intron 4 of the Cob (apocytochrome b) gene (bI4) promotes splicing of bI4 as well as aI4 introns [i.e., intron 4 of the Cox 1 (cytochrome oxidase subunit I) gene].

Furthermore, both in *Neurospora* and in yeast, nuclear mutations such as cyt-18, NAM-2 and CBP-2 can adversely affect the self-splicing activity of the group I intron RNA (Labouesse et al., 1985). In *Neurospora* a nuclear gene Cyt-18 encodes mitochondrial tyrosyl tRNA synthetase. The Cyt-18 gene product has been found to promote the splicing of pre-rRNA *in vitro* (Collins and Lambowitz, 1985). The nuclear-encoded yeast NAM-2 protein is a mitochondrial leucyl-tRNA synthetase (Herbert et al., 1988). Likewise, in *Neurospora*, mitochondrial tRNA synthetases are involved in RNA splicing; a tyrosyl tRNA synthetase was found to possess an unusual domain that was required for RNA splicing (Cherniak et al., 1990). It seems plausible that protein or protein–RNA interactions may be involved in the promotion of splicing by maturases (Cech, 1990). Another yeast CBP-2 protein encoded by a nuclear gene is required for the splicing of intron 5 of Cob pre-mRNA (Guernier-Takada et al., 1983). This protein is required for the *in vitro* splicing at lower Mg^{2+} concentration. It would be of interest to determine the CBP-2 sites involved in interaction with the intron RNA (Cech, 1990). The properties of a number of maturases are presented in Table 9.1.

F. Hammerhead RNA as Ribozyme

Certain plant RNA viruses and the satellite RNA, called *virusoids*, act as autocatalytic endoribonuclease ribozymes. These plant RNA viruses and the virusoids occur in linear or circular forms and replicate as rolling circles from which the genome size RNA are self-cleaved. Their self-cleavage generates RNA molecules with 5′-OH and 2′,3′-cyclic phosphodiester ends. The self-cleavage domain contains a consensus sequence with three stemmed secondary structure called *hammerhead* (Haseloff and Gerlach, 1988). The properties of a number of hammerhead ribozymes are summarized in Table 9.1. Based on the understanding of the structure of naturally occurring self-cleaving hammerhead ribozyme, it has been possible to synthesize hammerhead ribozymes *in vitro*. These have been found to cleave RNA substrate molecules (Fedor and Uhlenbec, 1990). It has further been possible to engineer such ribozymes to act as restriction endonucleases for the cleavage of specific target RNAs as discussed later.

G. Cis- and Trans-Acting Ribozyme Endonuclease

All self-cleaving RNA molecules are cis-acting ribozyme endonucleases. An understanding of the molecular nature of these self-cleaving ribozymes has provided clues to construct trans-acting ribozymes. Such trans-acting ribozymes make an endonucleolytic cut on a substrate RNA molecule other than itself (Branch and Robertson, 1991). In order to facilitate their trans cleavage actions, the substrate RNA molecule must be chosen to interact with the ribozyme to promote cleavage. The cis- and trans-acting ribozymes differ significantly in their molecular structures. The trans-acting ribozymes are much smaller in size because the site for the cleavage is carried by the substrate RNA molecules. Ribozyme has been used for directed cleavage of RNA (Shibahara et al., 1987). Characteristics of cis- and trans-acting ribozymes are presented in Table 9.1.

II. DNAZYMES

DNA has been always considered to be an informative molecule partly because of its stable double-stranded nature and lack of reactive group unlike RNA. This view was sustained by the fact that no DNA sequence with catalytic activity was found to occur in nature. This situation about the lack of catalytic DNA is in contrast to the fact that several ribozymes or RNA molecules with catalytic activity and biological role have been found to exist in nature. However, the development of combinatorial methods and an *in vitro* selection procedure for the creation of RNA molecules as ribozymes, along with the realization that the single-stranded DNA could assume

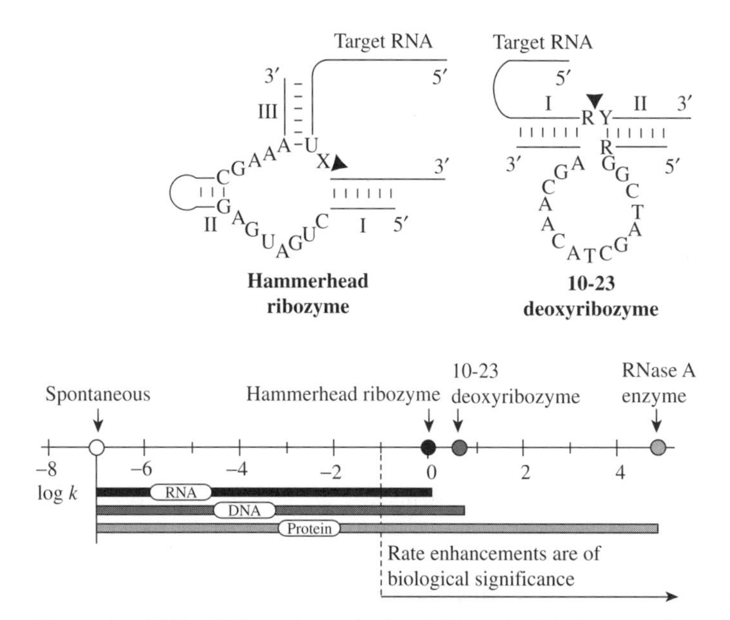

Figure 9.1 Comparing DNA, RNA, and protein. (**Top**) Structures of the natural hammerhead ribozyme (RNA enzyme) and the engineered 10-23 deoxyribozyme (DNA enzyme). The arrowheads identify the site of enzyme-catalyzed RNA cleavage. (**Bottom**) A comparison of the catalytic power of three RNA-cleaving enzymes that operate by an identical transesterification mechanism. The plot depicts the logarithms of the rate constants (min^{-1}) (Santaro and Joyce, 1997) for each enzyme. Bars depict the rate enhancements that are generated by RNA, DNA, and protein enzymes over the rate constant for the spontaneous cleavage of RNA. (Reproduced from Breaker, 2000, *Science*, with permission.)

secondary and folded structure, led to examine the possibility of artificially creating DNA sequences and examining their catalytic activity in cleaving RNA/DNA. Soon some DNA molecules, synthesized artificially, were identified to possess catalytic activity capable of cleaving RNA (Breaker and Joyce, 1994, 1995; Cuenoud and Szostak, 1995; Li and Sen, 1998) or DNA (Carmi and Breaker, 2001).

Of the many DNAzyme molecules, a particular one called 10–23 has been very well characterized (Santoro and Joyce, 1997). This DNAzyme consists of a sequence of 15 deoxynucleotides as the catalytic domain and possesses catalytic efficiency several orders of magnitude greater than that of ribozyme but somewhat less than the proteinzyme, RNaseA (Santoro and Joyce 1997; Breaker, 2000). The relative activities of different enzymes such as ribozyme, DNAzyme, and proteinzyme (RNaseA) on RNA substrate are presented in Figure 9.1.

A DNAzyme is an enzyme that can cleave RNA, preferably at an unpaired site with a purine followed by a pyrimidine. The DNAzyme usually recognizes the substrate RNA and forms a double-stranded structure on both the 5′ and 3′ ends of the substrate RNA, generating a loop of unpaired DNA sequence in between the two ends of the substrate molecules and causing cleavage at the junction of the unpaired region of the substrate molecule. The variety of DNAzymes that could be artificially created and their application is boundless (Breaker, 2000). DNAzymes are being evaluated for their ability to serve as therapeutics in order to treat several human diseases. However, unlike antisense RNA or ribozymes, the DNAzymes must be deliveried to the cell directly; a gene for DNAzyme could not be inserted into the cells. Nevertheless, like ribozymes, the delivery, stabilization, and physiological activity of the DNAzymes are equally complicated.

III. CHEMZYMES

A number of chemicals, metal ligand complexes, and oligopeptides act as nucleases by making a scission in the nucleic acid chain. These are described below. Even though DNA is known to exist in different forms, no nuclease is known which specifically cleaves one or the other form of DNA among DNA-A, DNA-B, and DNA-Z. In this respect, nuclease activity of certain chemzymes is particularly important because chemzymes can cleave one or the other form of DNA (Barton, 1986; Youngquist and Dervan, 1985). At present there is no evidence that chemzymes do indeed catalyze such cleavage of nucleic acids *in vivo*; however, it is known that reducing sugar makes cleavage in DNA in aging cells (Bucala et al., 1984); thus it is possible that chemzyme may play a role *in vivo*. A number of chemzymes which act as nucleases are listed in Table 9.2.

A. Chemicals and Metal Ligand Complexes as Nucleases

1. Piperidine. Piperidine (hexahydropyridine) can cause cleavage in the nucleic acid strand at the sites of nucleotide-base modification. Use of piperidine was made in the development of the chemical cleavage method of nucleotide sequencing originally described by Maxam and Gilbert (1977). In four separate reactions, base-specific modifications are created at G (by dimethylsulfate), G and A (by depurinization with acid), C (by hydrazine plus salt), and C and T (by hydrazine) sites in an end labeled DNA segment. The base modifications are followed by piperidine treatment at high temperature (90°C), yielding nucleic acid fragments that are analyzed by electrophoresis on a high-resolution polyacrylamide gel. The DNA fragments in the gel are visualized by autoradiography. The ladderlike cleavage pattern displayed by the autoradiography can be directly read into the nucleotide sequence. Piperidine can also cleave at purine sites modified by diethyl pyrocarbonate (Herr, 1985). Diethyl pyrocarbonate specifically modifies purines located in the DNA region that can readily assume the left-handed helical form of DNA, that is, DNA-Z (Herr, 1985). Thus the DNA strand scission following diethyl pyrocarbonate is a good method to detect the location of DNA-Z region in a DNA segment. Certain DNA intercalating agents such as echionomycin (Mendel and Dervan, 1987) can render certain base pairs more sensitive to diethyl pyrocarbonate. This has been used to identify Hoogsteen base pairs. Thus the ability of piperdine to cleave nucleic acid strands at the sites of base modification has been utilized to develop a technology for the nucleotide sequencing whose application has revolutionized the science of genetics and molecular biology and may lead to complete understanding of the nucleotide sequence of human genome.

2. Hydrogen Peroxide. DNA strand scission can also be brought about by treatment of DNA with hydrogen peroxide H_2O_2. Such treatment can cleave DNA at any base site and has been developed to generate a DNA footprinting (differential cleavage activity) (Tullius and Dombroski, 1986) as first described by Galas and Schmidtz (1978). The hydroxyl radical acts by attacking the sugar moiety of the nucleotide and is non-base-specific. This chemical footprinting has been used to map the DNA region protected by λ repressor and *cro* protein.

3. DNA Intercalating Agents. Most DNA intercalators, when attached to a metal chelator such as EDTA, can cause single-strand breaks in a DNA duplex in the presence of Fe(II) and O (Hertzberg and Dervan, 1982, 1984; Dervan, 1986; Wu et al., 1983). The different intercalators vary in their mode of cleavage. The DNA cleavage by methidiumpropyl–iron complex [MPE–Fe(III)] is almost independent of the base sequence. The sequence specificity is lower than that by DNaseI. Certain intercalators that specifically bind to AT- or GC-rich sequences cause affinity cleavage of DNA. Netropsin and distamycin are specific to AT-rich DNA segments; actinomycin and bleomycin are specific to GC-rich regions (Dervan, 1986). In contrast, phenanthroline causes cleavage specific to conformation (either B or Z form) of DNA without any specificity to the base sequence in DNA that might generate the conformation of the DNA. Antitumor drugs and other chemicals have been found to cleave double-stranded DNA in a site-specific manner (Zein et al., 1988; Schultz and Dervan, 1983; Sugiura et al., 1990) Bleomycin makes double-stranded break in DNA and creates mutational hot spots (Steighner and Povirk, 1990).

MPE–Fe(III) has been used extensively in the footprinting of the binding sites encompassing 4–5 base pairs of a number of other drugs such as actinomycin, netropsin, distamycin, chromomycin, mithramycin, and oliviomycin (VanDyke et al., 1982; VanDyke and Dervan, 1983). The MPE–Fe(III) most likely causes the oxidative degradation of the deoxyribose moiety (Hertzberg and Dervan, 1984).

4. Phenanthroline. The DNA cleavage action of trisphenanthroline chiral metal complexes has been described (Barton, 1986; Pope and Sigman, 1984; Mei and Barton, 1986, 1988; Barton and Raphael, 1985; Müller et al., 1987). A specific enantiomeric form of the metal–phenathroline complex has been found to cleave a specific form of DNA. The A-tris(9,7-diphenyl-1,10-phenathroline)–cobalt(III) can cleave only the Z form of DNA (Barton and Raphael, 1985; Müller et al., 1987). The 1,10-phenathroline–copper complex predominantly facilitates the cleavage of the B form of DNA. In contrast, the tris(3,4,7,8-tetramethyl phenalthroline)–ruthenium(II) is specific for the cleavage of A-form helices of DNA and RNA (Mei and Barton, 1986). It has been further shown that the 1,10-phenathroline–copper complex can actively cleave the poly(dG-dC) nucleotide in the B form at low salt concentration but is unable to cleave the same polynucleotide stabilized in Z forms at higher salt (3.0 M NaCl) concentration (Pope and Sigman, 1984).

The DNA-Z conformation-specific cleavage facilitated by the A-tris(4,7-diphenyl 1,10-phenanthroline)–cobalt(II) complex is of much interest because this complex seems to recognize secondary structures generated by certain regulatory elements that may act as a punctuation mark in a DNA segment (Barton et al., 1984; Barton and Raphael, 1985; Müller et al., 1987). The cleavage site in the secondary regions acted upon by the tris(4,7-diphenyl 1,10-phenanthroline)–cobalt complex appears to correspond to regions in the genome of pBR322 and SV40 which may be important for control of gene expression. In the plasmid pBR322, the cleavage is made at the end of each gene. This plasmid contains genes that confer resistance to ampicillin and to tetracycline, or act as origin of replication; thus, the three cleavages made by this chiral complex of phenanthroline–cobalt(II) correspond to the ends of the three genetic segments in the genome of pBR322. Likewise, the cleavages of SV40 also correspond to the ends of each genetic segment. Such cleavage by a phenanthroline–cobalt complex resembles the cleavage of DNA by mung bean nuclease (McCutchen et al., 1984). The mung bean nuclease has been shown to cleave at the beginning and at the end of a gene.

5. Factors Controlling the DNA Cleavage by Chemzymes. The DNA cleavage by ligand, chiral–metal complex, or intercalating agents is activated by generation of locally high concentrations of OH radicals as in the case of MPE–Fe(III) or Ca(Phen)^{2}distamycin–EDTA or by photoreductions as in the case of bleomycin, metals, and phenanthroline–cobalt complex (Sim and Lown, 1978; Sigman et al., 1979; Barton, 1986). Cleavage by several chemzymes is influenced by the primary and secondary structure of the substrate (Marshall et al., 1981).

B. Peptides

Oligopeptides with aromatic amino acid have been found to mimic the effect of the DNA damage specific endonucleases. The tripeptide lys–trp–lys has been found to bind specifically with the apurinic site in the DNA and then to make an incision on the apurinic site (Behmooras et al., 1981; Toulmé and Héléne, 1981). Another tripeptide, lys–gly–lys, did not possess any AP-endonuclease activity (Behmooras et al., 1981). The fact that the tripeptide lys–trp–lys has AP-endonuclease activity may provide an insight into the mechanism of action of the AP endonuclease. It is possible that AP endonuclease may contain the tripeptide at its active site. Furthermore, a number of peptides have been described which can cleave DNA in sequence-specific manor (Kozarich et al., 1989).

IV. DESIGNER NUCLEASE

An understanding of the nature of DNA cleaving and DNA binding molecules has been utilized in the design of nucleases that can be targeted to a specific site of DNA molecules for selective cleavage (Dervan, 1986). Such artificially designed molecules are called *designer nucleases* or *artificial nucleases.* They act very much like naturally occurring restriction endonucleases; however, because of their complex DNA recognition components, they can be made to cause much less frequent cleavage in DNA molecules than the known restriction endonucleases. Such attributes of the designer nucleases make them more suitable for this utilization in human gene mapping and cloning (Dervan, 1986). Some of the chemical approaches in the design of artificial nucleases are described below. A list of designer nucleases is presented in Table 9.3.

An oligonucleotide of a particular sequence can be tethered to an EDTA–Fe complex to generate an artificial nuclease that can cause site-specific cleavage

Table 9.3 Different designer nucleases with site-specific DNA cleavage activity

I. Site-specific oligonucleotide complexed with the following:
   A. EDTA–Fe (Chu and Orgel, 1985; Dryer and Dervan, 1985)
   B. Staphylococcal nuclease (Corey and Schultz, 1987)

II. Site-specific DNA binding protein attached to a DNA cleaving reagent or chemzyme
   A. Synthetic 52-mer residue of Hin-recombinase (139–190 AA) attached to EDTA–Fe at N terminus of the protein (Sluka et al., 1987)
   B. Yeast GCN-4 protein fused to EDTA–Fe at the N-terminus (Oakley and Dervan, 1990)
   C. *E.coli* CAP (catabolite activator protein) complexes with 1,10-phenanthroline at the tenth amino acid (Ebright et al., 1990)

III. Site-specific DNA binding drug attached to DNA cleaving reagent (chemzyme)
   A. Distamycin EDTA–Fe (Schultz et al., 1982)

(Dryer and Dervan, 1985). The oligonucleotide binds to complementary nucleotide sequence in DNA molecules, and then a cleavage is made by the attached EDTA–Fe complex. Likewise, phenanthroline–copper has been used to cause cleavage in a sequence-specific manner at a particular target (Chen and Sigman, 1986). Furthermore, it has been possible to construct a series of self-cleaving RNA duplexes using artificially synthesized 21-mer oligonucleotides (Koizumi et al., 1988). Similar site-specific cleavage can be facilitated by a nonspecific nuclease attached to an oligodeoxynucleotide. Corey and Shultz (1987) have used micrococcal nuclease attached to an oligonucleotide to target it to a specific region of a DNA molecule where a scission is made by the micrococcal nuclease component of the artificial enzyme. Alternatively, many site-specific DNA binding proteins (Schleif, 1988; Bruist et al., 1987) such as yeast GCN_4 proteins or *E. coli* CAP protein have been used to complex with a DNA cleaving reagent such as EDTA and Fe or phenanthroline, to generate a designer nuclease (Sluka et al., 1987; Kim et al., 1988; Ebright et al., 1990; Koob and Szybalski, 1990). Likewise, many drugs that bind with DNA molecules in a sequence-specific manner can be attached to a chelator to yield a designer nuclease (Schultz et al., 1982; Schultz 1988). Finally, a homopyrimidine oligonucleotide–chelator complex capable of forming triple helix has been used as a designer nuclease.

10

MOLECULES THAT INTERACT WITH NUCLEASES

In addition to nucleic acid molecules that are substrates for the enzyme, nucleases interact with a number of other molecules. These include (a) inhibitors that adversely influence the activity of nucleases, (b) certain proteins that interfere with the activity of nucleases by interacting with the substrates, and (c) certain nucleotide sequences in DNA which act as sites of recognition and/or cleavage by nucleases.

I. INHIBITORS

A. Proteins as Nuclease Inhibitors

Various inhibitors of nuclease have been described. These may include protein, RNA, or other molecules.

1. DNase Inhibitor—Protein. Proteins that strongly and specifically inhibit the activity of DNaseI have been described from various animal cells (Cooper et al., 1950). Protein from the pigeon crop gland which inhibited the activity of the pancreatic DNase has been described. Later, Lindberg (1967a,b) purified a protein from calf spleen which inhibited the activity of the pancreatic DNase. The inhibitor protein has a molecular weight of 42 kDa (Lindberg and Skoog, 1970) and requires Mn^{2+} for its activity. This inhibitor protein acted by binding with the DNaseI; such interaction between the inhibitor protein and DNaseI leads to a complex formation containing one molecule each of DNaseI and the inhibitor (Lindberg, 1967a,b). Similar proteins from calf thymus and rat serum have been extensively characterized and shown to possess properties of the spleen inhibitor. This inhibitor consists of a significant fraction of the cellular protein amounting to almost 5–10% of the total protein. Biochemical and immunological characterization have shown that this inhibitor protein is identical to actin (Lazarides and Lindberg, 1974; Lacks, 1981). The binding of actin or inhibitor protein to DNaseI on affinity chromatography using DNase-agarose has been shown to occur. Actin was also shown to inhibit the activity of DNaseI. The inhibitor protein and actin have a very close similarity in amino acid composition. It has been further shown that the mutant actin is unable to bind with DNaseI. Actin may play a role in the regulation of DNaseI. Actin exists in three forms. These are actin G (globular form), actin F (fibrous form), and actin P (paracrystalline form). DNaseI binds tightly to G-actin but not so tightly to actin F. But DNaseI depolymerizes actin F to form a 1 : 1 complex with actin. It has been further shown that DNaseI exists as a 1 : 1 complex with actin in the pancreatic juice (Rohr and Mannherz, 1979). DNaseI has been used to understand the dynamics of the interconversion of different forms of actin. Also, DNase binding to actin has been used as an immunoprecipitation assay for actin (Snabes et al., 1981). In such an assay, [^{21}I]-labeled actin complexed with DNaseI was assayed after immunoprecipitation with monospecific rabbit antibody to DNaseI. Thus the complex formation between actin and DNaseI can be used not only for the role of different proteins in the inhibition of DNase activity, but also for the understanding of the role of actin in cellular processes.

2. RNase-Inhibitor—Protein. An inhibitor of RNaseA has been characterized from human placenta. This protein, called PRI (placental RNase inhibitor), is ~50 kDa in size. This is a natural inhibitor of angiogenin (RNase5) and other members of the RNase superfamily but not onconase sialic acid-binding lectin-RNases. PRI also does not inhibit RNaseT1, RNase1, and S1 nuclease. PRI binds with RNaseA noncovalently in the ratio of 1 : 1 under a reducing environment. The crystal structure of the RNaseA and the inhibitor PRI shows that they have a similar structure and that PRI with its leucine-rich loops fits into RNase, thereby blocking the active site of the enzyme.

In addition to the RNaseA-PRI, the three-dimensional (3-D) structure of barnase and its inhibitor barstar is very well known.

B. RNA as Nuclease Inhibitors

The role of RNA in the inhibition of DNase has been described earlier. It was shown that the activity of DNase in the cell prepared from the extracts of *Escherichia coli* could be increased by a prior treatment with pancreatic ribonuclease. The ribonuclease was found to stimulate the DNase activity by destroying the RNA bound to DNase. Later, Lehman et al. (1962) showed that the inhibitory action of RNA was specific to a particular *E. coli* endonuclease. Both the amino acid acceptor tRNA and the ribosomal RNA from *E. coli* acted as inhibitors. Polyribonucleotides synthesized by polynucleotide phosphorylase were found inactive as inhibitors. The activity of other DNases (including pancreatic DNaseI, snake venom phosphodiesterases, and other *E. coli* DNases) was not adversely affected by the RNA. The inhibition of group A streptococcal deoxyribonuclease by ribonucleic acids from diverse sources is known to occur.

C. Other Molecules that Act as Nuclease Inhibitors

A number of cations, anions, polyamino acids, and chelators such as EDTA have been known to inhibit the activity of nucleases (Privat de Garilhe, 1967; Davidson, 1972; Cunningham et al., 1956; Lehman et al., 1962). Heparin has been shown to inhibit the activity of ribonuclease.

Also, vanadyl ribonucleoside complexes (VRC), 5′-diphosphoadenosine 3′-phosphate, 2′-CMP, 2′-UMP, Apu, 2′-GMP, and other nucleotides, as well as diethyl pyrocarbonate (DEPC) and bentonite, are potent inhibitors of RNases. In addition, a number of drugs such as coumermycin A, novobiocin, nalidixic acids, and oxolinic acid are inhibitors of topoisomerases (Gellert et al., 1976b, 1977; Higgins et al., 1978; Sugino et al., 1977). These drugs have been used to obtain mutants of bacterial gyrase or topoisomerase II. Mutants resistant to oxolinic acid and nalidixic acid inhibit the gyrase subunit-A, whereas the coumermycin A and novobiocin inhibit subunit B of the *E. coli* gyrase. A number of other drugs have been shown to inhibit the activity of nucleases (Scurlock and Miller, 1979).

II. PROTEINS THAT INTERFERE WITH THE ACTIVITY OF NUCLEASE BY INTERACTING WITH THE SUBSTRATE (NUCLEIC ACIDS)

In the studies described above, the inhibitor was thought to act by binding with DNase. However, there is some evidence that the inhibitors from some other sources may act by forming a complex with DNA and Mg^{2+} and thus block the activity of the inhibitors by not making the substrate available for the enzyme action. The most important proteins belonging to this class of inhibitors are nucleic-acid binding proteins. These proteins prevent the digestion of nucleic acid by binding with it such that nucleases cannot hydrolyze the phosphodiester linkages. Evidence for such role of DNA-binding proteins comes from the studies of the temperature-sensitive mutants of the bacteriophage T4 defective in gene 32 protein. These T4 mutants, when transferred to nonpermissive temperature, undergo rapid hydrolysis of its DNA; the T4 DNA, which is usually 600–1000S, is converted into small-sized DNA in the range of 30–80S. It has also been shown that a nuclease controlled by T4 gene causes this degradation of T4 DNA in the absence of the ssb protein, the product of T4 gene 32. These studies clearly indicate that the T4 gene 32 product, which is a single-stranded DNA-binding protein, protects ssDNA from nuclease digestion. It has further been shown that the T4 gene 32 product can protect ssDNA from deoxyribonuclease digestion *in vitro*. These studies have led to the following conclusions regarding the structure of T4 genome organization: (a) the genome length DNA of T4 is spaced with single-stranded DNA, and (b) the single-stranded regions are protected from endonuclease digestion by binding

Table 10.1 Properties of ssb protein from different organisms

Organize	Molecular Size	Other Features
T_4	301 AA, $M_r = 33,488$	Encoded by gene 32, 10,000 copies per *E. Coli* cell, tendency to aggregate
E. coli	177 AA, $M_r = 18,873$	Encoded by SSB gene, 300–350 copies per cell
Adenovirus	529 AA, $M_r = 59,000$	Genetic mutants known, tendency to aggregate
Yeast SSBI	$M_r = 50,000$	Nonessential gene; cloned
Yeast SSBII		Binds more tightly to ssDNA
Yeast SSBIII	$M_r = 20,000$	Mitochondrial ssb
Calf thymus UpI	$M_r = 24,000$	Very similar to prokaryotic ssb protein; proteolytic production (195 AA) of aHnRNA protein of HeLa cells

with ssb protein encoded by gene 32. The bacteriophage T4 ssb protein or gene 32 product contains 331 amino acids. Of these the region encompassing amino acid residues 22–253 is important for binding with the ssDNA. It is further shown that the tyrosine-rich region spanning amino acids 72–116 is important for ssDNA binding. These conclusions are based on a study involving the removal of N-terminal or C-terminal regions by proteolytic digestion or by mutation. The ssb protein organizes ssDNA in a manner similar to that of histones to DNA duplex in a nucleosome. Usually ssb protein binds with a stretch of 140 nucleotides of ssDNA.

The ssb proteins have been characterized from a number of organisms. The *E. coli* ssb protein is similar to the bacteriophage T4 gene 32 protein. The ssb protein from a variety of organisms has been very well characterized. Some of these characteristics are presented in Table 10.1.

III. DNA SEQUENCES THAT INTERACT WITH NUCLEASES

The site specificity of certain nucleases, particularly restriction endonucleases, is a well-known fact. Also, certain recombinases that participate in site-specific recombination are also known to recognize a specific sequence of nucleotides and cause cleavage within such sequence. This is very well shown by *Int* recombinase in prokaryotes, or by homing endonucleases in eukaryotes, or by recombinases that are involved in the rearrangement of the immunoglobin gene during the development of a mammal. Even certain other nucleases such as DNaseI, which was considered to be a general nuclease, have been shown to possess a preference for certain nucleotide sequence as substrate. This is true for certain topoisomerases as well.

However, among the different nucleotide sequences that interact with nucleases, chi elements are best described (Smith, 1983, 1987; Taylor et al., 1985). Chi elements are recognized by certain nucleases that participate in recombination. Such nucleases act preferentially at or near the chi sites. The chi site is an octamer with a nucleotide sequence 5′GCTGGTGG3′/3′CGACCACC5′ (Smith et al., 1981). This sequence occurs once every 5 kbp in the *E. coli* chromosome and seems to exert its influence on recombination up to a distance of 10 kbp on *E. coli* chromosome. This structure of chi elements has been described in λ phage, pBR322 plasmid, and *E. coli*. A mutant of chi element with nucleotide sequence 5′GCTAGTGG3′ has also been described; this mutant chi element has a partial activity.

There is good evidence which suggests the interaction of the recBCD enzyme with the chi elements (Smith, 1983; Taylor et al., 1985; Ponticelli et al., 1985). These include the facts that (a) chi stimulates the recBCD pathway and (b) the recBCD null mutants lack both the chi activity and the recBCD pathway of recombination. These observations suggest that chi interacts directly with the recBCD enzyme or any other protein of the recBCD pathway. However, it is ruled out that recA protein, which participates both in recBCD and recF pathway, interacts with chi because the latter is known not to influence the recF pathway. Different models of interactions of chi with recBCD enzyme have been considered (Smith, 1983, 1987), and the roles of ssb protein and rec protein have been invoked (Smith, 1987; Honigberg and Radding, 1988; Madiraju et al., 1988). It is postulated that recBCD enzyme produces a nick in one of the DNA chains of the chi which leads to formation of the Holliday junction; the latter upon resolution yields recombinant DNA molecules. The interaction of chi with recBCD enzyme during the process of recombination is depicted in Figure 10.1.

A. Chi-Like Elements in Eukaryotes

It is known that in fungi the frequency of gene conversion is higher at one end of the locus than at the other end. This polarity in gene conversion is considered to

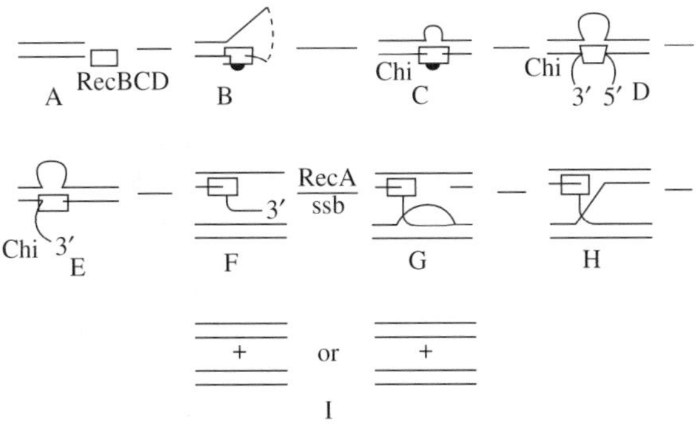

Figure 10.1. Interaction of chi element with recBCD enzyme during recombination in *E. coli*. [From Smith (1987), copyright Annual Reviews Inc., used with permission.]

indicate the presence of a chi-like element from which the recombination events are initiated as a result of the action of a recombinase. Such elements have been identified in *Neurospora crassa* and called *cog* (Catcheside, 1977). The *Neurospora cog* element is much like the chi element of *Escherichia coli* in that both act in cis and as dominant in heterozygotes. Certain other recombination-promoting sites in yeast and in *Sordaria* and the occurrence of chi element in other eukaryotes and their role in recombination have been described. The chi elements have also been shown to exist in the Ti plasmid of *Agrobacterium tumefacience* and in the mouse immunoglobulin genes.

IV. OTHER INHIBITOR MOLECULES

The 3′–5′ exonuclease activity of *E. coli* DNA pol I enzyme is inhibited by dNMP; the latter binds with the exonuclease active site and inhibits its activity without interfering with the polymerase activity (Que et al., 1978). Harmane (1-methyl-9*H*-pyrido[3,4-6]indole) inhibits the excision of thymine dimers by inhibiting the activities of APendonucleases with glycosylase activities (Warner et al., 1980a,b, 1981).

V. PROTEINS THAT INTERACT WITH DNA OR NUCLEASE TO ORCHESTRATE THE ACTIVITY OF NUCLEASES

There are a number of proteins that interact with DNA or nucleases to facilitate their function. RecA protein is one such protein. RecA protein has at least two functions that can influence the activity of nucleases (Williams et al., 1981). First, the RecA protein acts as synaptase and brings about the pairing of homologous DNA segments. Such pairing is a prelude to the process of genetic recombination (Holloman et al., 1975; McEntee et al., 1979; Shibata et al., 1982; Cunningham et al., 1981; Kmiec et al., 1986; Kmiec and Holloman, 1986; Maidaraju et al., 1988; Honigberg and Redding, 1988). The pairing of homologous DNA segments by RecA protein is preceded and followed by activities of a number of nucleases (recombinases). Second, the RecA protein acts as protease and hydrolyzes the product of LexA gene and thus controls the SOS response, which involves the activity of a larger number of nucleases. The properties of RecA protein and its various aspects in modulating the physiological role of nucleases have been discussed (Roca and Cox, 1990; Little and Mount, 1982; Burkhardt et al., 1988). RecA proteins or its counterpart have been found to exist universally in prokaryotes and eukaryotes, which may play roles during the process of DNA repair and DNA recombination and their SOS response. A larger number of other proteins, such as helicase, modulate the activity of nuclease involved in DNA repair (Hoeijmakers, 1993a,b). Some of these proteins facilitate the coupling of the process of transcription with DNA repair and thus facilitate the repair of the damage in the actively transcribing gene over non-transcribing genes (Hanawalt and Mellon, 1993). In human cells a helicase encoded by ERCC-6 gene is directly involved in linking transcription to DNA repair (Troelstra et al., 1992). Such factors that link transcription to DNA repair have been called *transcription–repair coupling factors* (TRCF). Human ERCC-3 and ERCC-6 and *E. coli* mdf gene products are thus TRCF. The role of *E. coli* mdf gene product as TRCF has been demonstrated *in vitro* (Selby and Sancar, 1990). Some of the TRCF are also transcription factors and thus promote transcription, but when they encounter a damaged DNA region, the transcription is halted and instead the TRCF commission the proteins and enzymes involved in DNA repair; thus the process of DNA repair gets a priority over transcription. Some aspects of relations betwen DNA transcription and DNA repair have been reviewed (Drapkin et al., 1994). In addition, the processes of DNA replication, repair, recombination, and transcription are interrelated, and nucleases play an important role; therefore, large numbers of proteins must interact with nucleases.

11

BIOLOGICAL FUNCTION OF NUCLEASES

Earlier nucleases were mostly considered as tools for biochemical studies. The only role ascribed to this group of enzymes included the salvage of DNA. Contrary to this view, nucleases are now known to play important roles in the different aspects of nucleic acid metabolism including DNA replication, repair, recombination, mutation and transcription; all of these genetic processes are related by participation of a number of common proteins including nucleases. Nucleases are also known to be involved in maintaining the superhelical structure of the chromatin. They determine the expression of a gene, indicate its transcriptional state and play an important role in the processing of a message. In this chapter the involvements of nucleases in the various genetic processes are discussed.

I. REPLICATION

The mechanism of DNA replication has been extensively discussed (Kornberg, 1980; Kornberg and Baker, 1992). DNA replication involves at least three steps such as initiation, elongation, and termination. Nucleases participate in all three steps of DNA replication to the extent that no step could possibly be completed without the action of this group of enzymes. Certain important nucleases involved in DNA replication are listed in Table 11.1.

A. Three Steps in DNA Replication

Briefly these steps are as described below.

1. Initiation. Replication of a helical DNA duplex proceeds with the untangling of its helical structure mediated by the action of a helicase or swivelase. The

Table 11.1 Nucleases involved in DNA replication

E. coli DNA pol I $5' \rightarrow 3'$ exonuclease in removal of RNA primers
E. coli DNA pol I $3' \rightarrow 5'$ exonuclease proofreading
E. coli DNA Q $3' \rightarrow 5'$ exonuclease proofreading
Mammalian DNaseV removal of RNA primers
DNA pol ε $3' \rightarrow 5'$ exonuclease proofreading and $5' \rightarrow 3'$ exonuclease removal of RNA primer and nick translation
DNA pol σ $3' \rightarrow 5'$ exonuclease proofreading
RNaseH /FEN-1/Dna2 nuclease removal of primers
Topoisomerases I release of torsional stress
Topoisomerase II separation of daughter DNA molecules after completion of replication

swivelase may act in a site-specific manner usually at the origin of replication. Topoisomerase may act as a swivel (Brill et al., 1987) or is required for DNA replication (Liu et al., 1979; Yang et al., 1985, 1987). Without such action of nuclease, DNA replication cannot begin.

Primer formation is the next thing that happens during the process of initiation of DNA replication. This is essential for the action of the enzyme DNA polymerase because it does not act *de novo*. Primer formation may require participation by a number of gene products (Kornberg and Baker, 1992). Primer is usually a stretch of ribonucleotides synthesized by RNA polymerase. Alternatively, primers may contain a mixture of ribo- and deoxyribonucleotides; such primers are synthesized by primase encoded by the dnaG gene in *E. coli*.

2. Elongation. During this process a primer is sequentially extended by DNA polymerase via the addition of deoxynucleotides complementary to the sequence of nucleotides in the template. In *E. coli* the nucleotides are added by DNA polymerase III to the $3'$ OH of the primer end. However, during this elongation process, error may occur due to the misincorporation of nucleotides that are noncomplementary to the nucleotide in the template. Nuclease action is required to remove the misincorporated nucleotides. In *E. coli* this $3' \rightarrow 5'$ proofreading activity is furnished by the E subunit of the DNA polymerase III homeoenzyme. The E subunit is encoded by dnaQ gene in *E. coli*. The fact that in *E. coli* a protein other than DNA polymerase III controls the proofreading functions is quite consistent with the observation that eukaryotic DNA polymerases also lack proofreading $3' \rightarrow 5'$ exonuclease activity and that such nuclease activity is provided by a distinct protein. Thus in eukaryotes as in prokaryotes the proofreading activity must be provided by a distinct protein that may act as an integral part of the replication machinery. In HeLa cells, the presence of such a protein with exonucleolytic activity has been demonstrated. In the replication machinery the two enzyme activities (i.e., the polymerizing and proofreading) must be coordinated because a change in ratio of these activities has been known to yield mutants of virus T4 which are proficient or deficient in the correction of misincorporated nucleotides; this aspect of the nuclease function is further discussed in the section on mutation (Section V) in this chapter. Properties of proofreading nuclease and other nuclease activities of DNA polymerase are listed in Table 11.2. Certain structure-specific nucleases as described in Chapter 3 are crucial in the removal of Okazaki fragments. These nucleases includes FEN-1 and Dna-2 nuclease in addition to RNaseH (see Figure 11.1b).

Table 11.2 Properties of proofreading and other nucleases associated with DNA polymerases

	I. Bacteriophage DNA Polymerases	
T4 gene 43	110 kDa	$3' \rightarrow 5'$ Exo (proofreading) activity
T7 gene 5	79.7 kDa	$3' \rightarrow 5'$ Exo activity ($5' \rightarrow 3'$ Exo activity controlled by gene 6)
Q29 gene 2	68 kDa	pol possesses both $3' \rightarrow 5'$ and $5' \rightarrow 3'$ Exo activities
T3	94.3 kDa	$3' \rightarrow 5'$ Exo activity
	II. Bacterial DNA Polymerases	
E. coli		
DNA pol I	PolA 103 kDa	$3' \rightarrow 5'$ Exo and $5' \rightarrow 3'$ Exo (DNase and RNaseH) *E. coli* polI $3' \rightarrow 5'$ intrinsic Exo 200 times more active than *E. coli* polI
DNA pol II	Pol B 90 kDa	Exo $5' \rightarrow 3'$. Exo activity present in N terminus.
DNA pol III	Pol C α subunit 130 kDa DNAQ α subunit 27.5 kDa	$3'$–$5'$-Exo activity PolC encodes polymerase activity DNA α (allelic to MutD) encodes $3' \rightarrow 5'$ Exo activity
B. subtilis		
DNA pol I	PolA	—
DNA pol II	PolB	—
DNA pol III	PolC 163 kDa	$3' \rightarrow 5'$ (N-terminus part) Exo activity intrinsic to peptide with polymerase activity
Streptococcus pneumoniae		
S.T.	Pol I-like *E. coli*	—
S.P.	Pol I	—
Micrococcus luteus		
	Pol II-like *E. coli*	
	Pol I	
	Pol III-like *E. coli*	
	Pol I	

Table 11.2 (*continued*)

	III. Animal Viruses	
DNA polymerases		
HSV	136 kDa	$3' \rightarrow 5'$ Exo activity $5' \rightarrow 3'$ RNaseH endonucleolytic
Vaccinia virus	110 kDa	$3' \rightarrow 5'$ Exo activity
Adenovirus	140 kDa	$3' \rightarrow 5'$ Exo activity
Bacculovirus	114 kDa	$3' \rightarrow 5'$ Exo activity
Reverse transcriptase		
Retroviral		
Avian		
A subunit	24 kDa	RNaseH N terminus of BSα
B subunit	32 kDa	Integrase C termini of B subunit
HIV		No Exo activity
Homodimer	66 kDa (subunit)	C-terminus RNaseH (15 kDa)
Integrase		32 kDa (in addition to RNaseH of the RT)
Hepatitis virus	90 kDa	RNaseH activity
	IV. Eukaryotic DNA polymerases	
α	170 kDa protein	No intrinsic Exo activity
σ	120 kDa protein	$3' \rightarrow 5'$ Exo activity
ε	170 kDa protein	$3' \rightarrow 5'$ Exo activity and $5' \rightarrow 3'$ Exo activity
β	36- to 38-kDa protein	Associated Exo activity (DNaseV)
τ	100- to 300-kDa protein	Exo activity Mismatch-specific No nuclease
	V. Telomerase	
Ribozyme		
	VI. RNA Polymerase	
Influenza virus		$3' \rightarrow 5'$ Exo activity
RNA polymerase II Yeast and *Drosophila*		RNase (barnase) domain present

3. Termination. This step of DNA replication involves (a) the removal of the RNA primer, (b) filling of the gap left by removal of the primer, and (c) the ligation of the newly synthesized DNA strands. Nucleases play a key role in the removal of RNA primer (see Figures 11.1a and 11.1b). In *E. coli* the $5' \rightarrow 3'$ exonucleolytic function of the DNA polymerase I is responsible for the removal of RNA primer. Without this exonucleolytic function of the DNA pol I, the process of DNA replication cannot be completed. Mutants of *E. coli* defective in the 5–3-exonucleolytic function are also defective in DNA replication. This exonucleolytic function of the enzyme acts in concert with its polymerizing activity; the removal of primer is maximal when the gaps are filled by the polymerizing activity of this enzyme. The exonucleolytic activity of DNA polymerase I can remove RNA or a mixture of ribonucleotides and deoxyribonucleotides in a primer. This property of DNA pol I is rather important because a primer could be purely RNA or a mixture of ribonucleotides and deoxyribonucleotides. The nature of primer may vary with the kind of viral or bacterial DNA being replicated (Kornberg, 1980). The removal of RNA primer may be achieved by the action of RNaseH such as in *E. coli*, ColE plasmid, and animal viruses. In bacteriophage T4, the removal of RNA primer is facilitated by the nuclease encoded by the viral gene 6 (Kornberg, 1980). It is possible to replicate a viral DNA *in vitro*. Using this *in vitro* DNA replication system, it has been shown that different viruses differ in their protein requirements essential for DNA replication. However, they uniformly require the same kinds of nucleases in each step of DNA replication. The involvement of nucleases in DNA replication is shown in Figures 11.1a and 11.1b. The role of nucleases in proofreading is crucial for faithful DNA replication (Kunkel et al., 1981b, 1987; Kunkel 1988; Echols and Goodman, 1991; Loeb and Kunkel, 1982).

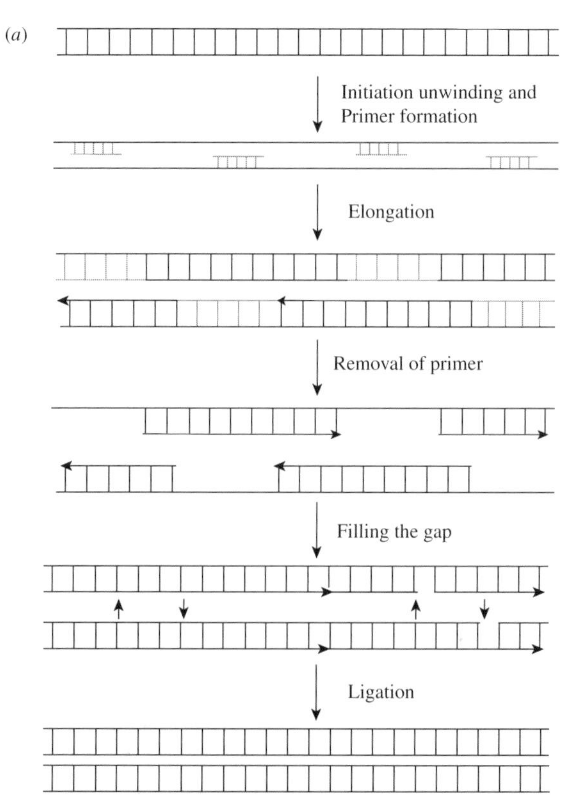

Figure 11.1a. Participation of nucleases in different steps of replication.

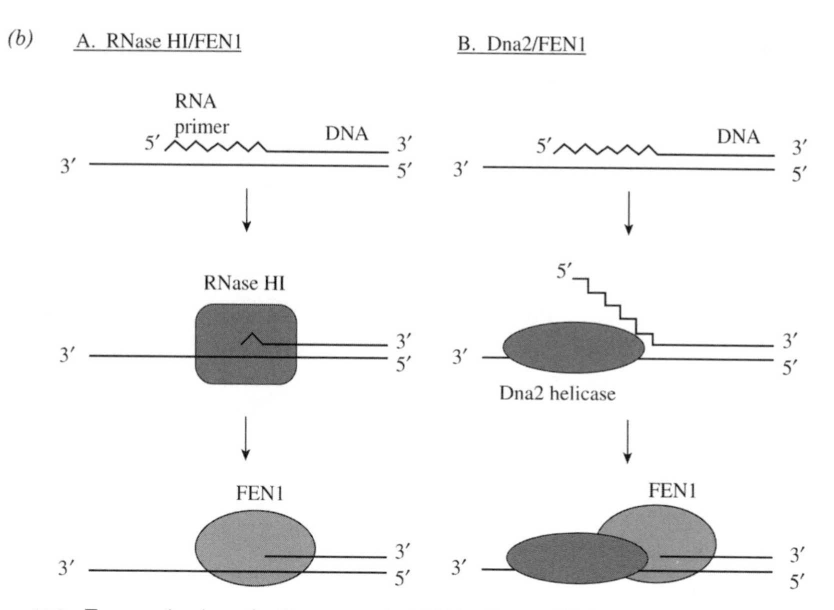

Figure 11.1b. Two mechanisms for the removal of RNA primers. (A) RNase HI cleaves the RNA segment attached to the 5′-end of the Okazaki fragment, leaving a single ribonucleotide adjacent to the RNA–DNA junction. FEN1 then removes the remaining ribonucleotide. (B) Dna2 helicase displaces the RNA segment (or RNA–DNA). FEN1 then cleaves endonucleolyticaly the branch point, releasing the displaced RNA (or RNA–DNA). Although the Dna2/FEN1-dependent mechanism has not been proved, recent biochemical and genetic studies strongly support this model (see Waga and Stillman, 1998). Reproduced with permission *Annu. Rev. Biochem.*, 1998.

B. Role of Viral Nuclease in the Degradation of Host DNA

Besides their role in different steps of DNA replication, the viral nucleases play important roles in the unfolding of the host supercoiled DNA, disruption of nucleoid, and its final degradations to nucleotides. The nucleotides released by the degradation of the host DNA may be reutilized in the synthesis of viral DNA. In bacteriophage T4 an extensive breakdown of the host DNA is achieved by the joint actions of the viral endonuclease and exonuclease encoded by the viral gene 3 and gene 6, respectively. However, a number of viruses—for example, T1 and T3—do not utilize the nucleotides obtained after degradation of host DNA.

C. Involvement of Nuclease During the Separation of Daughter Helices at the End of Replication

Nuclease must act to facilitate the separation of daughter helices at the end of replication of a circular DNA molecule. This may be brought about by the action of topoisomerase, perhaps in a site-specific manner at the point of termination. Topoisomerases have been shown to catalyze such reactions leading to separation of knotted DNA circles *in vitro*.

D. Involvement of Nucleases in the Rolling Circle Mechanism of DNA Replication

Most viruses with circular DNA as their genetic material replicate by the mechanism of rolling circle first described by Gilbert and Dressler (1969). In this mode of replication, replication starts with a nick caused by a nuclease which creates a free 3′-OH terminus and a phosphate group at the 5′ end. The 3′-OH group is then acted upon by DNA polymerase to elongate the chain using the template generated by the rolled-out strand with 5′-phosphate group. The cycle of replication goes on as the circular template rotates on its axis releasing several unit lengths of DNA at the 3′ end. The continuous DNA strand released as a "tail" must be acted upon by site-specific nucleases to generate unit lengths of DNA. The free ends of the genome size DNA then may be covalently or noncovalently joined to generate the circular virions, which are then packaged inside the viral head. Replication of phage P2, λ, φX174, and many other viruses occurs in this manner. The φX174 A protein and fd gene 2 protein are examples of nucleases that make the initial cut in DNA to start the replication, while λ terminase protein is an example of the nuclease required to produce the genome size DNA from the "tail" of rolling circle.

E. Involvement of Nuclease in the Replication of Linear DNA

Replication of linear DNA such as in virus T7 poses a problem during the termination step due to removal of primer. The gap created by the removal of primer at 5′ end of the strand cannot be readily filled in since DNA polymerase cannot synthesize new DNA chain *de novo*. This problem is solved by the formation of concatemers; this provides the 3-OH region, which is then extended by the DNA polymerase. The gap is thus filled and the adjacent nucleotides are joined by ligase to generate a concatemeric form of DNA with multiple genome size DNA joined end to end in a linear array. Later the concatemer is acted upon by nuclease to generate genome-size DNA chains that are packaged into a viral head.

F. Involvement of Nuclease in the Replication of Chromosome in Eukaryotes

Eukaryotic chromosomes are unique in containing an array of replicons that are 10–100 times smaller than a bacterial replicon. It is implied that nucleases must be involved in the initiation, elongation, and termination steps of DNA replication in eukaryotes. No nuclease mutants defective in specific steps of DNA replication in eukaryote have been identified so far. The topoisomerase mutants of yeast do seem to affect the separation of chromosomes at the end of mitosis.

Telomers of eukaryotic chromosome and that of vaccinia virus are organized alike. Therefore the replication of vaccinia virus can provide an insight in the replication of eukaryotic chromosome or at least the replication of the telomeric region and the involvement of nucleases in this process. Endonucleases are required to cause nicks in the circular templates, leading to the formation of linear duplex structures. Subsequent nibbling of the palindromic terminus of the duplex structure causes the formation of hairpins with internal discontinuities; the latter, when sealed by ligase, yields the circular genome of the pox virus.

II. DNA REPAIR

Among the different DNA transactions of replication, repair, and recombination, the role of nucleases in DNA repair is most extensively documented (Bernstein, 1981; Collins et al., 1987; Friedberg, 1985; Hanawalt et al., 1979; Howard-Flanders, 1981; Sancar and Sancar, 1988; Linn, 1982a; Lindahl, 1979; Hoeijmakers, 1993a,b; Wang et al., 1993; Cleaver, 1994). Nucleases are required to remove the damaged lesions in the DNA chain during the process of repair. There are several DNA repair pathways depending upon the nature of DNA damage to be repaired. Not all DNA repair pathways require the involvement of nucleases; certain DNA damages can be repaired without the participation of nucleases (Lindahl, 1979). Repair of thymine dimers by the mechanism of photoreactivation is an example of DNA repair without the involvement of nucleases. However, photolyase seems to be missing in humans; therefore, the role of DNA repair pathways involving nucleases is of particular significance in relation to their role in humans (Li et al., 1994; Ley, 1993), particularly in view of the increased incidence of UV radiation due to depletion of ozone layer in the stratosphere (Blaustein et al., 1994).

Damage to DNA may occur due to a variety of sources. This may include tautomerism of the bases in the DNA chain, an error in DNA replication due to a change in the fidelity of DNA polymerase or lack of proofreading mechanisms, free radicals produced during chemical reactions within the cell, and other changes in the environment. Damage to DNA may occur under physiological conditions. Damage to DNA may include the following types.

A. Baseless Sites

It has been estimated that the mammalian cells may lose up to 5000–10,000 purines and 200–500 pyrimidines during a period of one cell generation time due to hydrolysis of basis from the DNA chain (Lindahl, 1979). Such loss of bases may be caused by exposure to chemicals (especially alkalating agents) and to radiation. Altered bases are principally removed by the action of the enzyme DNA glycosylase. The apurinic and apyrimidinic sites must be repaired in a series of steps involving the participation of nucleases. An *in vitro* system for apurinic mutagenesis has been described (Hevroni and Livneh, 1988).

B. Sites with Altered Base or Incorrect Base

Altered bases of nucleoside may occur in a DNA chain due to the action of ionizing radiation and alkaling agents. Presence of base analog at the time of replication may cause the substitution of normal base (nucleoside). An incorrect base may be incorporated into a DNA chain due to a change in the fidelity of DNA polymerase or of the proofreading activity of exonuclease associated with the replication machinery. The fidelity of DNA polymerase may change due to an intrinsic alteration in the DNA polymerase per se or due to change in the ionic environment such as substitution of Mg^{2+} by Mn^{2+}. Deamination of cytosine and adenine leading to the formation of uracil and hypoxanthine respectively may lead to improper base pairing.

C. Cross-Linking and Other Damages

Inter- and intrastrand cross-links may arise as a result of the action of radiation, free radicals, alkalating agents, and light activated psoralen compounds. Intrastrand cross-links such as pyrimidine dimers are much less difficult to repair than the interstrand cross-links. Other DNA damages may include deletion or duplication, strand breaks, and formation of bulky adducts and oxidative damages. The precise nature of some of these damages is not yet known.

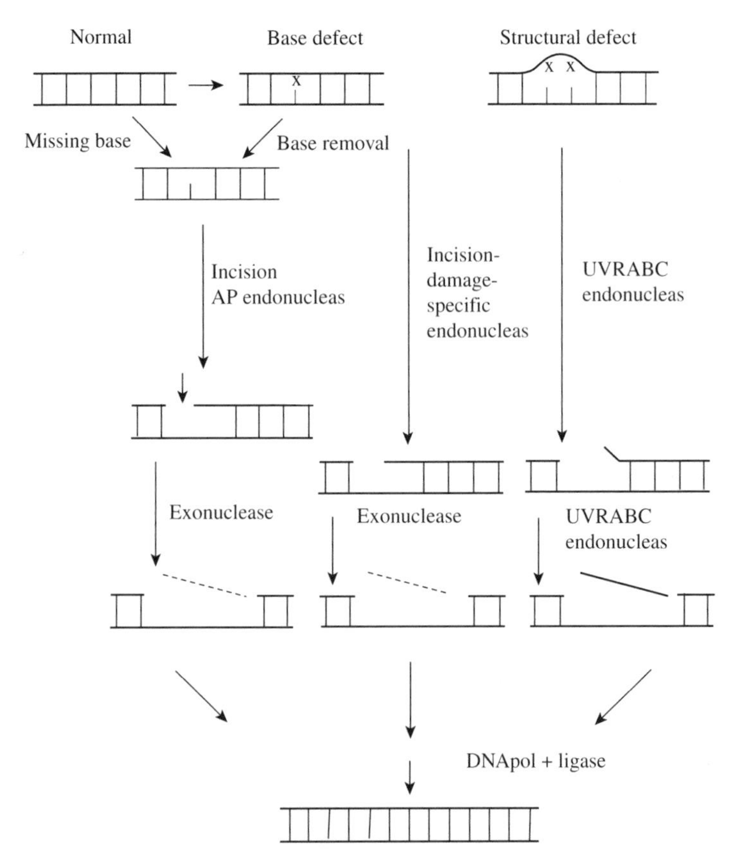

Figure 11.2. Different DNA repair excision pathway. [From Hanawalt et al. (1979), copyright Annual Reviews, Inc., used with permission.]

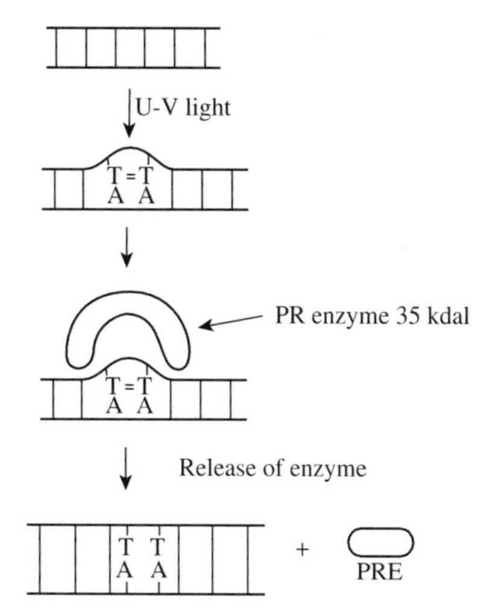

Figure 11.3. DNA repair pathway without involvement of nucleases. (Courtsey of Professor P. C. Hanawalt.)

D. DNA Repair Mechanisms

A variety of repair mechanisms has been found to occur in a cell to ensure the repairability of a multitude of DNA damages. These are described below and depicted in Figures 11.2 to 11.6, and the involvement of different nucleases and genes encoding them are listed in Table 11.3.

E. Excision Repair

Perhaps this is the most common repair pathway that repairs baseless site or sites with thymine dimer. The repair machinery, which involves nucleases, DNA polymerase, and ligase, recognizes a distortion in the DNA helix and then repairs the damaged site. The repair process proceeds in several steps, which include:

1. An incision by an endonuclease in the DNA chain containing the damaged site. The incision is usually made by an endonuclease on the 5 side of the damaged lesion in the DNA chain.
2. An excision of nucleotides encompassing the damaged region. This is carried by an exonuclease (usually a $5' \rightarrow 3'$ exonuclease) that may remove only up to 30 nucleotides (short patch) or up to 1000 nucleotides (long patch).
3. Repair DNA synthesis by DNA polymerase to fill in the gap. In *E. coli* this step is usually catalyzed by DNA polymerase I. The extent of DNA synthesis may vary depending upon the extent of nucleotides removed by exonuclease in the second step. These are generally called short-patch and long-patch

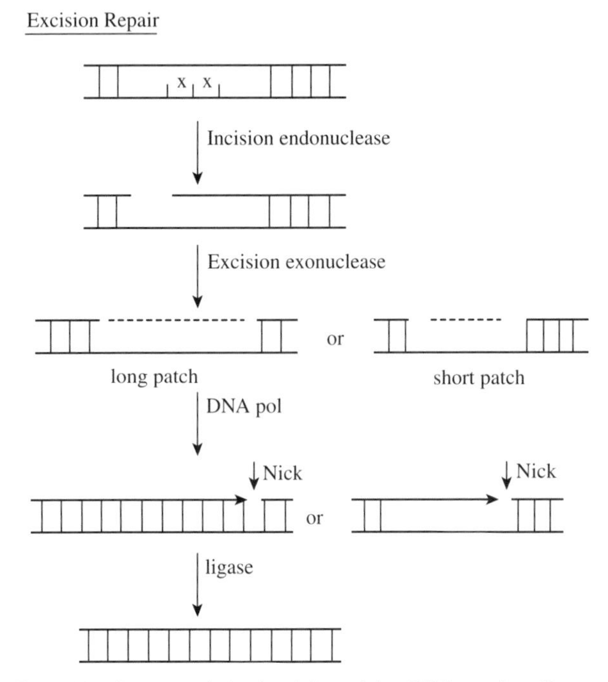

Figure 11.4. Long- and short-patch excision DNA repair pathway.

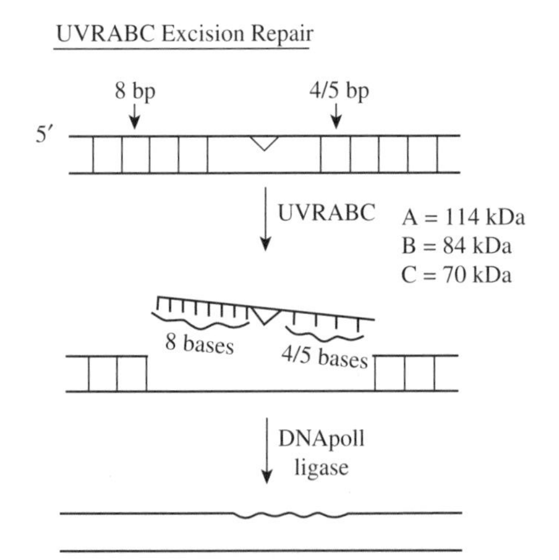

Figure 11.5. UVR excision repair pathway. [From Sancar and Rupp (1983), copyright Cell Press, used with permission.]

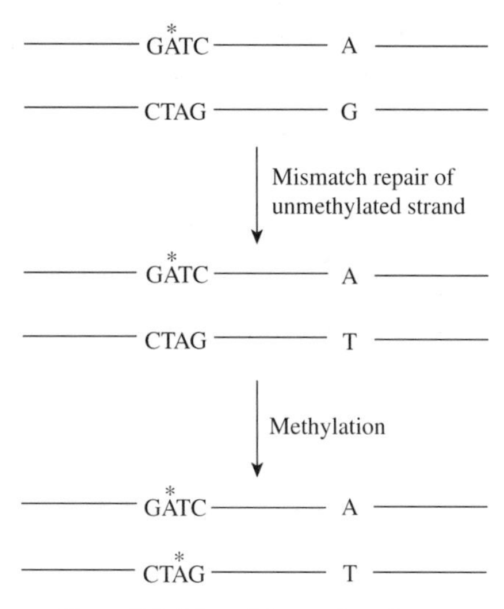

Figure 11.6. Mismatch repair pathway.

repair depending upon whether 10–30 nucleotides (short patch) or 100 to over 1000 nucleotides (long patch) are synthesized during the repair of the damaged lesion.

4. Ligation of newly synthesized DNA chain to the previously existing DNA chain. This is carried by the enzyme ligase to seal in the nick between adjacent nucleotides at the 3′ end of the newly synthesized DNA and the 5′ end of the old DNA chain by forming a phosphodiester bond.

The evidence for the role of nuclease in the excision repair pathway is based on the finding that nuclease-deficient mutants (a) are mutagen-sensitive, (b) retain

Table 11.3 Genes and nucleases involved in *E. coli* excision repair

Gene	Nuclease
Xth	Exonuclease III
	Endonuclease IV
	Endonuclease III
	Endonuclease VII
XSE	Exonuclease VII
uvrA	uvrA protein
uvrB	uvrB protein
uvrC	uvrC protein
uvrD	Helicase
PolA	Exonuclease of DNA polI
DNAQ	E subunit of DNA polIII

extensive amounts of thymine dimer per DNA molecule after exposure to UV light, and (c) contain less breaks in DNA chain after exposure to UV light when DNA is analyzed by alkaline sucrose gradient.

In the excision pathway (as depicted in Figure 11.2) the incision and excision steps may be catalyzed by the same nuclease or by distinct nucleases. If the damaged region contains an apurinic or apyrimidinic site, the chain may be acted upon by an AP endonuclease in concert with glycosylase or by an AP endonuclease with associated glycosylase activity (characteristics of such enzymes are described in Chapter 5). The role of AP endonuclease is depicted in Figure 11.2.

Another version of the excision pathway has been recently described as depicted in Figure 11.5. In this excision repair pathway, the damaged lesion is precisely removed by a cleavage to the 5′ end and another to the 3′ end of the damaged site. Such cleavages are made by the concerted action of the uvrABC endonuclease. Evidence for the presence of such repair pathway is based on the removal of thymine dimers by uvrABC endonuclease *in vitro*. It has been shown that uvrABC endonuclease introduces a nick in the DNA chain precisely eight nucleotides upstream to the 5′ end of the thymine dimer. Another nick three or four nucleotides downstream to the 3′ end of the thymine dimer is shown in Figure 11.5. Because of these nicks a stretch of nucleotides (about 13 nucleotides in length) containing the thymine dimer is released. Thus the damaged lesion is removed and the gap can be filled by the action of DNA polymerase I, and finally the newly synthesized DNA can be joined to the preexisting DNA chain by the action of ligase. DNA damages causing a distortion in DNA chain such as pyrimidine dimers and bulky adduct formation are most likely repaired by this excision repair pathway involving uvrABC endonuclease.

F. Bypass Repair Pathways

Excision repair pathways cannot possibly remove all the damages in DNA, particularly if the DNA enters replication immediately following the occurrence of damage. Thus those damages must be repaired by other pathways. These are appropriately called postreplication pathways or bypass repair pathways. There are two pathways for postreplication repair of DNA. These are described below and depicted in Figure 11.2.

G. Recombinational Repair Pathway

When replication ensues in a DNA chain containing thymine dimer or other damage, replication is usually stalled when polymerases encounter the damage region. However, replication soon starts beyond the damaged region, thus creating a gap in the newly synthesized DNA chain. The DNA helix containing both the thymine dimer and a gap (opposite the thymine dimer in the new strand of DNA) is destined to doom. However, this situation is usually averted via the distribution of these defective lesions to different DNA helices by recombination of DNA strands.

Once the defective lesions are distributed by the mechanism of recombination, these can be repaired via the mechanism of excision repair. Because this pathway requires the occurrence of recombination between DNA strands, this mechanism has been called the recombinational repair pathway.

H. Inducible and Error-Prone Repair Pathway

This pathway is both inducible and error-prone, resulting in high frequency of mutation among survivors of DNA damage. It is known that DNA damages inflicted to DNA may themselves induce this repair pathway. Presence of a DNA polymerase with low fidelity or of an insensitive proofreading nuclease may promote this repair pathway. Gene 42 T4 mutants with altered exonuclease proofreading ability produces a spectrum of mutator and antimutators. Thus the DNA polymerase, when it encounters a thymine dimer or other damage lesion, may bypass it and continue the synthesis of DNA chain, leaving errors in the DNA chain; the latter in turn provide the reservoir for mutations among the survivors (Hall and Mount, 1981). It is possible that in a particular niche these mutations may not be expressed or may not be lethal and may contribute to the survival of organism. Later these errors in DNA chain may be corrected among the survivors. Thus both postreplication repair pathways provide means to postpone the repair of damaged region to a convenient time when it can be removed by the mechanism of excision repair pathways. The involvement of nucleases in these pathways is limited to their participation in the mechanism of recombination itself and their involvement in subsequent excision repair pathways. In *E. coli* the postreplication repair pathways as well as the long-patch excision repair pathway are dependent on recA protein. This may reflect an interaction between recA gene product and the nucleases. This aspect is discussed in Chapter 10.

In the recombinational repair pathway, involvement of recBCD nuclease in bacteria and of single-strand-specific fungal nucleases is evidenced by the fact that mutants deficient in these nucleases are defective in both DNA repair and recombination. It has recently been shown that the recA gene, among the multitudes of things that it does within *E. coli* cells, does also interact with the recBCD nuclease (Williams et al., 1981) and with the editing nuclease subunit (E) of DNA polymerase III (Lu et al., 1986) This aspect is further discussed in the section on mutation (Section V).

I. Mismatch Repair

Repair of mismatch base pairs is crucial for the occurrence of the phenomenon of "conversion" that has been described as an important source of genetic recombination in phages and in fungi (Stahl, 1980; Fishel et al., 1986). Mismatch may also occur due to failure of the editing nuclease to remove the misincorporated nucleotide during the process of replication. Although no specific nuclease has been identified to play a role in mismatch repair, it is obvious that a region of mismatch is repaired via the excision repair pathway. It is also obvious that the distortion in DNA structure caused by the extent of mismatch itself may serve as an identifying landmark for the action of nucleases and associated proteins (that modify the activity of nucleases). If the extent of mismatch is limited to a single or to a few base pairs, it may not be noticed by the nucleases and may remain unrepaired, leading to what is called postmeiotic segregation of markers in fungi (Whitehouse, 1982). Mismatch of base pairs in the DNA helix due to misincorporation of nucleotide during replication may be identified by a transient lack of methylation in the misincorporated nucleotide which may then be removed by excision repair pathway. There are at least two mismatch repair pathways in *E. coli*: One is methylation-dependent and requires dam, mut H, mut L, mut S, and mut U gene products and ATP (Langle-Ronault et al., 1987; Modrich, 1987), and another is independent of dam-methylation but requires mut Y and PolA gene products (Tsai-wu et al., 1992). Human genes homologous to *E. coli* mut L and mut S have been identified and characterized, and their role in cancer of the colon has been elucidated (Service, 1994).

J. Mismatch Repair in Mammalian Cells

Presence of a mismatch repair mechanism in mammalian cells has been very well documented. A biological selection method was used to identify conversion of heteroduplexes into homoduplexes. Two distinct variants of polyoma virus designated as TS-A and CR differ in their ability to form plaque at higher temperature; the CR strain can form plaque at 39°C, but the TS-A strain cannot. Mutation at a site called A controls the ability of the CR strain for plaque formation at higher temperature. These two strains of polyoma also varied at three other loci (B, C, and D); differences at these could be detected by restriction endonuclease analysis. DNA from these two polyoma variants was used to generate heteroduplex. Mouse cells were transfected with heteroduplex DNA and then examined for plaque formation at 39°C. Under these conditions, plaque formation could occur only if a homoduplex is generated as a result of the mismatch repair of the heteroduplex. Plaque formation at 39°C by mouse cells after infection with heteroduplex clearly indicated the occurrence of mismatch repair. Correction of mismatch at other sites (B, C, and D) was also documented in the progeny virus by restriction endonuclease analysis. These results clearly indicate the presence of a mismatch repair system in mouse cells. It is imperative that this repair system must involve participation by specific endonuclease. Using such transfection of mouse cells by heteroduplex polyoma DNA, it has been shown that sites which are separated by less than 90 nucleotides were corrected simultaneously and therefore belonged to the same excision tract.

K. Incision of Damaged DNA is a Complex Process Involving Several Proteins

Nucleases whose specific roles in DNA repair have been established are described in Chapter 5. These nucleases act in concert with each other or with other proteins that modify or facilitate their nuclease activity. Genetic evidence suggests involvement of a number of proteins in the incision of damaged DNA chain during the excision repair in *E. coli*, *Neurospora*, yeast, *Drosophila*, and human cells. In *E. coli*, uvrA protein exhibits endonuclease activity only in concert with the uvrB and uvrC gene products; also the activity of uvrABC endonuclease is promoted by the uvrD gene product. In yeast there are at least 50 loci that are involved in DNA repair, which clearly indicates the participation of several gene products in a specific repair step. In *Drosophila* there are at least 30 genes that are involved in DNA repair mechanisms. Likewise, in mammalian cells a large number of genes control DNA repair. Participation of at least eight gene products in the excision repair of DNA in human cells is inferred from the occurrence of different complementation groups among the cells from individuals with xeroderma pigmentosum (XP). The complexity of the situation is further shown by the fact that each complementation group defective or deficient in the excision repair of thymine dimers can overcome this difficulty by the provision of a single UV endonuclease from the virus T4 or the bacteria *Micrococcus luteus*. The human XGP protein participates in the DNA repair in certain complementation groups of human XP patients as described in Chapter 3.

L. Excision Repair Mutants of *Neurospora*

Over a dozen loci controlling mutagen sensitivity in *Neurospora* have been identified (Mishra, 1991). Among them, UVS-2 is the only one that has been shown to be defective in excision repair pathways. However, the role of nuclease in this defect has not yet been demonstrated. UVS-2 mutants are unable to excise UV-induced pyrimidine dimers. This allele (UVS-2) appears to be defective in the incision step, like the uvrA and uvrB allele of *E. coli* and *S. typhimurium*. It is therefore defective in incision endonuclease or in protein that controls the activity of incision endonuclease. An mms-sensitive mutant of *Neurospora* may be deficient in AP endonuclease (Mishra, 1986). Certain nuclease-deficient mutants of *Neurospora* such as nuc-3 (Mishra and Forsthoefel, 1983) and nuc-2 (Ishikawa et al., 1969) seem to be defective in DNA repair pathways because these mutants are sensitive to UV radiation. The DNA-repair-defective mutants of *Neurospora*, like the yeast mutants, belong to three different groups (Mishra, 1991).

M. Excision Repair Mutants of Yeast

Genetic analysis of the radiation-sensitive mutants of yeast shows the presence of at least three epistatic groups with respect to their ability to incise UV-induced lesions in DNA *in vivo* (McCardy et al., 1987; Hoeijmakers, 1993a,b). One group, which includes Rad1, Rad2, Rad3, Rad4, and Rad10, completely lacks the ability to incise UV-induced lesions in DNA. Another group of mutants, which includes Rad7, Rad14, Rad16, and Rad23, is somewhat deficient but does not completely lack the ability for the incision of UV-induced damage in DNA *in vivo* (Friedberg, 1985). The ability to incise UV-induced dimers *in vivo* was measured by comparing the extent of strand breaks in the wild-type and mutant cells exposed to UV light and then was analyzed by sedimentation in alkaline sucrose gradient. In such experiments, when the percent of strand breaks for the wild type was considered 100, the comparative number of strand breaks in mutant cells was 0 (first group of rad mutants) or 29 (rad14) or 63 (rad16). These data clearly indicated that the first group of mutants completely lacked the ability to incise UV-damaged lesions in DNA. Thus in yeast there are at least 5 genes (Rad1, Rad2, Rad3, Rad4, and Rad10) that are required for the incision of damaged DNA. Even though these rad mutants were defective in *in vivo* incisions of UV-damaged lesions, they were not defective *in vitro*.

Thus in yeast, as in human cells, at least a number of genes are required for the incision of damaged DNA lesions. However, none of these mutants have been demonstrated to lack a specific nuclease. Thus it seems that the products of these genes may be involved in modifying the activity of an incision endonuclease *in vivo*. This aspect is further discussed later in this section.

N. Excision Repair Mutants of *Drosophila*

In *Drosophila*, there are at least 30 loci that control sensitivity to mutagens. These genes are localized on chromosome 2, 3, and X of *Drosophila* (Boyd et al., 1987). These mutants are also defective for the ability for excision repair as shown by the number of residual strand breaks after exposure to UV light analyzed by alkaline sucrose gradient. Here again the mutants seem to fall into two categories. One group of mutants represented by mei-9 and mus(2)201 completely lacks the ability for incision of UV-induced dimers, whereas the other group of mutants showed only a partial deficiency in excision-repair ability; the extent of excision repair varied from 24% (mus(3)308) to 72% (mus(3)302).

O. Excision Repair Mutants of Mammalian Cells

Mutants of mammalian (Chinese hamster and mouse) cells deficient in DNA repair have been characterized (Thompson et al., 1980). Among the mutants of Chinese hamster cells defective for DNA repair, five complementation groups have been found to exist (Thompson, 1998; Thompson et al., 1980, 1985). These mutants, when examined for that ability to incise UV-induced damaged lesions in DNA, were found to be defective; the mutants showed less than 10% strand breaks. Furthermore, a particular Chinese hamster mutant (designated as UV-20) belonging to the CHO complementation group 2 was found to complement the DNA repair defects of the human XP cells belonging to group A and C (Thompson et al., 1980). The hybrid between the CHO UV20 and human XP cells was found to be much less sensitive to killing by exposure to UV radiation (Thompson et al., 1983). These data clearly showed that the defects in these cell lines (i.e., CHO UV20, XP-A, and XP-C) are due to mutation in genes that control distinct functions essential for the incision of UV damage in DNA. The human XPG nuclease is very well characterized; this nuclease or its homolog in yeast RAD2 protein participate in the base excision repair. The characteristics of these nucleases and the gene(S) encoding them are discussed in Chapter 3.

The fact that a number of gene products are involved in the incision of a damaged lesion in DNA can possibly be explained on the basis of three-dimensional organization of chromatin structure in eukaryotes. In human cells it has been shown that the DNA synthesized during DNA repair is more sensitive to digestion by micrococcal nuclease, suggesting a nucleosomal control of excision DNA repair. It is possible that among the number of proteins required to facilitate the incision of damaged DNA, some may be required for the dissociation of nucleosomal structure so that incision endonuclease can freely act on the damaged DNA, while other proteins may participate in the reorganization of the nucleosomal structure.

III. RECOMBINATION

Recombination involves the exchange of genetic information, which is the most basic aspect of all living systems. It generates variation among organisms and thus provides the raw material for the process of evolution. In recent years it has been shown to play important roles in the control of gene expression. Above all it has been used as the main tool for genetic analysis and manipulation in Mendelian and molecular genetics.

The phenomenon of recombination was first described in the beginning of this century. However, only in the last few years has it been possible to determine the biochemical steps underlying the mechanism of this genetic process. Several hypotheses have been proposed (Holliday, 1964; Meselson and Radding, 1975; Szostak et al., 1983; West, 1992, 1994) to explain the different aspects of recombination. The essence of these recombination models lies in explaining the phenomena of reciprocal recombination, gene conversion, high negative interference, and asymmetry in the direction of gene conversion. Several excellent treatises are available on the genetic and biochemical aspects of the recombination mechanisms (Stahl, 1979, 1980; Whitehouse, 1982). However, none of them deals with the specific role of nucleases in genetic recombination, which is the subject of discussion in this section of this book.

A. Different Kinds of Genetic Recombination

Clark (1971) first defined the process of genetic recombination as the interaction of nucleic acid molecules leading to changes in the linkage relationship between genes or their parts. Two kinds of genetic recombination have been identified. These are homologous and nonhomologous recombinations.

Homologous recombination can further be described in two ways. These are general and site-specific recombinations. General recombination involves the interaction among any homologous sequences of DNA, whereas the site-specific recombination involves interaction between specific stretches of DNA, both leading to a new linkage relationship. However, it is pertinent to mention that even in general recombination, all interactions of nucleic acids are initiated at specific stretches of DNA. There is evidence that in *Escherichia coli* as well as in other organisms, general recombination is initiated at specific sites called *chi sites*. Thus all homologous recombinations are site-specific. Therefore, any classification of homologous recombination into general and site specific recombinations is just a matter of convenience.

Nonhomologous recombination (or illegitimate recombination as it is sometimes called) involves interactions between DNAs of little homology. Initially, nonhomologous recombination was so named because at that time it was just not possible to detect the involvement of a small stretch of homologous DNA involved in this process (Franklin, 1971). Illegitimate recombination usually leads to deletion, duplication, substitution, and insertion of DNA segment. Integration of viral DNA such as SV40 into resident host chromosome is also mediated by illegitimate recombination. All transpositional recombinations are mediated by illegitimate recombination. This process of genetic recombination might be involved in the gene amplification. Although the precise mechanism of illegitimate recombination is not known, it is certain that it must involve breakage and reunion of chromosomal fragments mediated by nucleases such as topoisomerases and transposases. The role of DNA gyrase (topoisomerase II) in illegitimate recombination of bacteria has been elucidated.

Over the years, several models of genetic recombination have been developed to account for the bulk of genetic and biochemical data gathered in different organisms, particularly fungi and bacteriophages. The majority of these models represent a variation of the Holliday model (Holliday, 1964). An account of this model is presented below. Knowledge of the steps involved in recombination is essential for the understanding of the roles that the different nucleases play in this genetic process.

B. Recombination Mechanisms and Nucleases

The mechanism of recombination involves several steps and requires participation by several nucleases. For the sake of convenience, the steps in recombination may be broadly described under three major stages, such as initiation, progression, and completion. These events involved in recombination are schematically presented in Figure 11.7a.

1. Initiation includes events that precede the formation of the Holliday structure. This begins with a nick introduced into a polynucleotide chain followed by strand displacement and strand aggression (see Figure 11.7a).
2. Progression includes the formation of Holliday structure and involves strand assimilation, branch migration, and heteroduplex formation (see Figure 11.7a). This stage of recombination has been designated here as *progression*.
3. Completion involves the maturation and resolution of Holliday intermediates into recombinant DNA molecules. This is marked by isomerization of

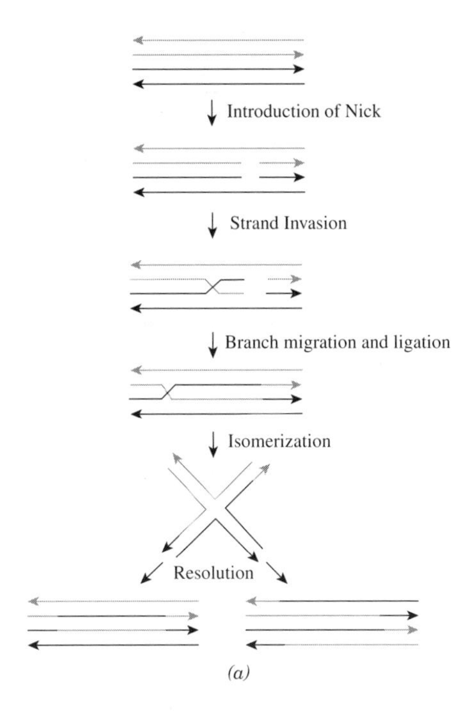

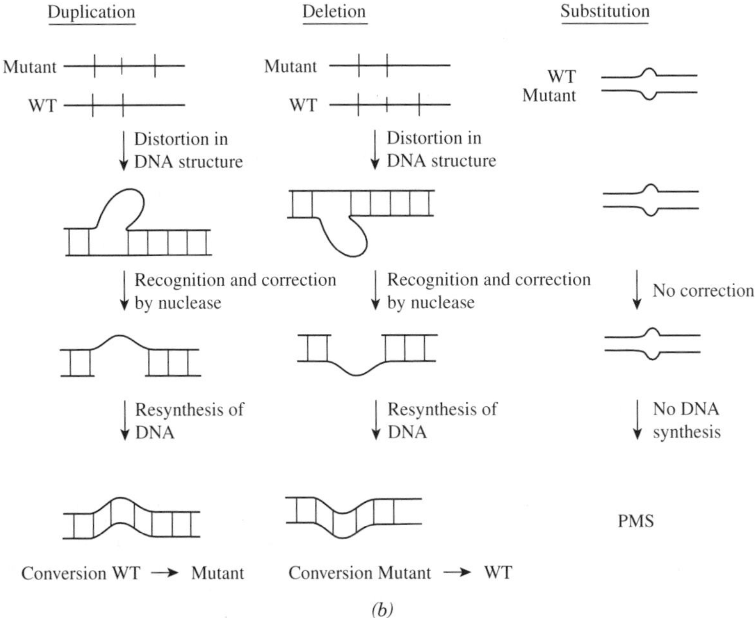

Figure 11.7. (a) Holliday model of DNA recombination with formation of heteroduplex regions including both DNA helices. [Based on Holliday (1964), copyright Cambridge University Press, used with permission.] (b) Extent of mismatch in the heteroduplex region caused by different mutants and their correction leading to gene conversion and postmeiotic segregation (PMS).

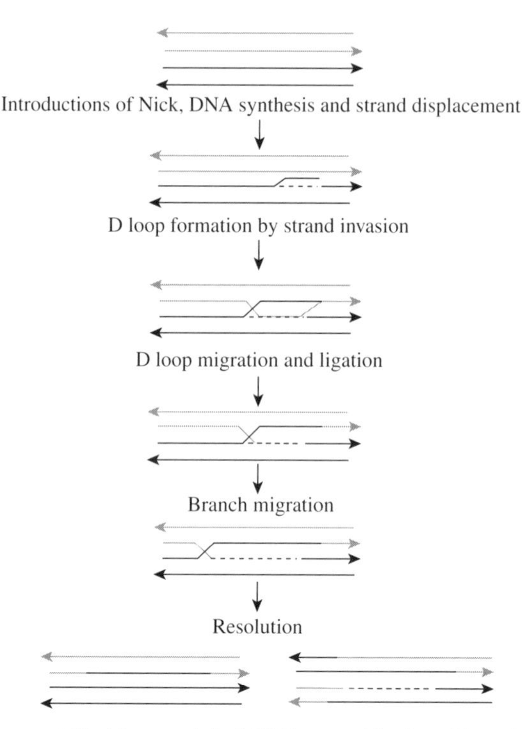

Figure 11.8. Meselson and Radding model of DNA recombination. [Based on Meselson and Radding (1975), used with permission.]

polynucleotide chains and resolution of Holliday structure by transnicking and ligation. This is followed by repair of the mismatched region in the heteroduplex structure which later yields the conversion of genetic markers. This stage of recombination has been designated here as *completion*.

The models of recombination (see Figures 11.7, 11.8, and 11.9) as originally proposed (Holliday, 1964) and later modified (Meselson and Radding, 1975) invoked simple strand breaks in DNA duplex. However, evidence from recent studies of gene conversion events underlying mating type switch in yeast suggested a model based on double-strand breaks (Szostak et al., 1983). The idea of double-strand break during recombination is further supported by the fact that the cleavage of plasmid by restriction enzyme increases the recombinogenic activity during yeast transformation. The double-strand break model of recombination eliminates the need for the occurrence of mismatch repair as the source of gene conversion. The role of nucleases in the process of recombination via different mechanisms as implicit in different models of recombination is shown in Figures 11.8 and 11.9.

All these major steps in recombination involve participation by nucleases. The initial nick(s) must be made by an endonuclease. The strand assimilation is usually accompanied by degradation of the polynucleotide chain(s) ahead of the delinquent

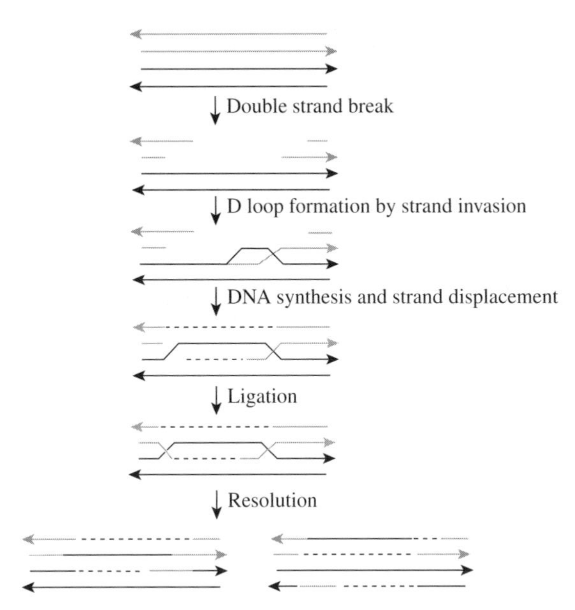

Figure 11.9. Double-strand break model of DNA recombination. [Based on Szostak et al. (1983), copyright Cell Press, used with permission.]

DNA chain(s); such degradation is mediated by exonucleases. The transnicking and mismatch repair during the completion of recombination also require participation by nucleases.

The role of nucleases in specific steps of recombination has been elucidated by the biochemical analysis of genetic mutants deficient in recombination first initiated by Clark (1971). Some of the roles of nucleases in recombination have now been confirmed by *in vitro* demonstration of the involvement of a particular enzyme in a specific step of recombination. The role of nucleases in the formation of Holliday structure, its resolution, and subsequent repair of mismatch region in heteroduplex leading to gene conversion has been demonstrated by *in vitro* studies. Different nucleases that mediate a specific step in recombination are listed in Tables 7.1, 7.2, 7.3, and 11.4.

C. Gene Conversion and Postmeiotic Segregation

Aberrant segregation ratios of the mutant and wild-type phenotypes among the progeny of crosses between wild-type and mutant parents have been described as phenomena of gene conversion and postmeiotic segregation. In yeast and other organisms with four-spored asci gene, conversion has been identified as showing segregation of wild-type and mutant phenotypes as 3 : 1 or 1 : 3 ratios. In *Neurospora* and other fungi with eight spored asci, the gene conversion can be identified as asci with 6 : 2 or 2 : 6 ratios of the wild-type and mutant spores. The phenomenon

Table 11.4 Nucleases involved in homologous recombination

Enzyme	Gene	Mode of Action/Possible Involvement in Recombination Stage
	I. Endonucleases	
1. T7 Endonuclease I	T7 gene 3	Initiation of recombination by introduction of nick or resolution of Holliday structure
2. T4 endonuclease VII	T gene 49	Resolution of recombination intermediate
3. Yeast cruciform endotransnicking enzyme	—	Resolution of Holliday structure
4. *E. coli* mismatch repair	—	Repair of mismatch in heteroduplex
5. Yeast mismatch repair enzyme	—	Rapair of mismatch in heteroduplex
	II. Exonuclease	
ATP-independent exonuclease		
1. λ exonuclease	red α gene	Acts processively on duplex DNA's role in strand assimilation
2. *E. coli* Exo VIII	Rec E gene	Same as λ exonuclease
3. T Exonuclease	T gene 6	Formation of recombination intermediates
ATP-dependent exonuclease		
RecBCD nuclease[a] *E. coli* Exo V	RecB RecC RecD	Role in initiation of recombination and strand transfer

[a] RecBCD-type nucleases have been described from a number of organism that play roles during recombination.

of postmeiotic segregation (PMS) can give rise to asci with 5 : 3 or 3 : 5 wild-type and mutant spores in the asci. The nature of gene conversion and PMS mostly depends on the extent of distortion caused by the mismatch in the heteroduplex structure formed during the process of recombination involving wild-type and mutant parents. Duplication and deletion mutants can cause severe distortion in DNA structure which can be recognized by the mismatch repair endonuclease and then corrected after removal of the DNA strand opposite the distorted DNA as shown in Figure 11.7b. Usually, duplication and deletion mutants are corrected and lead to gene conversion. A duplication mutant necessarily leads to conversion of the wild-type allele to mutant, whereas a deletion mutant leads to a conversion of the mutant allele to wild type. Substitution mutants involving one or a few base-pair mismatches that do not cause extensive distortion of the

Table 11.5 List of recombination systems characterized *in vitro*

λ integrate system	Nash (1981), Nash and Robertson (1981)
Hin-mediated site-specific recombination	Nash et al. (1987)
P recombinase	Johnson et al. (1986, 1987, 1988)
Cin recombinase	Johnson and Simon (1985)
γδ resolvase	Hoess and Abremski (1984, 1985)
Bacteriophage T4 system	Iida (1984)
Bacteriophage T7	Reed and Gridley (1981),
E. coli mismatch repair	Reed (1981a)
Yeast FLP system	Sadowski and Vetter (1976)
Yeast "resolvase"	Sadowski, (1971, 1977, 1986), Roeder and
Yeast mismatch repair enzyme	Sadowski (1979)
Endonuclease VDJ	Kemper and Brown (1976)
E. coli ruvC gene product	Symington et al. (1983)
	Kolodner (1980)
	Kondo et al. (1984)
	Connolly et al. (1991)

heteroduplex remain undetected by the mismatch repair enzyme and later appear as PMS asci as seen in Figure 11.7b.

D. *In Vitro* Recombination System

The *in vitro* recombination system has been used in a number of instances to elucidate the role of particular protein in a specific step of general, site-specific, or illegitimate recombinations. The different *in vitro* systems studied are listed in Table 11.5. As mentioned before, the intrinsic aspect of all recombination models (Holliday, 1964; Meselson and Radding, 1975; Szostak et al., 1983) is the formation of the recombination intermediates, or Holliday structures as they are called. Both the formation of Holliday structures and their resolution by specific enzymes have now been demonstrated (West, 1992, 1994) by *in vitro* studies. The physical presence of Holliday structures has been shown through the formation of DNA with the figure eight by the *E. coli* enzyme (Kolodner et al., 1980, 1987, Kolodner, 1980). In these studies, the fact that two circular plasmid DNA molecules were joined into the figure eight could be demonstrated by gel electrophoresis analysis and/or by electron microscopy. The resolution of Holliday structures into recombinant molecules was demonstrated by the inclusion of resolvase from a variety of sources. These include the T4 endonuclease 7 (Kemp et al., 1984), T7 endonuclease I (de Massy and Weisberg, 1987; Sadowski and Vetter, 1976) *E. coli* ruvC and yeast enzyme (Symington et al., 1983). In these studies, Holliday structures such as chi-form DNA generated from recombinant plasmid with figure-eight structure or other chi-form DNA or cruciform DNA were treated with a specific enzyme. Size measurement or electron microscopy of the DNA after enzyme treatment revealed the cleavage of the Holliday structure. The fact that T4 endonuclease 7 and T7 endonuclease I catalyze the same reaction (i.e., the resolution of the Holliday structure) has been demonstrated by their *in vivo* complementation. A plasmid carrying T7 gene 3 (which encodes endonuclease I) was found to complement T4 gene-49 mutation. The involvement of different enzymes in the site-specific recombination elucidated by genetic analysis was decidedly confirmed by *in vitro* studies (Kitts et al., 1982). The breakage and reunion of DNA during λ integration occurs by the topoisomerase-like activity of the λ int protein (Kitts et al., 1982). The role of V(D)J endonuclease in site-specific cleavage of immunoglobin genes has been confirmed by *in vitro* studies (Kondo et al., 1984). Likewise, the role of *E. coli* gyrase in illegitimate recombination has been elucidated by analysis of the recombination product after treatment with the enzyme in an *in vitro* study.

E. Fungal Recombination Nucleases

A particular nuclease of *Ustilago maydis* seems to be involved in DNA repair and recombination. This enzyme deficiency seems to impair repair and recombination in *Ustilago* (Ahmed et al., 1975; Holliday et al., 1974). This enzyme is known to be involved in the repair of mismatched bases. A *Neurospora* enzyme with endo–exonucleolytic activity has been described (Chow and Fraser, 1983). This enzyme is a proteolytic product of a native enzyme of higher molecular weight (90 kDa). A *Neurospora* mutant called UVS-3, sensitive to UV light and ionizing radiation and meiotically impaired, lacks a proteolytic activity leading to reduced level of the endo–exonuclease activity. The wild-type yeast has been shown to possess a nuclease activity that is antigenically similar to *Neurospora* endo–exonuclease as described above. However, the *Rad-52* mutants of yeast have been shown to lack this nuclease both at the log growth phase as well as during meiosis (Resnick et al., 1984). This finding suggests that a functional RAD52 gene is required for this nuclease activity which may play a crucial role in meiotic recombination.

F. Mismatch Repairs During Recombination

Most organisms seem to possess a mismatch repair process capable of recognizing and repairing incorrect base pair in DNA. The process of mismatch repair has been considered as the basis for (a) gene conversion (Holliday, 1964), (b) phenomenon of high negative interference, (c) correction of replication error, and (d) marker discrimination during pneumococcal transformation (Lacks et al., 1982). The mismatch repair may be facilitated by an endonuclease similar to S_1 nuclease. Different aspects of mismatch repair have been discussed (Claverys and Lacks 1986; Claverys et al., 1983). Mismatch repair may play a role in providing a barrier to recombination between different species. It has been shown that such a barrier to recombination between *Escherichia coli* and *Salmonella typhimurium* may be removed in a mutant deficient in mismatch repair (Rayssigvier et al., 1989).

In *Escherichia coli* there seem to exist at least three mismatch repair pathways. Of these, one is DNA-methylation-dependent while others are DNA-methylation-independent (Fishel et al., 1981, 1986). Additionally, these pathways differ in the extent of excision tracts and in their requirement of additional gene products. In

E. coli the mismatch repair is controlled by a mut system which encodes for proteins with nuclease activity [see Claverys and Lacks (1986) for discussion]. In *Streptococcus*, the mismatch repair is controlled by *hex* genes that presumably encode for nucleases. In *E. coli* the DNA-methylation-dependent pathways require a dam gene product while other pathways require uvrD and RecF gene products. However, both the DNA-methylation-dependent and -independent mismatch repair pathways require a nuclease induced by recJ gene (Fishel et al., 1986). Furthermore, the mismatch repair in *E. coli* suggests that the strand identification is dictated by the methylation of the base. Both in *E. coli* and in yeast, the presence of a protein that can repair a mismatch in a heteroduplex has been identified by *in vitro* assay (Lu et al., 1983; Kolodner, 1980; Kolodner et al., 1980, 1987). The enzyme has been identified by its ability to restore a restriction site by the repair of a mismatch in the heteroduplex region of a plasmid (see Figure 7.4). The restoration of a restriction site was detected by the analysis of DNA fragments generated after treatment with appropriate restriction endonuclease(s). Further characterization of such enzyme activity in *E. coli* and yeast will elucidate the nature of the enzyme. Ahmed et al. (1975) have described a single-strand-specific nuclease from *Ustilago* which can act on a mismatch region in DNA. In addition to these, topoisomerase may be involved in nonhomologous recombination (Bullock et al., 1988). Mismatch repair specific enzymes have been identified in nuclear extracts of human and *Drosophila* cells (Holmes et al., 1990). Genes and enzymes involved in mismatch repair are listed in Table 11.6.

G. Recombination Pathways

In bacteria alternative pathways for homologous recombination have been identified (Clark, 1991; Smith, 1988, 1989a,b). This is based on their dependence on the product of recA genes and the nature of nucleases involved during recombination. At least four major pathways have been identified. These are listed below.

1. RecBCD Pathway. This pathway is dependent on the products of recA genes and of the recBCD genes. RecBCD genes have been shown to code for the components of the enzyme exoV. This recombination pathway is responsible for 99% of chromosomal recombination in Hfr crosses and is stimulated by chi sequences (Stahl, 1980).

2. RecFJ Pathway. This pathway is also dependent on the recA gene product. In addition, it requires the product of the recF and recJ genes. The activity of this pathway is increased extensively (over 100-fold) in the presence of SbcB mutations. The SbcB gene product(s) are repressors of exonuclease I. This pathway is insensitive to chi sequences. RecF and recJ gene products are nucleases that have not yet been characterized.

3. RecE Pathway. This pathway is independent of the recA gene product. This is mediated by the recE gene product, which is exonuclease VIII. In addition, it may

Table 11.6 Genes and enzymes involved in mismatch repair

Gene	Protein	Remarks
E. coli		
mutH	25 kDa	Endonuclease must have evolved from DPNII
mutL	70 kDa	Orchestrate nuclease action; identify the distortion in the DNA helix due to mismatch
mutS	98 kDa *Salmonella* 97-kDa *E. coli*	
mutU	73–75 kDa	Helicase
*hex*A	90 kDa	Corresponds to mutS (98 kDa)
*hex*B	83 kDa	Corresponds to mutL (70 kDa)
dam		Must have evolved from DPNII methylase
mutY		39.1-kDa nuclease
DNA polI		Fill in the gap created by the action of nuclease to remove the mismatch region
Yeast		
PMS-1		Equivalent to *E. coli* mutL
Human[a]		
hMLH1		Located on human chromosome 3; is equivalent to *E. coli* mutL
hMLS2		Located on human chromosome 2; is equivalent to *E. coli* *mu*S

[a]Mutation in these gens leads to 90% hereditary nonpolyposis colon cancer (HNPCC) in humans. A third gene involved in the remaining 10% of HNPCC has not yet been identified. Mutations in these human genes lead to failure of repair of DNA mismatch and consequently cause colon cancer.

involve activities other than exonuclease VIII. This may not facilitate recombination of bacterial chromosome and has been shown to be insensitive to chi sequences. Also, the recombination of plasmids in *E. coli* is independent of the recA gene product (Fishel et al., 1981).

4. Red Pathway. This actually involves the recombination of bacteriophage λ. This system is independent of recA gene product and is insensitive to chi sequences. This system is controlled by λβ nuclease and λ exonuclease.

H. Recombinational Control of Gene Expression

Since the elucidation of the control of *lac* operon in *Escherichia coli*, interaction between DNA and protein has been considered as the hallmark of the regulation of gene expression. Such a regulatory mechanism does not involve any change in the structural organization of the gene. However, in recent years it has become obvious that recombination between different segments of genome in an organism provides additional avenues for control of gene expression (Zieg et al., 1978). Thus by genome reorganization, the genetic process of recombination provides not only variations essential for evolution but also new avenues for genetic control of regulatory mechanisms.

Iino (1984) has summarized three different ways of genetic recombination leading to control of gene expression. These are excision, exchange, or inversion of genetic segment(s). The excision type of regulatory system has been found to occur during mammalian immunodifferentiation where a functional immunoglobin gene is produced by the excision of an intervening DNA sequence. The exchange type causes the activation of gene by translocational exchange of DNA segments. This type includes the mating-type conversion in yeast and antigen diversity in trypanosomes (Borst and Cross, 1982; Van der Ploeg et al., 1984). The inversion type affects the activity of gene by inversion of a genetic segment. The G inversion in bacteriophage Mu (Hsu and Landy, 1984) and C inversion in bacteriophage P1 and P0 inversions in *Salmonella* are the most common examples of the inversion type of recombinational regulatory mechanism.

All types of recombinational regulatory mechanisms involve site-specific recombination mediated by recombinase. The excision type is mediated by V(D)J endonuclease, the exchange type is mediated by HO endonuclease, and the inversion type is mediated by a number of invertases. These recombinases identify a specific sequence in DNA and then facilitate site-specific recombination. Some of the cleavage site for recombinase is similar to that found in transposon, which suggests that recombination control systems may have evolved from transposable systems (Iino, 1984).

Other examples of the role of nucleases in gene expression include gene amplification and gene sequestering. Gene amplification for ribosomal RNA in *Xenopus* or for chorion gene in *Drosophila* has been very well elucidated. The process of gene sequestering is seen during the development of nucleus in somatic cells of certain organisms. In *Tetrahymena*, the chromosomal segments are reorganized for expression in macronucleus, and virtually all repeat sequences are lost. Likewise, in certain nematodes and in other multicellular invertebrates, the chromosomes in somatic cells are reorganized and during this process almost all repeat sequences are lost; thus the chromosomes in somatic cells are markedly reduced in their size in comparison to the chromosomes in the nucleus of the germ cells. Nucleases are also involved in the loss of mitochondrial DNA leading to uniparental transmission of mitochondrial genes in most organisms. Likewise, in certain wasps the development of the male is accompanied by the elimination of an entire set of chromosomes leading to haploidy. Nucleases play an important role in destruction of chromosomes during such sexual development. In addition, nucleases can influence the expression of a gene by bringing an enhancer DNA sequence or a promoter closer to a gene or by degradation of mRNA upon translation.

I. Role of Recombinase in Mammalian Antibody Diversity, Allelic Exclusion, and Class Switch

In mammals, the antibody molecule synthesized by B cells consists of light-chain and heavy-chain polypeptides. Each polypeptide chain contains an N-terminal variable region and C-terminal constant region. The DNA segments encoding variable and constant regions are situated at distant locations in the mammalian chromosomes, and they are brought together during the development of pre-B cells. The DNA segments encoding kappa light-chain variable and constant regions are located on human chromosome 2. The lambda light-chain variable and constant regions are encoded by DNA segments on human chromosome 22. The heavy-chain variable and constant regions are encoded by DNA segments on human chromosome 14. During the assembly of gene encoding light chain, the variable region (V) and joining region (J) adjacent to the constant region (C) are brought together. Likewise, the assembly of genes encoding heavy chain involves the joining of variable region (V), diversity region (D), joining region (J), and the adjacent constant region (C). Such assembly of the components of the heavy and light chains is mediated by the process of intramolecular site-specific recombination controlled by the product of *rag-1* and *rag-2* genes. The assembly of antibody genes during the differentiation of the mammalian pre-B cells is very complex but highly ordered, and regulated sets of reactions in which recombinases play a great role. The assembly of heavy chain on one of the homologs of chromosome 14 in human is initiated first. If such assembly fails, then the DNA segment encoding components of heavy chain on another homolog of the human chromosome 14 is attempted. When the assembly of heavy chain is successful, the assembly of light chain is initiated first on chromosome 2. If this assembly of kappa light chain is successful, then the alleles of kappa light chain on another homolog of chromosome 2 is inactivated. If and when the assembly of kappa light-chain gene is completed, the lambda light-chain assembly is not attempted. However, if kappa light-chain assembly on both homologs of human chromosome 2 is unsuccessful, the assembly of lambda light chain on one of the homologs of chromosome 22 is attempted. The sequence of events in the assembly of genes for heavy and light antibody chain and their allelic exclusion is indeed very orderly. This suggests a great precision on the part of nucleases involved in recombination of these DNA segments. Further complexity of the process of recombination is indicated by the phenomenon of class switch in which a particular B cell synthesizing a specific immunoglobulin such as IgG switches to synthesize another immunoglobulin such as IgA or IgE. Such switch is facilitated by substitution of one constant region in the heavy chain by another constant region. The constant region of heavy chain is encoded by either of the DNA segments μ, δ, γ, ε, and α occurring in tandem; of these, each segment encoding a constant region of heavy chain occurs only in one copy—except for the γ segment, which occurs as multiple copy. This class switch is mediated by intramolecular homologous recombination controlled by recombinases in response to another antigen. Recombinases involved in assembly of immunoglobulin genes and sequence of events involved therein are listed in Table 11.7.

Table 11.7 Role of recombination in rearrangement of antibody genes: characteristics of human antibody protein genes

I. Light-chain genes
  A. Kappa (κ)-chain component on chromosome 2
    $V_\kappa + J_\kappa + C_\kappa$
    (300) (5) (1)
  B. Lambda (λ)-chain component on chromosome 22
    $V_\lambda + J_\lambda + C_\lambda$
    (300) (5) (1)
II. Heavy-chain gene(s) on chromosome 14
    $V_H + D_H + J_H + C(\mu, \delta, \gamma, \varepsilon, \text{and } \alpha)$
      μ, δ, ε, and α each occurs as a single copy, whereas γ occurs as multiple copies
III. Sequence of events in antibody gene assembly
  1. Heavy-chain gene rearrangement must occur first in one of the homologs of human chromosome 14. If that does not result in successful assembly, then gene rearrangement is initiated on the other homolog of chromosome 14.
  2. A successful assembly of H-chain gene triggers signal for the assembly of light (κ)-chain gene on chromosome 2.
  3. Failure in assembly of kappa light-chain gene triggers signal for the assembly of the components of the lambda light-chain gene on chromosome 22.
  4. Once the arrangements of the components of gene for heavy chain and light chain are successful on one of the chromosomes, the assembly of such components on another homolog is stopped by the mechanism of allelic exclusion. The mechanism of allelic exclusion involving precise actions by recombinases is not yet known. Also, the mechanism of class-switch involving assembly of heavy-chain gene involving DNA segment encoding a particular constant region is not known.

J. T-Cell Surface Receptor

T lymphocyte, which differentiates in thymus, mediates the cellular and humoral immunity via the T-cell surface receptor protein. The T-cell receptor protein consists of α and β polypeptides; occasionally, it is also comprised of a γδ heterodimer. The molecular size of the α, β, and δ chains about 45 kDa, whereas the γ chain is about 35 kDa. Each chain consists of N-terminus variable and C-terminus constant region. The DNA segments encoding the V, D, J, and C components of the β chain are located on human chromosome 7, whereas the DNA segments of α chain are located on human chromosome 14. The assembly of these components is facilitated by the recombinases controlling the assembly of antibody genes in pre-B cells. The rearrangement of the components of β chains precedes the assembly of the components of α chain, and their successful rearrangement on one chromosome leads to allelic exclusion.

K. Application of Recombinases: Engineered Expression of Genes

A number of animal, bacterial, and plant genes have been expressed *in vivo* using the cloned recombinase gene and their specific sites (see Chapter 7). The expression of a silent reporter gene (e.g., β-galactosidase gene on Xgal medium) is triggered by the action of a site-specific recombinase (Sauer and Henderson, 1988; O'Gorman et al., 1991). Such expression of the silent reporter gene via site-specific recombination can be facilitated in response to environmental factor(s) or in tissue-specific or temporal manner.

IV. DNA TRANSFECTION OR TRANSFORMATION

Transformation is the transfer of genetic character(s) mediated by naked DNA molecule(s). This process must have evolved to promote genetic exchanges among organisms in nature. This conclusion is based on the observation that restriction systems which degrade the viral DNA upon entry into a bacterial cell leave the donor DNA unharmed during transformation. Nucleases play a great role during the process of transformation (Strauss, 1962). Historically, their negative roles have been emphasized because nucleases were seen to interfere with the process of transformation by hydrolyzing the donor DNA (Avery et al., 1944). The development of competence during bacterial transformation is designed to avoid the negative effects of nucleases. Similar objectives are achieved in *Hemophilus* by the identification of homologous donor DNA, which is prevented from degradation by nucleases and allowed entry into the interior of a recipient cell while the heterologous DNA is degraded in the absence of their recognition during the process of transformation.

The evidence for the positive roles of nucleases during transformation is based on several facts: (a) The duplex donor DNA is degraded to yield single-stranded DNA that enters into the interior of the recipient cells. Both the uptake and the degradation of the complementary strand of DNA require the development of competence (Lacks, 1970; Lacks and Neuberger, 1975; Lacks and Greenberg, 1973; Lacks et al., 1967, 1975). (b) The DNase activity in the culture filtrates of the transformable strains of *Pneumococcus* and *Streptococcus* is increased during the process of transformation. The addition of RNA inhibits both the DNase activity and the transformation frequency in bacteria. (c) Nuclease-deficient mutants have reduced frequency of transformation (Lacks et al., 1975; Mishra, 1979).

Transformation as a process for genetic exchange has been best studied in *Streptococcus pneumoniae*, *Streptococcus sanguis*, and *Bacillus subtilis* among the gram-positive bacteria and in *Hemophilus influenzae* among the gram-negative bacteria (Kooistra and Venema, 1970, 1973, 1974, 1976). In gram-positive bacteria, a duplex donor DNA becomes single-stranded during the process of entry into a recipient cell; the single-stranded donor DNA is then integrated into resident chromosome. The situation in gram-negative bacteria is somewhat different because the uptake of DNA is in the form of a duplex that is rendered single-stranded immediately before integration [see Claverys and Lacks (1986) for discussion]. Moreover, the uptake of DNA is donor-specific. Thus the nucleases are involved in two steps during transformation in bacteria. These steps include the transport of the donor DNA into the interior of a cell and the integration of donor DNA into the host chromosome. During the transport of the donor DNA, nucleases first act to generate nicks that act as sites for binding with the cell surface. Later nucleases act to generate single-stranded DNA capable of entry into the interior of a cell. Concomitant with the entry of the single-stranded DNA, equivalent amounts of nucleotides are produced outside the recipient cell by the nuclease action. Thus during the transport of donor DNA, nucleases act like DNA translocase. In certain bacteria, specialized structures called *transformasomes* are developed to facilitate the uptake of the donor DNA.

A number of nucleases have been known to exist in *Diplococcus pneumoniae*; of these, only one nuclease that is membrane-bound is known to be involved in the uptake of the donor DNA (Lacks et al., 1975). A particular nuclease-deficient mutant of *Diplococcus pneumoniae*, called *noz-1*, is blocked in the entry of the donor DNA into the recipient cell, causing a drastic reduction in the transformation frequency. This finding is consistent with the model of DNA entry which entails that nucleases act as DNA translocase during transformation such that one strand of DNA is pulled into the interior of a cell while the other strand is degraded into nucleotides by the nucleases outside the host cells (Lacks et al., 1967). Certain ATP-dependent DNase-deficient mutants of *Diplococcus pneumoniae*, *Bacillus subtilis*, and *Hemophilus influenzae* have been shown to have reduced frequency of transformation (Vivis, 1973; Wilcox and Smith, 1975, 1976).

The evidence for involvement of nuclease in the second step of transformation (i.e., integration of the donor DNA into the recipient chromosome) comes from the study of the marker effect on transformation frequency in *Streptococcus pneumoniae*. The nature of the donor–recipient heteroduplex and subsequent excision and correction of the mismatch DNA are controlled by the *hex* gene(s) (Lacks et al., 1982) The streptococcal *hex* system is very much similar to the mut system of *Escherichia coli*. The *hex* and mut genes have been cloned, and their products have been characterized. Based on such characterization, it is suggested that the *hex* (*hexA* and *hexB*) and mut (mutL and mutS) genes encode for proteins capable of recognition of a mismatch region and its excision (Claverys and Lacks, 1986). The yeast PMS-1 gene has been characterized after cloning and has been found to be similar to the *E. coli* mutL gene, which encodes a protein involved in recognition of mismatch region in DNA before its repair.

A system similar to bacterial *hex* and mut systems seems to exist in eukaryotes (Williamson et al., 1985; White et al., 1985). The role of nuclease in eukaryotic transformation has been investigated in *Neurospora crassa* (Mishra, 1979). A particular *Neurospora* nuclease-deficient mutant called *nuc-2* (Hasunuma and Ishikawa, 1967) has been shown to be deficient in the uptake of donor DNA as well as in the frequency of transformation as compared to the wild-type recipient cells (Mishra, 1979). A number of nuclease-deficient mutants are now known in yeast, *Neurospora* and *Ustilago*; however, their role in transformation remains to be elucidated.

V. MUTATION

Mutation may result from the nonremoval of errors in DNA. Mismatch in DNA is one of the major sources of error. Misincorporation of bases during DNA replication, exposure of DNA to mutagens, or reorganization of DNA duplex during recombination may lead to the formation of mismatch in DNA. The mismatch may involve a single base pair or a number of base pairs. Single base-pair mismatch may include transitions (A/C and G/T) or transversions (G/G, C/C, A/A, T/T, A/G, and C/T). Different mismatches are corrected with different efficiencies (Kramer et al., 1984). Single base-pair mismatch and small deletion and insertion mismatch usually originate due to the failure of the accuracy of DNA replication machinery, while mismatch involving a large stretch of DNA (i.e., deletion and insertion) arises during recombination. The mismatch in DNA is usually recognized by the cellular replication or repair machinery and has been corrected *in vitro* (Lu and Chang, 1988). A simple mismatch may be removed by the editing function of a nuclease usually associated with the DNA polymerase. This type of nuclease may be an integral part of the DNA polymerase as seen in *E. coli* DNA polymerase I or an associated protein as in *E. coli* DNA polymerase III. *E. coli* mutants defective for DNA Q gene encoding the proofreading nuclease component of DNA polymerase III show a 1000-fold increase in spontaneous mutation. Additionally, the mismatch is removed by nucleases encoded by mut and *hex* genes of *E. coli* and *S. pneumoniae*, respectively. The *hex* and mut systems are very much similar in the recognition of the mismatch spectrum, acting efficiently on transition mispairs and some transversion and frame shift mismatches but not on long deletions and insertions. Thus nucleases play a great role in mutation avoidance by removal of errors caused by the infidelity of DNA replication machinery.

In phage T4 as well as in *E. coli*, mutation in the proofreading component of DNA polymerase has been identified as mutator or antimutator phenotypes. The mutator nuclease has a deficient proofreading activity; therefore, a number of mismatches created during replication are not repaired, leading to high frequency of mutation. In contrast the antimutator nuclease is rendered more proficient in proofreading due to changes in the enzyme structure as a result of genetic mutation; consequently, the antimutator nuclease can repair most of the mismatches such that the mutation frequency is much lower in the mutant as compared to the wild type. Proofreading seems to be the foremost function of a replication machinery. This conclusion is based on the fact that T4 DNA polymerase, when provided with a mismatch primer template under conditions suitable for DNA synthesis, engages in the removal of mismatch prior to DNA synthesis. Thus in T4, among the different functions of DNA polymerase (a protein that has three functions, i.e., DNA polymerase activity, mismatch removal, and nuclease activity), the mismatch removal activity takes the precedence over other functions. The general nuclease function is expressed only under conditions when DNA synthesis is not possible. Similar antimutator mutants of mammalian cells have now been shown to possess proficient proofreading 3′-exonuclease activities; the 3′-exonuclease activity of the DNA pol δ was found to be six times higher in the mutant than in the wild type; likewise, the 3′-exoactivity of the DNA pol ε was found to be three times higher in the mutant than in the wild type (Feher and Mishra, 1994, 1995).

Both *hex* and mut play a role in mutation avoidance by correcting errors in replication. Both mut and *hex* mutants have been shown to increase severalfold in spontaneous mutation rates. It is suggested that the *hex* system may have nuclease activity similar to type I restriction enzyme or to the *E. coli* recBCD enzyme and may function as a DNA translocase. The mut and *hex* genes have been cloned from *E. coli* or *Salmonella* or both. The protein products and their possible nuclease functions have been characterized. These data are summarized in Table 11.4. It is suggested that these proteins act as a complex *in vivo*. Based on the study of the repair of λ heteroduplexes after transfection of *E. coli*, there seem to be some differences in the efficiency of repair of the transition and transversion mismatches by the mut system. All transition mismatches are repaired with equal efficiency regardless of their location, whereas the repair of the transversion mismatches is influenced by adjoining nucleotide sequences (Jones et al., 1987). The mut system recognizes the mismatch region by methylation site, while the *hex* system recognizes the mismatch region by a single-strand region produced by other means. The methylation in *E. coli* is carried out by the dam gene product, which is responsible for the methylation of GATC sites. The mutH gene product has been known to attack the unmethylated sites (Claverys and Lacks, 1986; Lahuc et al., 1987; Laengle-Ranault et al., 1986). The dam-dependent mismatchers are corrected with different efficiencies. It has been further suggested that dam and mutH genes might have evolved from the *S. pneumoniae* DPNII methylase and endonuclease, respectively (Claverys and Lacks, 1986).

In addition to errors in DNA replication, mutations may arise as a result of a mistake during DNA repair and recombination, including transposition. Transposition of a DNA segment causes mutations by disruption of a gene or by imprecise exit of a transposon. Site-specific recombination may also lead to deletion, insertion and base substitution mutation (Leong et al., 1985a,b; Almasan and Mishra, 1988, 1991). As a matter of fact, transposition is the major source of spontaneous mutations both in prokaryotes and in eukaryotes, particularly shown in bacteria, yeast and *Drosophila* (Boeke and Corces, 1989). Mutations due to mistakes in these sources may overburden an organism. In this sense, nucleases play a great role in mutation avoidance. The main job of nucleases during these DNA transactions is to ensure the removal of mispaired or damaged DNA. Defective nuclease activity in DNA replication, repair, and recombination leads to increased mutation frequency. Defects in editing nuclease or nuclease encoded by mut genes or by *hex* genes have been known to increase the mutation frequency of bacteria by 1000-fold.

Nucleases thus play a significant role to modulate the mutation rate of an organism required to sustain optimal growth in a given environment. Any drastic change in the environment may require a change in mutation rate. A higher mutation rate can be attained by a mutation in nucleases belonging to *hex*, mut, or DNA polymerase editing function. Preponderance of *hex* or mut mutation has been shown to occur when bacteria are continuously grown in the presence of MNG. Mutation in mut or *hex*, in turn, increases the adaptability of an organism by causing mutation at other loci. Thus, *hex* and mut systems can turn an organism to demands of high and low mutation rates for growth in a particular environment. This role of nucleases is consistent with the notion of punctuated evolution (Gould and Eldridge, 1977) as discussed in the last chapter of this book. Nucleases also play a role in the maintenance of homogeneity within a multigene family via the control of gene conversion events. Recent studies show that in eukaryotes a gene conversion event is controlled by a system similar to those of *hex* and mut (White et al., 1985; Williamson et al., 1985). In addition, transposon-like Tn element and P elements have been used for insertional mutagenesis (Kleckner, 1981; Cooley et al., 1988). Some of these aspects of the role of nucleases in evolution have been discussed by Claverys and Lacks (1986).

VI. DNA SUPERCOILING AND MAINTENANCE OF CHROMOSOME STRUCTURE

DNA exists as supercoiled molecules inside a cell. The extent of DNA supercoiling depends on the dynamic balance between the two groups of enzymes that are mutually antagonistic. These enzymes are topoisomerase I, which relaxes supercoiled DNA, and topoisomerase II (gyrase), which introduces supercoil into DNA. Vinograd et al. (1965) first demonstrated the existence of polyoma DNA in supercoiled form. The different forms of DNA and their degree of supercoiling can be resolved by ultracentrifugation (Vinograd et al., 1965) or by agarose gel electrophoresis (Aaij and Borst, 1972) and visualized by electron microscopy. The role of topoisomerase in DNA supercoiling has been discussed (Gellert, 1981; Drlica, 1984; Cozzarelli, 1980; Brillan & Sternglanz, 1988). Supercoiling is now considered as the property of all DNA which is readily visualized in the closed circular molecules. It has been suggested that genomic DNA may be organized into discrete regions of supercoiling in order to facilitate different DNA transactions. The *E. coli* DNA seems to be distributed over 50 domains of supercoiling (Sinden and Pettijohn, 1981).

The fact that DNA exists as supercoiled molecules and that topoisomerases control and maintain this supercoiling has been established in *E. coli* by the examination of the psoralen-binding property of DNA. Supercoiled DNA was found to bind much more psoralen than the nicked DNA (Sinden et al., 1980). *E. coli* DNA nicked by exposure to gamma rays or upon infection of the bacterial cells by bacteriophage T4 showed a markedly reduced level of psoralen binding. Furthermore, the DNA from *E. coli* cells treated with coumermycin, which inhibits gyrase, was found to show a marked reduction in the level of psoralen binding. Similar loss of psoralen binding was seen in DNA obtained from temperature-sensitive gyrase mutants of *E. coli* grown at the restrictive temperature. In contrast to gyrase mutant, the top A mutant of *E. coli* or *Salmonella typhimurium* showed a higher level of psoralen binding as compared to the isogenic topA cells. These data are interpreted to suggest that in the *topA* mutant, the DNA is much more supercoiled due to the lack of the relaxing activity of topoisomerase I which is missing in the top A mutants. It has also been shown that *E. coli* mutants with complete deletion of the top A gene usually acquire a compensatory mutation in the gyrase A or gyrase B genes. These data led to the conclusion that *in vivo* the enzymes topoisomerase I and topoisomerase II (gyrase) act in opposite ways to attain a particular level of supercoiling needed for several DNA-dependent functions such as DNA replication, transcription, recombination, transposition, and encapsidation. Unlike the bacterial chromosome, the eukaryotic chromosome is under much less torsional stress (Lilley, 1983). In eukaryotes, most of the supercoiling is due to DNA wrapping of histone octamers in nucleosomes. Furthermore, the activity of topoisomerase avoids the positive supercoiling in nonnucleosomal DNA. The conclusion that eukaryotic DNA is mostly free of torsional stress is supported by the lack of psoralen binding by *Drosophila* and HeLa DNA (Sinden and Pettijohn, 1981). However, it is still possible that a small fraction of eukaryotic DNA may exist as domains of supercoiling which may not be detected by the psorlen binding studies of Sinden and Pettijohn (1981). Further evidence for the role of topoisomerase in maintaining the superhelical structure of eukaryotic chromosome comes from the effect of novobiocin, an inhibitor of topoisomerase II, on the DNaseI sensitivity of an active globin gene. It is established that active genes in eukaryotes exist in a nucleosomal form that renders them more sensitive to DNaseI. It has been shown that when chicken erythrocytes are treated with novobiocin the DNaseI sensitivity of the β-globin gene is markedly reduced, suggesting the role of topoisomerase in maintaining the superhelicity of the active genes (Fleishman et al., 1984; Villeponteau et al., 1984). Also, the above view that at least a small fraction of eukaryotic DNA must exist as supercoiled domain is supported by the study of Simian virus 40 minichromosomes. It has been shown that the SV40 minichromosomes isolated from the infected monkey cells can be further relaxed by the activity of topoisomerase I. Furthermore, it has been found that novobiocin interferes with the heat shock response in *Drosophila*, suggesting the role of topoisomerase and DNA supercoiling in gene activity (Han et al., 1985). The level of supercoiling of DNA segment in the chromosome determines its level of transcription (Kunuic et al., 1986).

VII. TRANSCRIPTION

Topoisomerases play an important role in determining the expression of genes by controlling the superhelicity of DNA (Yang et al., 1979; Fisher, 1981, 1984; Gellert, 1981; Stirdivant et al., 1985). As a matter of fact, the expression of the *E. coli* gyrase gene is controlled by the topological state of the DNA. Gellert and others (1976a,b, 1977) have demonstrated the activation of gyrase synthesis by conditions that reduce gyrase activity. In these studies the net synthesis of gyrase protein was determined by precipitation with antibodies prepared against purified gyrase A and gyrase B proteins in bacterial cells treated with gyrase inhibitors such as coumermycin and novobiocin. The wild-type *E. coli* cells treated with gyrase inhibitors showed a 10-fold increase in the amount of gyrase protein synthesis as compared to the untreated wild-type cells. The gyrase mutants resistant to novobiocin or other inhibitors showed no increase in the gyrase synthesis when treated with the inhibitors. The results of these studies indicate that the synthesis of gyrA and gyrB proteins is promoted by a loss in the superhelicity of DNA due to the presence of gyrase inhibitor. This conclusion is further supported by *in vitro* transcription and translation studies using relaxed and supercoiled forms of the cloned gyrase A and gyrase B genes. In these studies, cloned gyrase gene(s) obtained in the relaxed form from bacteria grown in the presence of novobiocin were found to provide up to 100 times more gyrase protein(s) compared to the supercoiled form of the cloned gyrase genes obtained from the bacteria grown in the absence of gyrase inhibitor. Furthermore, the supercoiled form of the cloned gyrase DNA treated with *E. coli* topoisomerase I *in vitro* showed similar enhancement in the production of gyrase protein.

Originally, the topoisomerase mutation ($SupX^-$ or *Top*A^-) was identified as suppressor of leucine auxotrophs (Leu 500) in *Salmonella* (Dubnau and Margolin, 1972; Dubnair et al., 1973). The results of the study of suppression of leucine auxotrophy in *Salmonella* led to the conclusion that *topA* mutation affected the transcription of genes involved in leucine biosynthesis by changes in the superhelicity of its promotor region. Likewise, it was found that the *bgl* locus of *E. coli* which controls the utilization of β glucoside and is normally inactive becomes operative in *topA* strains with compensatory mutations in gyrase. In bacteria the expression of a number of genes is influenced by a change in the superhelicity of DNA; these mostly include the catabolite-sensitive genes such as maltose, lactose, and galactose operons and tryptophanase gene.

It has been further shown that the loss of topoisomerase I in *Salmonella* alters many cellular activities. In *Salmonella topA* mutants, the recBCD (exonuclease V) activity is increased severalfold. Also, the *topA* mutants of *Salmonella* showed almost a total loss of mutagenesis by alkalating agents. The activity of topoisomerase I is necessary for the expression of genes required for the growth of *Salmonella* under anaerobic conditions, whereas the activity of gyrase is necessary for the expression of genes required for the growth of bacteria under aerobic conditions (Yamamoto and Droffner, 1985). The expression of *his* operon is also regulated by the wild type level of supercoiling in *Salmonella typhimurium* and *E. coli* (Rudd and Menzel, 1987). Thus the expression of a number of genes is influenced by changes in supercoiling of DNA due to chemical or genetic perturbation of topoisomerase genes.

In eukaryotes, the effects of DNA supercoiling on transcription have been examined with the use of cloned herpes simplex virus thymidine kinase and adenovirus major late promoters injected into *Xenopus* oocytes. A 500- to 1000-fold increase in transcription was seen when the cloned genes were injected in supercoiled form as opposed to their linear forms. The linear form of genes was generated *in ovo* by injection of restriction endonucleases. In another study, the reactivation of a silent yeast mating-type locus (HML2) was found to depend on the superhelicity of plasmid containing the HML α locus. Further involvement of toposiomerase I in transcription is indicated by the association of this enzyme with the DNA-hypersensitive sites. The fact that topoisomerase I activity as visualized by immunofluorescence is associated with active genes in *Drosophila* supports the view that topoisomerases are involved in transcription of genes in eukaryotes (Fleischmann

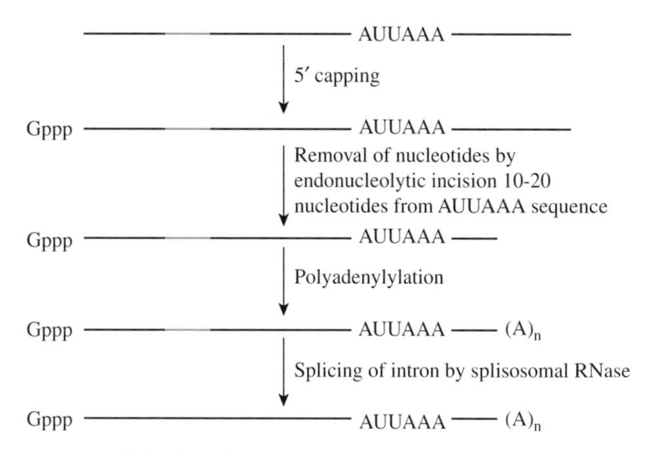

Figure 11.10. Role of nucleases in the processing of nuclear transcript.

et al., 1984). Brill and Sternglanz (1988) have demonstrated that the supercoiling of DNA is dependent on transcription in certain yeast mutant. Topoisomerase has been shown to interact with the transcribed region in *Drosophila* (Gilmour et al., 1986).

VIII. RNA PROCESSING

RNA processing involving several steps leads to maturation of RNA such that it can participate in the process of translation or becomes translatable. Nucleases participate in the trimming, splicing, and editing of RNA as shown in Figures 11.10 to 11.15.

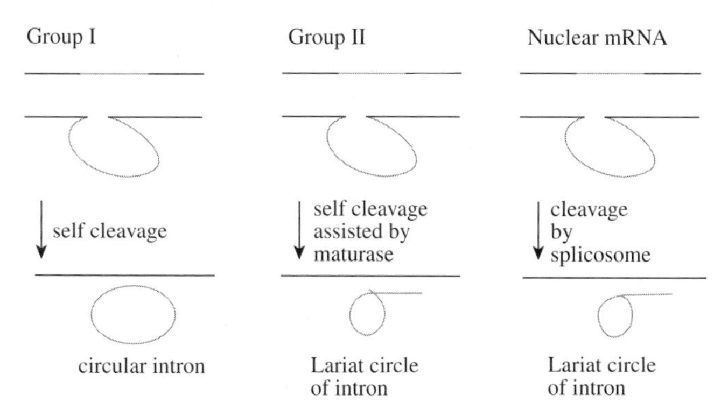

Figure 11.11. Role of nucleases in intron splicing. [Based on Ceech (1990), copyright *Annual Review of Biochemistry*, used with permission.]

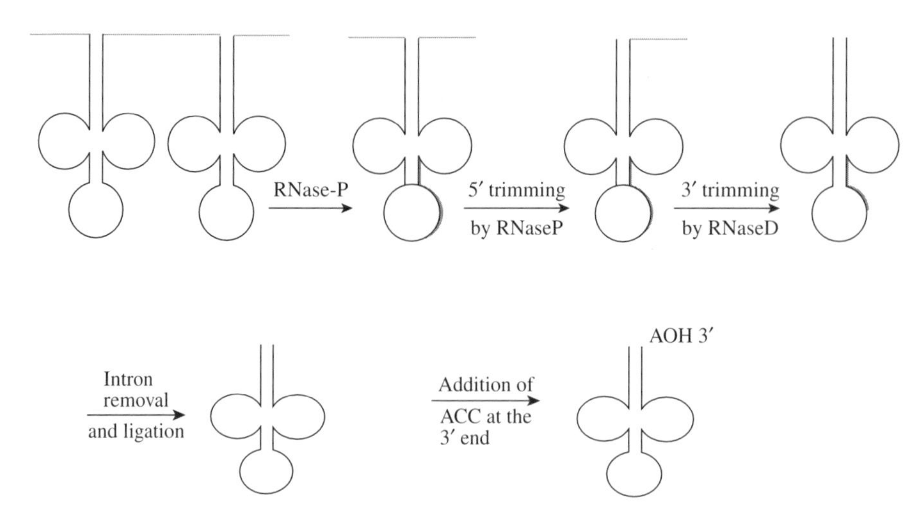

Figure 11.12. Role of nucleases in tRNA processing. [Based on Cech (1990), copyright *Annual Review of Biochemistry*, used with permission.]

A. RNA Trimming

A number of RNA molecules (such as tRNA) undergo trimming or removal of nucleotides from its 3′ or 5′ ends (Deutscher, 1984). Ribonucleases involved in such trimming are described in Chapter 2. In addition, a stretch of nucleotide in the pre-mRNA at the 3′ end is removed by endonucleolytic cut at the polyadenylation sites before the polyadenylation of the 3′ mRNA (Shaw and Kamen, 1986). The detail analysis of the endonuclease involved in this process is lacking; however, a 64-kDa protein that acts as on RNA-binding protein and facilitates the recognition of polyadenylylation signal sequence—and, thereby, recognition of the cleavage site—has been known.

B. RNA Splicing

It is known that most genes in eukaryotes contain exons and introns that are transcribed into a primary transcript or HnRNA or pre-mRNA. However, only exon

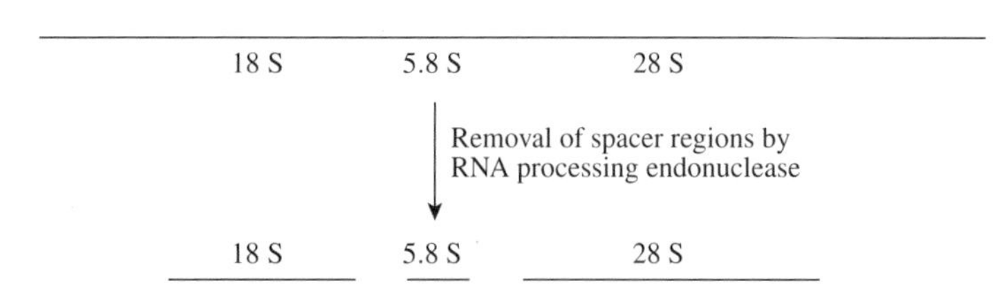

Figure 11.13. Role of nucleases in rRNA processing.

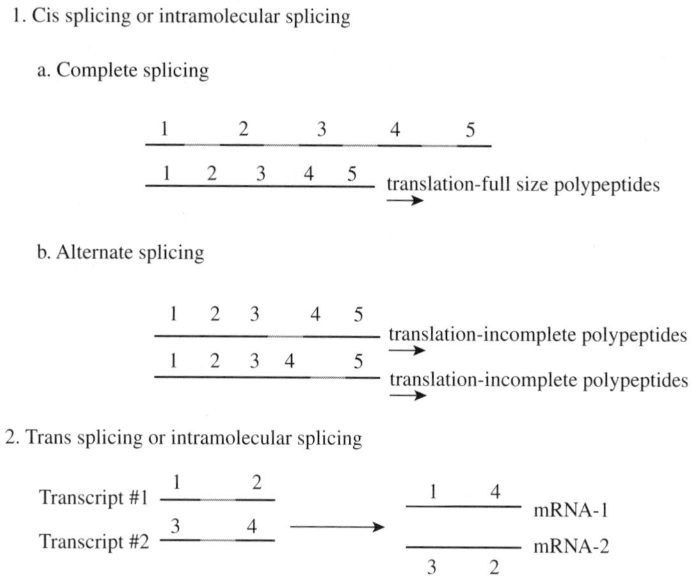

1. Cis splicing or intramolecular splicing

a. Complete splicing

1 2 3 4 5

1 2 3 4 5 translation-full size polypeptides

b. Alternate splicing

1 2 3 4 5 translation-incomplete polypeptides

1 2 3 4 5 translation-incomplete polypeptides

2. Trans splicing or intramolecular splicing

Transcript #1 1 2 → 1 4 mRNA-1

Transcript #2 3 4 → 3 2 mRNA-2

Figure 11.14. Role of nucleases in nuclear pre-mRNA splicing.

sequences are translated, whereas intron sequences are removed by the process of splicing during the processing of the primary transcript. The exons are then brought together in mRNA by the process of splicing. In addition to mRNA, certain tRNAs also undergo splicing.

Ribonucleases involved in splicing of bacterial and yeast tRNA have been studied in detail, and their properties are described in Chapter 2. The removal of certain introns in eukaryotes and in certain bacterial viruses are catalyzed by ribozymes. In group I introns, the intron is removed by its own ribozyme activities, producing mature translatable mRNA and circular intronic RNAs. In group II introns, the ribozyme activity of the intron in conjunction with other proteins acting as helpers (called maturases) removes the intrinsic region in the form of a lariat structure. In addition, amino acyl tRNA synthetase and several other proteins may be involved in the splicing of the group I and group II introns (Lambowitz and Perlman, 1990). However, the intron(s) of eukaryotic transcript are removed by a very complex organellar substructure, called *splicosomes*, which contain five

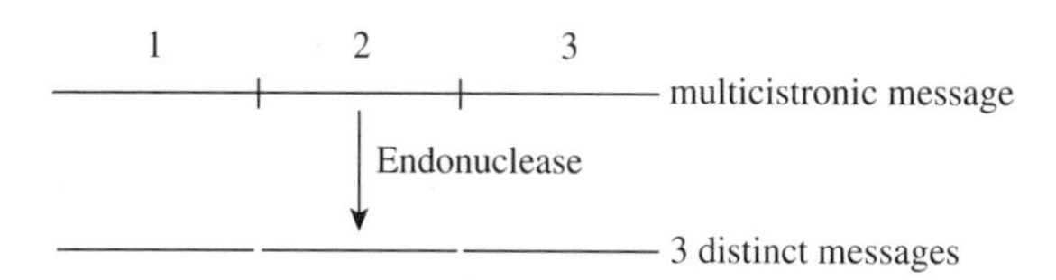

Figure 11.15. Role of nucleases in processing of multicistronic RNA.

snRNAs and a large number of proteins. Mutational analysis of snRNAs indicates that the splicing activity of the splicosome is indeed due to the ribozyme activity of the snRNAs per se in conjunction with the other protein components of splicosomes. Thus the protein components of splicosomes are functionally equivalent to the maturases participating in the splicing of group II intron by ribozyme. The introns from eukaryotic transcripts are also removed by the splicosomes in the form of a lariat RNA. A gene may contain 0–40 or more introns. Introns may be removed in cis to join the exon sequences present in a particular transcript. Alternatively, introns may be spliced out in trans-joining exon sequences from two different transcripts (Kress et al., 1983; Solnick, 1985). Thus a translated protein in eukaryote does not necessarily correspond to the genetic information contained in a particular gene. Nucleases must play a precise role in facilitating such cis and trans splicing of transcripts in eukaryotes. The processes of alternate cis splicing and trans splicing may be considered as a remnant of the process of shuffling of exons. Exons are considered to correspond to domains in an enzyme. It has been suggested that genes for eukaryotic proteins evolved by shuffling of a small number of exons (estimated to be 1000–7000) via recombination mediated by introns. The role of nucleases in exon shuffling is discussed Chapter 15.

The mechanism of trans-splicing is a major source of protein diversity, which plays an important role during the differentiation of multicellular organisms including human (Leff et al., 1986; Breitbart et al., 1987). The fact that sexual differentiation in *Drosophila* is mediated by alternate splicing of the same message is very well established (Baker, 1989).

In addition, an intron may encode for a site specific endonuclease that facilitates intron mobility or may encode a maturase that facilitates intron removal during the process of splicing. Introns may also harbor certain signals for translation because it has been shown that genes from which introns have been removed (and cloned in the form of cDNA) may not be expressed in a cell. Also, the introns removed by nucleases are tissue-specific. Thus the process of splicing and the role of nucleases therein play a significant role in the gene expression and developmental biology of an organism.

C. RNA Editing

In addition to RNA trimming and RNA splicing, nucleases may facilitate RNA editing by insertion, deletion, and substitution of ribonucleotide in a transcript (Eisen, 1988). A number of mitochondrial DNA transcripts in trypanosomes and in plants have been shown to undergo the process of RNA editing (Covello and Gray, 1989; Araya et al., 1995). At least two human transcripts have been shown to undergo the process of RNA editing such that these human proteins contain an amino acid at certain positions different from what is encoded by the gene for that protein. Thus RNA editing defies the concept that the primary structure of a protein is encoded by nucleotide sequence in a gene. Such RNA editing of human neuronal transmembrane receptor protein has been shown to be useful in the control of ion transport. Thus RNA editing adds another dimension to the process of control mechanisms involved in eukaryotic gene expression. RNA editing may also be considered as reminiscent of the process that ancient RNA might have undergone to produce a translatable RNA. The nuclease involved in RNA editing has not yet been identified and characterized.

IX. CONTROL OF TRANSLATION

Nucleases control translation by controlling the amount of translatable mRNA via the maturation of transcript into mRNA. In addition, nucleases play a role in the turnover of mRNA. Certain mRNA with a short half-life may be more prone to degradation by nucleases. Most mRNA with polyA stretches at the 3 end are not readily digested by the nucleases. However, over the long range, all mRNA are degraded by nucleases. A nuclease does not necessarily always control the level of translation by determining the degradation of mRNA. Instead, it may expose certain regions of mRNA by specific binding leading to conformational changes in the mRNA, promoting its translation, as has been shown in the case of phage C III genes by RNaseIII (Altuvia et al., 1987).

X. VIRAL MATURATION AND ENCAPSIDATION

Most viral DNAs are synthesized in concatemeric forms: The multigenomic DNAs are packaged in genomic size by the action of nucleases during the process of encapsidation. The packing of λ DNA is facilitated by terminase protein with part of the viral body. With the help of terminase, genomic-size DNA are cut at the cohesive (*cos*) site of λ DNA and then encapsulated. Similar nucleases are involved in the encapsidation of other viral genomic-size DNA. In addition, certain viruses such as T4 DNA during the process of replication produced branched DNA molecules that resemble the Holliday structures. Such DNA molecules are first resolved by a T4 gene 49 product before encapsidation. The properties of this T4 recombinase (gene 47 product) have been described in Chapter 7.

The λ terminase is a heterodimer between a small subunit (gpNaI) and large subunit (Spa) encoded by λ phage. The N terminus of the gpNaI interacts with *cos* region, and its C terminus interacts with the gpA. Terminase alone can cleave λ DNA at *cos* to some extent, but cleavage is promoted by interaction with head protein (Becker and Gold, 1978; Feiss, 1986). The maturation of HSV has been shown to require two separate cleavage reactions (Varmouza and Smiley, 1985).

XI. NUCLEASE IN DEFENSE MECHANISM

Host range specificity of a virus is based on its ability to escape digestion by restriction endonucleases produced in the host cell. Most of the time it depends on the methylase enzyme to acquire special marking in its DNA so that its DNA is not digested by the host nuclease upon entry into host cells. In certain other instances, virus encodes proteins that inactivate the host nucleases. RNaseL is an important part of defense system against viruses in humans, as shown in Figure 11.16. Likewise, nucleases like dicer and RISC are crucial for gene silencing in a variety of organisms including plants, fungi fruit flies, and worms. Genetic engineering of these RNases can be harnessed to maintain the good health of these organisms

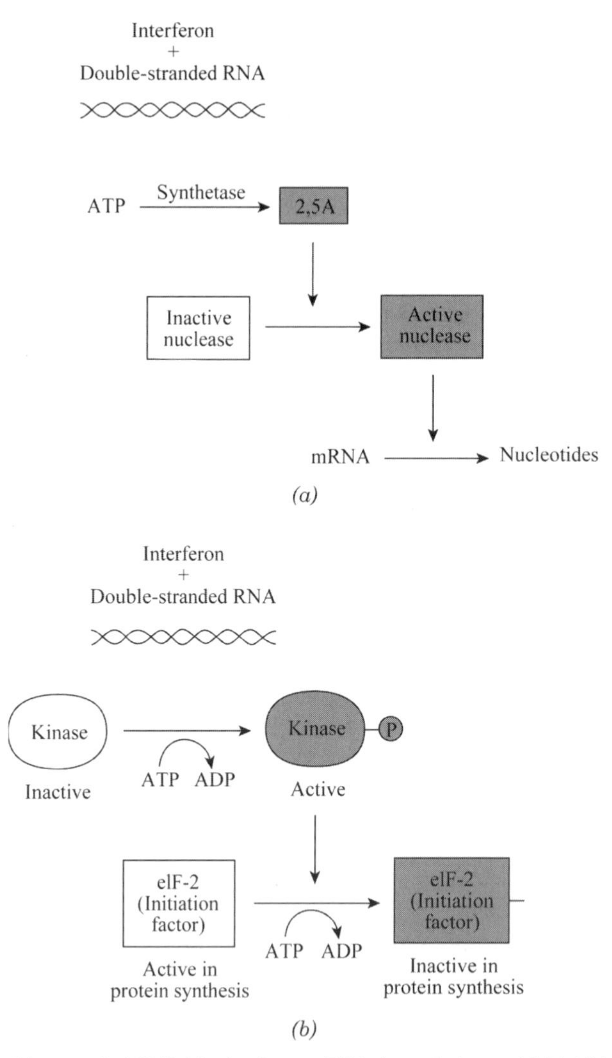

Figure 11.16. Double-stranded RNA (a) stimulates mRNA degradation and (b) inhibits the initiation of protein synthesis in interferon-treated cells. [From Stryer (1982), copyright W. H. Freeman, used with permission.]

and can be exploited to control the expression of certain desired genes for their biotechnological manipulation.

XII. APOPTOSIS AND NUCLEIC ACID SALVAGE

Apoptosis is the programmed cell death that is leashed out during the development of multicellular organisms under physiological conditions.

Apoptosis always leads to cleavage of chromosomes into nucleosomal fragments, which is carried out by several nucleases such as DNaseI and DNaseII. Apoptosis is required in order to kill cancerous cells and is also required in the organization of neuronal cells during normal human brain development.

Finally, nucleases are involved in the digestion of unwanted nucleic acid molecules. Nuclease is a major component of the pancreatic secretion which enables digestion of DNA in the food samples ingested by the organism such as cows and humans. Without the help of pancreatic nuclease, digestion of nucleic acids present in our food would cause immense problems. The utilization of protein-rich yeast cells as a major source of protein for cattle is hindered by the overabundance of nucleic acids that cannot be readily digested by the pancreatic nucleases of the animal.

12

NUCLEASES AND HUMAN DISEASES: BASIS FOR APPLICATION

Nucleases are important in the understanding of human diseases from different points of view. First, nucleases are the key enzymes in the DNA replication, repair, and recombination pathways, and a number of human diseases are caused by dysfunction of these enzymes. Second, nucleases are important tools in reverse genetic methodology to identify and characterize genes controlling inherited disease of mankind. Third, a number of antitumor drugs are inhibitors of nuclease activities such as topoisomerases (Felix, 2001). Fourth, the knowledge regarding the nature of certain nucleases can be used to control cancer and other diseases including those caused by viruses. Fifth, a number of nucleases are involved in the process of DNA replication, repair, recombination, transcription, and cell cycle control; these function as a member of multiprotein complex, and their function is affected even if there is change in any of the proteins other than nucleases. Also, DNA repair is coupled to transcription; such coupling of DNA repair to transcription is called TCR (Hanawalt and Mellon, 1993). TCR may influence the DNA repair process involving the activity of nucleases via a change in the activity of a transcription factor. Also, several nucleases possess activities other than nuclease activity such as helicase activity, and a mutation in motiffs other than nuclease could also lead to a defective nuclease function causing human diseases. Finally, DNases and RNases including ribozymes have been or are being used to treat many human diseases. A defective DNA repair can cause mutagenesis leading to several human diseases including cancer, arteriosclerosis, and aging. A number of human diseases defective in DNA repair and recombination pathways have been known as listed in Tables 12.1 and 12.2. Some of these diseases are genetically inherited as autosomal recessive and are characterized by a defective DNA repair accompanied by mutagen–carcinogen hypersensitivity and chromosome instability (Friedberg, 1985; Collins et al., 1987, Lehmann, 1982). Among different human diseases with DNA

Table 12.1 Human diseases defective in DNA repair

Disease	Mode of Inheritance	Frequency of Occurrence	Biochemical Defect
1. Xeroderma pigmentosum (XP)	Autosomal recessive	1 in 40,000–200,000	AP-endonuclease
2. Ataxia telangiectasia (AT)	Autosomal recessive	1 in 40,000	Topoisomerase
3. Bloom syndrome (BS)	Autosomal recessive[a]	Very low but occurs 1 in 48,000 in Ashkenazi Jews	Ligase
4. Finconi's anemia (FA)	Autosomal recessive		Not known
5. Cockayne syndrome (CS)	Autosomal recessive		Not known
6. Werner syndrome	Autosomal recessive		3′–5′Exonuclease

[a] The fact that a higher number of males than females are diagnosed to have BS suggests a sex-linked inheritance.

repair defects, only two seem to be associated with a defective or missing nuclease or a protein influencing the activity of a nuclease. These include xeroderma pigmentosum (XP), which seems to lack AP endonuclease, and ataxia telangiectasia (AT), which seems to lack a topoisomerase II activity or a protein that influences the activity of this enzyme. Bloom syndrome, another DNA repair defective disease, has now been shown to be defective in ligase and not in nuclease. The biochemical basis of other DNA repair defective enzyme such as Finconi's anemia (FA) and Cockayne syndrome (CS) is not yet established. It is now known that persons suffering from Werner syndrome, a disease characterized by early onset of aging, have altered 3′–5′-exonuclease activity (Machwe et al., 2000) associated with the Werner protein (WRNp) (Huang et al., 1998). A number of other human diseases may be caused by DNA recombination and/or transposition in which recombinases may play an important role. Some of these diseases are listed in

Table 12.2 Role of nuclease in human disease and development

1. Carcinogenesis and cancer
2. Hypercholesterolemia
3. Thalassemia
4. Adenosine deaminase deficiency
5. Hemophilia
6. Aging—Werner syndrome
7. Neuronal development and neurological disorders
8. Development of immune system and related immunological disorder

Table 12.2. Nucleases are tools in reverse genetic analysis of several diseases leading to the identification of human genes and their products involved in all kinds of human birth defects such as Huntington's chorea, cystic fibrosis, muscular dystrophy, a number of neurological disorders, and many other human disease conditions (Caskey, 1987). A number of human diseases in which nucleases are identified to be defective are discussed below.

I. INVOLVEMENT OF NUCLEASES IN HUMAN DISEASE

A. Xeroderma Pigmentosum

This was the first human disease that was shown to lack excision repair of DNA damages induced by UV light. Following this discovery, an extensive genetic and biochemical analysis of the disease has been undertaken. The disease is shown to be inherited as autosomal recessive and is clinically manifested as skin cancer and neurological disorder. The disease occurs with different frequency in European (1 in 200,000) and in Japanese (1 in 40,000) populations. The phenotypic abnormalities of XP have recently been summarized and are presented in Table 12.3. Analysis of XP cells collected from different patients by somatic cell fusion indicated the existence of more than eight complementation groups. This suggested the role of more than eight genes in excision repair of UV-induced damage in humans. The characterization of the different complementation group has recently been summarized (Friedberg, 1985) and is presented in Table 12.4.

Data presented in the table show that XP cells in comparison to normal cells varied significantly in the range of UV sensitivity, which is determined as the percent survival of the cells after exposure to UV, extent of repair DNA synthesis, and loss of sites sensitive to pyrimidine-dimer-specific enzyme. These data clearly show that XP cells are defective in endonucleases or in proteins that influence the activity of such damage-specific endonucleases (Martelmans et al., 1976). In view of such

Table 12.3 Phenotypic characteristics of XP cells

1. Increased sensitivity to UV and UV-mimetic chemicals and, frequently, to ionizing radiation.
2. Defective host cell reactivation of UV-irradiated viruses.
3. Defective removal of thymidine dimer from DNA.
4. Defective repair of synthesis of DNA.
5. Increased sister chromatid exchanges following exposure to UV radiation.
6. Enhanced mutagenesis after exposure to mutagenesis.
7. Enhanced susceptibility to neoplastic transformation.
8. Defective AP-endonuclease activity in D group of XP cells.
9. Lower levels of DNA photolyase activity.

Source: Friedberg (1985), copyright by W. H. Freeman, used with permission.

Table 12.4 Comparative parameters of DNA repair in XP cells

Complementation Group	Repair Synthesis (% of Normal)	Loss of ESS[a] in 32 Hours (% of Normal)	Relative UV Sensitivity
A	0–5	0	++++
B	3–7	10	+++
C	10–15	20	++
D	10–40	20	++++
E	40–60	50	+
F	0–10	60	+
G	10–15	0	++++
H	30	?	?

[a] ESS refers to endonuclease-sensitive sites.

Source: Friedberg (1985), copyright by W. H. Freeman, used with permission.

expectations, it is important that XP cells have been shown to lack an AP endonuclease (Kuhnlein et al., 1978). This finding that the primary biochemical defect of XP cells results from a missing AP endonuclease is further supported by the fact that XP cells can overcome their defect when provided with the bacteriophage T4 specific AP endonuclease (Smith and Hanawalt, 1978; Tanaka et al., 1977). However, results of the studies show that XP cells of E group are deficient in a nuclear factor that binds to the damaged region (Chu and Chang, 1988; Kemey et al., 1994). Furthermore, it has been shown that the human excision repair gene involved in complementation group A of xeroderma pigmentosum encodes a zinc finger protein (Tanaka et al., 1990).

The most intriguing aspect of xeroderma pigmentosum is the involvement of a large number of genes in the process of DNA repair which can be overcome by the supply of a single endonuclease of the bacteriophage. The UV-induced damage in the mammalian cells seems to be controlled by at least 15 gene products. This is based on the identification of a total of at least 15 complementation groups (9 complementation groups in XP cells and 6 complementation groups in mouse cells). This situation is comparable to repair of similar damage in yeast where again a large number of genes are involved in DNA repair. It is quite possible that a large number of proteins may have a role in regulation or that *in vivo* the UV-induced damages are repaired by a multienzyme complex. The precise role of these large number of genes involved in simple DNA repair in diverse organisms such as yeast and humans must await the identification and characterization of these gene products. A number of yeast genes have been recently characterized; among them, none except Rad52 encodes for a nuclease. There is evidence that suggests that the Rad52 gene product encodes a nuclease or a protein that controls a nuclease in yeast. The nuclease missing in the yeast in Rad52 mutant has been shown to cross-react immunologically to a *Neurospora* nuclease. A human XP patient belonging to complementation group G has been very well characterized and was found to be defective for XPG protein, which is a structure-specific nuclease. This is an acidic protein of 133 kDa which is essential for base excision repair in humans and in rodents (see Chapter 3).

B. Ataxia Telangiectasia

This is also known as Louis–Bar Syndrome. This is inherited as autosomal recessive and appears in quite significant frequency among live births (1 in 40,000). The disease is characterized by disorders in nervous system, immune system, and skin, leading to progressive degeneration and the dilation of small vessels. The discrepancy of immune system leads to repeated infection and occurrence of neoplasms. Among ataxia telangiectasia (AT) individuals, the neoplasms occur at the age of 20, which is much earlier than that reported for other neoplasms; the latter generally appear at the age of 55 (Friedberg, 1985). It has also been shown that individuals heterozygous for AT are more likely to die of cancer than the individuals lacking the gene for AT.

At the cellular level, the AT cells are characterized by sensitivity to x-radiation and x-ray mimetic chemicals. The AT cells are not sensitive to UV light or UV-mimetic drugs (see Friedberg, 1985). In this way, AT cells resemble certain mutagen-sensitive mutants of *Neurospora crassa* which are also sensitive to x-ray but not to UV light (Delange and Mishra, 1982). AT cells exposed to x-ray show double-strand breaks in chromosomes. The AT cells do not seem to differ quantitatively from the normal cells with regard to the repair of the double-strand breaks (Lehmann, 1982) but are characterized by the misrepair of double-strand breaks (Debenham et al., 1987). The fidelity of repair during the joining of the double-strand breaks of chromosomes in AT cells has been examined ingeniously by Debenham and his colleagues. These workers have used a chimeric plasmid (pPM17 or pPMHSV16) containing two selectable markers such as G418R (resistance to G418) and gpt (ability to grow in XHATM medium). The chimeric plasmid contains a unique restriction site KpnI within its *gpt* gene. The fidelity of repair of the double-strand break was examined by comparing the expression of the *gpt* gene among G418R transformants of the normal and AT cells treated with chimeric plasmid linearized by KpnI restriction. These data showed that the AT cells had reduced fidelity of the repair during joining of the ends of the linear plasmid. Similar misrepair during the joining of the linear plasmid was also seen in certain mutants (A4) of hamster cells. However, other hamster mutant cells such as C11 and Xrs-1 (Kemp et al., 1984) showed no difference in the rejoining of the KpnI-digested chimeric plasmid. Also, the XP cell showed a somewhat reduced efficiency of the faithful rejoining of the linearized plasmid compared to the normal cell; their efficiencies, 34% and 56%, respectively, were much greater than that of the AT cells with an efficiency of only 26%. These studies of mammalian cells showed that there are several groups of mutants with respect to their ability to carry out faithful repair during the joining of the double-strand breaks in chromosomes. Comparable mutants have been described in *E. coli*. Among *E. coli* mutants, rorA and recN are capable of faithfully rejoining the restriction-enzyme-generated double-strand breaks in plasmid DNA molecules (Debenham et al., 1987), whereas the rorB mutants rejoin such double-strand breaks with low fidelity. Further study of the rorB mutant provided an insight into the biochemical basis of this defect. It has been shown that the *E. coli* rorB mutant is sensitive to comermycin A1 but not to nalidixic acid. This observation suggests that there has been a specific change in the topoisomerase II (gyrase B subunit) of this mutant. It is also shown that the change in rorB mutant cannot be complemented by a cloned gyrB gene. Results of such complementation analysis clearly indicated that rorB gene is not the structural gene for topoisomerase II but instead codes for some factor that influences the topoisomerase activity. Results of these studies prompted a similar analysis of the AT cells. The AT cells were found to be sensitive to novobiocin, which is known to inhibit the activity of eukaryotic topoisomerase II. These results do indicate a change in the topoisomerase II of AT cells (Debenham et al., 1987). The significance of possible change in the topoisomerase II of AT cells has been discussed by Debenham et al. (1987). These workers have explained the role of altered topoisomerase in facilitating the DNA synthesis in AT cells without a delay after exposure to x-ray. The gene products of some of these genes in humans may soon be identified using a gene transfer system. Recently a PCR-based analysis has shown an increase in V-J-D recombination in the AT patients.

The most intriguing aspect of these human diseases is the correlation between the deficient DNA repair and neurological disorders. The abundance of DNA polymerase β and lack of DNA polymerase α in nerve cells is consistent with the idea that nerve cell DNAs do not replicate but require consistent DNA repair for its maintenance. Therefore it is plausible that a lack of nuclease leading to faulty DNA repair may cause neurological problems in human patients. It would be of interest to examine whether or not a chimeric mouse containing DNA repair defective cells can cause neurological disorders (Sun et al., 2001).

C. Cockayne Syndrome

Another human disease in which a defective DNA repair may be involved is Cockayne syndrome (CS), which is inherited due to an autosomal recessive gene. This disease is clinically manifested as growth retardation, neurological disorder, and severe photosensitivity. The removal of thymine dimer following exposure to UV radiation seems to be normal in the cells of the patients with Cockayne syndrome. However, when the extent of the repair of the transcriptionally active and inactive genes is compared, it becomes obvious that CS patients are defective in the repair of transcriptionally active genes (Venema et al., 1990). It is of interest to mention here that in a UV-irradiated cell, the transcriptionally active genes are repaired preferentially over the transcriptionally inactive genes (Mellon et al., 1986; Kantor et al., 1990). It is further known that transcriptionally active genes are situated in nuclear-matrix-associated DNA, whereas the transcriptionally inactive genes are contained in the DNA loop. Even though the role of nucleases in the DNA repair in CS patients is not yet established, the fact that DNA repair is defective certainly implies the involvement of nucleases or factors associated with these enzymes.

D. Cancer

Cancer is known to be caused by mutation in oncogenes controlling the promotion of cell growth or in antioncogenes or tumor suppressor genes causing the suppression of cell proliferation. Nucleases may play a role in causing such mutation directly or by causing rearrangement in chromosomes (Croce, 1987; Duesberg, 1987; Pardue, 1991, Ward et al., 2001). Cancer may be related to fragile sites in chromosomes, presumably caused by transposable elements.

Cancer is also caused by mutation in genes controlling mismatch repair (MMR); for example, colorectal cancer is caused by mutation in MMR gene.

Also, the instability of chromosomes associated with cancer is due to mutation in repair nucleases; for example, mutation in FEN-1 leads to chromosomal abnormality or loss of heterozygosity (LOH).

One of the major causes of cancer involves mutation in tumor suppressor genes.

The role of nucleases in cancer can be envisioned in terms of the fact that the occurrence of cancer is almost always associated with chromosomal rearrangements and mutations and that nucleases are known to participate in events leading to chromosomal rearrangements and mutations. Such chromosomal rearrangement and mutation have been shown to cause dysfunctional genes in humans. In certain human lymphomas, occurrence of recombination between chromosome 14 (antibody heavy-chain region) and chromosome 8 (c-myc region) leads to dysfunctional oncogene c-myc by putting its transcription under the control of a powerful enhancer of the immunoglobulin gene. Likewise, in chronic myelogenous leukemia a recombination between chromosome 22 (region containing the tumor suppressor gene bcr) and chromosome 9 (region containing another tumor suppressor gene c-abl) produces a fusion gene, bcr-c-abl; the product of this fusion gene causes the leukemia. The translocation involving chromosomes 9 and 22 in the chronic myelogenous leukemia is known as Philadelphia chromosome. Likewise, other chromosomal abnormalities, such as the amplification of a certain chromosome region that is identified as homogeneously strain region (HSR) or double minute chromosome (DMS), may cause the dysfunction of certain oncogenes as seen in certain neuroblastomas where many copies of the myc gene have been known to exist.

In other instances such as human bladder carcinoma, point mutations in the ras oncogene can lead to its activation. The dysfunction of oncogenes is very important in causation of cancer because oncogenes are known to encode proteins that function as growth factors (e.g., sis) or growth factor receptors (e.g., erb-B) or intracellular signal transducers (e.g., src, abl, and ras) or transcription factors (e.g., myc, fos, and jun).

Likewise, chromosomal rearrangement and mutation may cause dysfunction of tumor suppressor genes (also called antioncogenes). The RB gene mutation that causes retinoblastoma is an example of the tumor suppressor gene. Likewise,

FAP and p53 genes are tumor suppressor genes. Mutation in FAP gene leads to a precancerous polyp formation on the colon, called *familial adenomatous polyposis* (FAP). The loss of tumor suppressor genes such as FAP gene on chromosome 5q and p53 gene on human chromosome 17p is one of the factors in development of colorectal cancer. Mutation in p53 gene also leads to inherited susceptibility to cancer and has been attributed to the cause of Li–Fraumeni syndrome. It is pertinent to mention that active mutant oncogenes behave as dominant over the wild-type oncogenes, whereas the mutant antioncogene is recessive to its wild-type allele. In addition to mutation in oncogenes and antioncogenes, development of cancer in terms of metastasis and tumor formation is promoted by a number of other mutations. A mutation in a gene that controls the cell adhesion leads to metastasis, whereas tumor formation requires mutations leading to overproduction of angiogenesis factors required to provide blood vessels to tumors. It should be noted that angiogenin itself has RNase activity. However, at this time it is not yet obvious how its RNase activity plays a role in its angiogenic function. Thus, at least from the academic point of view, the bottom-line question is why certain chromosomal rearrangements take place and what recombinases or nucleases mediate such processes. What is the nature of these recombinases and how are their actions controlled? As discussed earlier in Chapter 7, a number of colon cancers are caused by mutation in a gene or genes that orchestrate the role of nuclease and other protein in the repair of mismatch base in DNA (Service, 1994).

In addition to cancer, a number of other human diseases involve genetic recombination. Some examples are as follows: (a) familial hypercholesterolemia in which recombination between Alu sequences in the DNA segment containing the gene for LDL receptor deletes an exon (the deletion of exon in mutant gene leads to the secretion of its gene product instead being retained in its place in plasma membrane); (b) muscular dystrophy in which CGT sequences are amplified in the 3′ untranslated region of the gene; and (c) likewise in fragile X syndrome, the CAC region is amplified. The mechanism of such amplification is not yet known. It is possible that these sequences are amplified due to slippage in replication involving the synthesis of a lagging strand of DNA or amplifed as homogeneously staining region in chromosomes. Whatever the mechanism is, the question is why they are not identified and removed by the DNA repair mechanisms.

Nucleases are further important as therapeutic agents against a number of cancer because of their tumirocidal effects. A number of nucleases are being evaluated for their therapeutic efficacy in clinical trials (Schein 1997, 2001; I. Lee et al., 2000).

E. Aging—Werner Syndrome

The aging process may result from accumulation of mutations leading to faulty proteins in the cells. Mutations may be attributed to nucleases due to defective proofreading mechanisms or to failure of the removal of damages to DNA. The proofreading activities of DNA polymerases in aging human fibroblasts have been shown to be defective. Now it is shown that Werner syndrome, characterized by the early onset of the symptoms of the process of aging, involves a defective WRN protein; this protein has both 3′–5′-exonuclease and a 3′–5′-helicase activities. A point mutation in the N′ terminus has been found to completely abolish this 3′–5′-exonuclease activity of the WRN protein (Huang et al., 1998).

The 3′–5′-exonuclease activity of WRN protein is also altered with respect to its inability to act on DNA containing AP sites, 8-oxyguanine, and cholesterol adducts (Machwe et. al. 2000). The WRN protein has a 3′–5′-exonuclease activity in its N′-terminal region, whereas the central region of this protein contains motifs that are homologous to the RecQ family of helicases (Yu et al., 1996). Both these activities are required for DNA repair and for other aspects of DNA metabolism. The understanding of the nature of changes in the WRN protein may help find avenues to help these patients.

F. Immunological Diseases

Several immunological diseases may be caused by the failure of V(D)J recombinase and other associated recombinases, leading to faulty assembly of genes encoding antibody, which, in turn, leads to subsequent defective or deficient antibody production. Scid is known to be caused by the failure of the V(D)J recombinase (Shuler et al., 1986; Maylynn et al., 1988; McCune et al., 1988; Moiser et al., 1988). A defective immune system may lead to several human diseases including cancer, infectious diseases and other diseases.

The illegitimate activity of V(D)J recombinase can cause disruption of several genes affecting hematopoiesis and other development in human (Aplan et al., 1990).

G. Nucleases and Neurological Disorders

The skin cancer xeroderma pigmentosum is known to lead to neurological disorders. It is still not known how and why a deficiency in AP endonuclease can cause neurological disorder. Furthermore, intramolecular recombination in mitochondrial DNA is known to lead to several neurological disorders (Martin, 1987; Wallace, 1989). This must involve nucleases because nucleases are the key enzymes involved in the process of intramolecular recombination. However, the role of nucleases and other recombinases in the recombination of mitochondrial DNA leading to a number of diseases remains to be elucidated. It is further tempting to speculate that the genetic process of recombination and the associated recombinases may play a role in the normal development of human brain. It is of interest to mention that the active development of brain in humans occurs at the same time that the immune system is developing. Therefore it is not unimaginable that recombination may play a major part in human development. Some support of this view is provided by the fact that the products of RAG genes are known to play a role in neuronal development in the mouse.

H. Human Diseases Involving Defective Protein Folding

A number of human diseases results from improper or defective folding of secretory or other membrane proteins. Some of these devastating human diseases include α_1-antitrypsin deficiency, cystic fibrosis, and osteogenesis imperfecta (Kuznetsov and Nigam, 1980). These proteins are folded and undergo several modifications in the lumen of endoplasmic reticulum (ER) facilitated by certain specialized ER-resident proteins (Shamu et al., 1994). Accumulation of improperly folded proteins leads to unfolded protein response (UPR), which is mediated by Ire1p protein in yeast or its homolog RNaseL in humans. Both of these proteins possess kinase and endoribonuclease activities, and their oligomeriation activates the RNase activity. The yeast Ire1p protein cleaves and splices the transcript for mRNA encoding the UPR-specific transcription factor Hac1p. The Hac1p upregulates the transcription of a number of ER-resident proteins. Thus Ire1p in yeast and presumably RNaseL in human controls the folding of proteins in the lumen of ER. The IR1p protein carries out the splicing of the Hac1p transcript by itself, only involving tRNA ligase and not as a member of the splicosome containing a multitude of proteins and RNAs (Sidrauski and Walter, 1997). An understanding of the RNaseL and its role in humans is crucial to the control of these human diseases.

I. Other Human Diseases

In lupus erythematosus and related rheumatic diseases, antibody against RNaseP has been found to exist in the blood serum (Gold et al., 1988). The basis for the presence of such anti-RNaseP antibody is not yet known. Likewise in several autoimmune diseases, antibody against topoisomerase I or autoantibodies with DNase activity has been shown to occur (Shuster et al., 1992). An understanding of the mechanism of these processes can enhance our ability to control these diseases in the future. In addition, some of the hemoglobin disorders, such as β-thalassemia, are known to be caused by improper mRNA splicing. Increase in the level of eosinophilic cationic protein has been associated with hyperbronchial activity and asthma in certain patients allergic to pollens (Di Gioacchino et al., 2000).

II. REVERSE GENETICS, HUMAN DISEASES, AND NUCLEASES

Methods of reverse genetics have been very useful in identification of genes and their products controlling several human diseases (Caskey, 1987). Without the use of nucleases, the methods of reverse genetics cannot be utilized for such understanding of human diseases. A large number of human genes and their alleles causing inherited diseases have been identified by detection of restriction fragment length polymorphisms (RFLPs) during pedigree analysis of human genomic DNA blots. In such analysis the role of specific restriction endonuclease is critical.

III. USE OF NUCLEASES IN CONTROL OF HUMAN DISEASE

In the past, nucleases have been used to control meningitis. Human DNaseI synthesized by the methods of recombinant DNA technology has been used to degrade the DNA contained in the mucus of the patients suffering from cystic fibrosis (Shak et al., 1990). The chemotherapy of certain kinds of cancer is based on the utilization of inhibitors of topoisomerases (Bodley and Liu, 1988). Certain RNases belonging to the family of RNaseA (see Chapter 2) such as onconase (a frog RNase) and BS-RNase (bovine seminal RNase) show anticancer activity as well as antiviral activity against HSV-1 in H9 leukemia cells (Youle et al., 1994).

Several diseases could also be controlled by developing the inhibitors targeted against the nucleases involved in the replication or integration of virus into human genome. Drugs can be developed to inhibit the activity of the enzyme RNaseH or viral integrase in several retroviruses or DNA virus; these enzymes are required for the viral replication and subsequent integration of the viral genome into the host eukaryotic cells. One such example is the RNaseH of human immunodeficiency virus (HIV) because of its integration into human chromosome which may be inhibited by certain drugs and prevent the replication of HIV and may thus prevent HIV infection among AIDS patients in humans (Bushman et al., 1990). Certain inhibitors of integrase is being explored for this purpose. Ribozymes have been engineered to suppress the activity of specific gene in monkey cells; therefore it is plausible that ribozymes may be used in therapy of human diseases by specific suppression of genes that play key roles in causation of these diseases (Cameron and Jennings, 1989).

13

NUCLEASES AS TOOLS

Nucleases have been widely used as tools in biological investigation. Pancreatic deoxyribonuclease was first used as a tool by Avery et al. (1944) to establish the chemical nature of the "transforming principle" during the transformation of pneumococcus by DNA. Nucleases were also used by Kornberg and his colleagues (Kornberg, 1961) to establish the complementary nature of DNA synthesized by DNA polymerase. They used nuclease to carry out the nearest-neighborhood analysis of nucleotides synthesized *in vitro*. Use of the *Neurospora* single-strand-specific nuclease was crucial in the first isolation of a gene (Shapiro et al., 1969). Discovery of restriction endonucleases has been the major factor in the development of recombinant DNA technology (including molecular cloning of a gene and the determination of the nucleotide sequence of a DNA segment) and in providing a new and rapid method for linkage analysis particularly useful for organisms not amenable to analysis by classical method of Mendelian genetics (Southern, 1975, 1982). Applications of nucleases have been useful in determining the organization of DNA into chromosome and in probing into the nature of gene as to whether or not it is transcriptionally active. Nucleases have been also useful in providing the picture of the internal organization of a eukaryotic gene into stretches of coding (exons) and noncoding sequences (introns). Analysis of eukaryotic DNA at different stages of development by restriction endonucleases led to the establishment of the idea that genetic recombination is a dynamic biological process that is occurring at all stages of development and not limited to events involved in sexual reproduction. Some of these aspects of the application of nucleases are described below.

I. NATURE OF "TRANSFORMING PRINCIPLE" AS DNA

Avery et al. (1944) showed that the chemical basis of heredity was DNA. In their experiments with pneumococcus, they showed that the transforming ability of DNA was abolished when treated with pancreatic DNase. A similar treatment with RNase or protease was ineffective in abolishing the transforming ability of the nucleic acid preparations. Results of these studies conclusively established the chemical nature of transforming factor as DNA.

II. ISOLATION OF DNA AND RNA

As an extension of observation that DNase effectively destroyed DNA during pneumococcal transformation, nucleases have been effectively used for the purification of DNA or RNA from a mixture of nucleic acids preparation. Customarily, a mixture of nucleic acids is treated with DNase during the purification of RNA and with RNase for the isolation of DNA. The DNase and RNase used during such isolation procedures must be of the highest purity and completely free of any contaminating activity. Ordinarily an RNase preparation is boiled at 80°C for 15 min to get rid of contaminating DNase. DNase completely free of RNase is available commercially.

III. NEAREST-NEIGHBORHOOD ANALYSIS

This method determines which two nucleotides are adjacent to each other in a DNA chain. This is based on the transfer of a PO group from the 5 position of a nucleotide to the 3 position of a neighboring nucleotide because of the manner in which these P–O bonds are attacked by a nuclease (e.g., micrococcal nuclease). Thus during the synthesis of a DNA chain by DNA polymerase a nucleotide (say dATP carrying ^{32}P-labeled α-PO attached to the fifth carbon of the sugar moiety of the nucleotide) can be linked to the third carbon of the sugar of another nucleotide as shown in Figure 13.1. Now when this DNA chain is digested to mononucleotides by micrococcal nuclease, then the nucleotide (dCMP) adjacent to dATP in the DNA chain can be easily identified because dCMP has acquired the radioactive phosphorus (^{32}P).

This kind of analysis was used to establish the antiparallel nature of the DNA strands in a duplex structure (Kornberg, 1961). Such knowledge of the antiparallel nature of the strands in a DNA duplex has profound consequences in the understanding of basic genetic processes of DNA replication, repair, recombination, and transcription.

IV. ISOLATION OF A GENE

Even before the coming of recombinant DNA technology, Shapiro et al. (1969) were successful in isolation of the lac operon of *E. coli*. Utilization of *Neurospora*

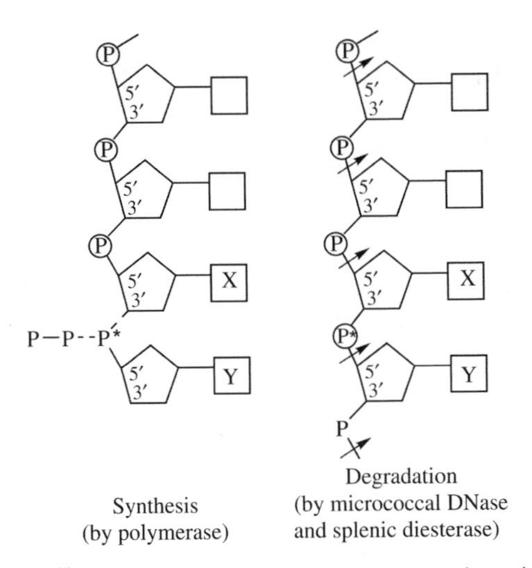

Figure 13.1. Synthesis of ^{32}P-labeled DNA and its subsequent degradation by micrococcal nuclease to 3′ deoxyribonucleotides. Arrow indicates the position of attack by nuclease. The net effect of nuclease digestion is the identification of nucleotide (dCMP), which was adjacent to dAMP in the DNA chain. [From Kornberg (1961), used with permission.]

crassa single-strand-specific nuclease was crucial for the success of their strategy to isolate the lac operon of *E. coli*. The workers first constructed transducing λ phages carrying *E. coli* lac operon region in the normal and in inverted orientations. The transducing phage DNAs were separated on CsCl gradient as heavy (H) and light (L) strands. The heavy DNA strands were found to contain the coding and noncoding sequences of the lac region. The lac region was regenerated by hybridization of the heavy DNA strands of the transducing phages. During such hybridization, only the *lac* region could hybridize while the λ DNAs were left as single-stranded DNAs that were then removed by digestion with *Neurospora* nuclease to yield intact *lac* operon region of *E. coli*.

V. UNIPARENTAL TRANSMISSION DURING CYTOPLASMIC INHERITANCE

Besides the nucleus, cytoplasm harbors several reservoirs of genetic information in the form of either organelles (mitochondria and chloroplast) or plasmid DNA. In fungi there is good genetic evidence that shows the uniparental transmission of these cytoplasmic determinants. However, such demonstration of uniparental transmission of cytoplasmic determinants in higher organisms was not possible until the discovery of restriction endonucleases. As soon as the cleavage map of mitochondrial DNA was available, it became obvious that the cleavage map of mitochondrial DNA is characteristic of a species. Hutchinson et al. (1974) showed that horses and donkeys possess characteristic but different mitochondrial DNA cleavage maps when treated with restriction endonuclease HaeIII. They also showed that the HaeIII cleavage map of the mule mitochondrial DNA resembled the pattern of the maternal parent, horse. Likewise, the HaeIII cleavage map of the hinney mitochondrial DNA resembled that of the maternal parent, donkey. Thus in both cases of mule and hinney the mitochondrial cleavage map resembled their respective maternal parent, suggesting that they inherited mitochondrial DNA from their respective maternal parents. Similar analysis of mitochondrial DNA has revealed the differences in the cleavage map of mitochondrial DNA from the normal and male sterile strains of corn (Dewey et al., 1986). Results of this study suggest that male sterility of corn is controlled by mitochondrial DNA. Such correlation would not have been possible without the use of restriction endonucleases.

VI. PHYSICAL MAPPING OF DNA

A physical map of DNA can be constructed using restriction sites as markers. This method was first used by Nathans and his colleagues (Danna and Nathans, 1971; Lai and Nathans, 1974, 1975) to generate a map of the virus SV40. First, the genomic DNA, after complete digestion with a particular restriction endonuclease, is electrophoresed on a gel and then the mobility pattern of the DNA segment is visualized. The DNA bands are usually numbered as A, B, C, D, E, and so on, according to their decreasing size (or increasing mobility). In this way the definite number of fragments generated by restriction endonuclease is determined. Second, the order of these bands in the DNA is determined by the analysis of DNA after partial digestion by the restriction endonuclease. The products of partial digestion are obtained from the gel and individually subjected to complete digestion by the enzyme and then reanalyzed by gel electrophoresis to reveal their linkage.

Restriction fragments of DNA generated by different restriction endonucleases can be correlated to a physical map of the chromosome (Danna and Nathans, 1971). For such correlation a particular DNA fragment among products of the first enzyme digestion is treated with the second enzyme and then identified by their correspondence to complete digestion of the genomic DNA by the second enzyme. In this manner one can find out which fragment of the first enzyme product yields a fragment corresponding to the second enzyme product.

Such correlation can be easily obtained by the application of the method of Southern hybridization (Southern, 1975). This method is an extension of the discovery that single-stranded DNA immobilized in a nitrocellulose filter can retain its ability to hybridize with complementary sequence. In this method, restriction fragments of DNA fractionated by gel electrophoresis are denatured *in situ* and then transformed onto nitrocellulose filter generating a replica of the original gel pattern of the DNA fragments. The homology of DNA segments on nitrocellulose filter can be determined by hybridization to a radioactive probe followed by autoradiography of the filter. The method of Southern hybridization has been used for isolation of a DNA fragment containing a eukaryotic gene or its part, for linkage analysis via chromosome walk and restriction fragment length polymorphism (RFLP). The latter in conjunction with Southern hybridization has been used for identification of mutations responsible for certain human diseases. RFLP may result due to a mutation in DNA sequence (Kan and Dozy, 1978) or due to changes in the number of tandem repeats in chromosomes (Namanura et al., 1987).

VII. USE OF NUCLEASE IN THE DEVELOPMENT OF RECOMBINANT DNA TECHNOLOGY AND THE MOLECULAR CLONING OF A GENE

Discovery of restriction endonucleases has been crucial for the development of recombinant DNA technology. The latter provides a powerful tool for the cloning and amplification of a particular gene or DNA segment and its physical and biochemical genetic characterization. Restriction endonucleases have been used to obtain fragments of genomic DNA to be cloned and for the specific cleavage of vector DNAs before they are joined by ligase to generate recombinant plasmid. The presence of a recombinant molecule can be identified by (a) its ability to complement a defective function in an appropriate host, (b) its ability to confer a new phenotype such as drug resistance, (c) its hybridization to a specific probe such as RNA transcript which is readily available, or (d) immunological methods. These aspects of molecular cloning have been reviewed (Mishra, 1985). A number of prokaryotic and eukaryotic genes have been cloned; over 3000 human genes and a number of other mammalian genes have been cloned and characterized (Olson, 1994). Restriction endonucleases have been also used for the subsequent physical mapping and nucleotide sequence analysis of a cloned DNA segment or gene.

Recombinant DNA technology can be used for the following purposes: (1) the cloning of a desired sequence of DNA from a complex mixture of DNA molecules present in the eukaryotic cell and subsequent amplification in large amounts for biochemical characterization to ascertain its complete nucleotide sequence and to determine the role of different regions of nucleotide sequences as controlling elements; (2) the cloning of an artificially synthesized nucleotide sequence to examine its expression in eukaryotic cells; (3) the modification of a desired nucleotide sequence to examine its effect on the expression of a gene; and (4) the synthesis of a large number of peptides and chemicals or organisms of interest to science, medicine, commerce, and environmental preservation.

VIII. CONSTRUCTION OF AN ARTIFICIAL CHROMOSOME

Molecular cloning of genes could provide a better understanding of the structure organization and function of the genome. A particular DNA sequence can be identified by different methods and then enriched in an appropriate host organism. A large number of the genes have been cloned in this manner. Over 3000 human genes have been cloned, and the nucleotide sequences of a number of them are now available. Besides the isolation of genes, a number of promoters, autonomously replicating and enhancer sequences, centromeres, and telomeres have been isolated and characterized by the methods of molecular cloning. One of the highlights of recombinant DNA methodology is the ability to construct biologically active artificial chromosomes (Dani and Zakian, 1983; Murray and Szostak, 1983).

Eukaryotic chromosomes are highly organized organelles containing several elements needed for their maintenance and inheritance during cell division. Typically, a eukaryotic chromosome contains (1) a *structural gene*, encoding a protein (such gene can be cloned by complementation of a defective function in the recipient cell); (2) *ars*, a DNA segment containing an origin of replication and which can be cloned by its ability to maintain high copy of a recombinant plasmid; (3) a *centromere* required for chromosome movement during cell division (this genetic element can be cloned by its ability to endow stability to the recombinant plasmid carrying it); and (4) a *telomere*, which maintains the stability of a linear chromosome after transfection into a host. Using these cloned components and λ DNA segment as fillers, it has been possible to construct artificial yeast chromosomes called YAC (Dani and Zakian, 1983; Murray and Szostak, 1983; Schlessinger, 1990). The stability of chromosomes during cell division is a function of the size of the artificial chromosomes; larger-size chromosomes (50 kbp or greater) were as stable as the naturally occurring yeast chromosomes. The development of YAC led to the construction of BAC (bacterial artificial chromosome) and PAC. Both BAC and PAC have been extensively used in genomic analyses. Finally, these developments led to the construction of human artificial chromosome (HAC) (Henning et al., 1999). It is therefore possible to introduce a new set of artificial chromosomes in humans at the time of *in vitro* fertilization or immediately before implantation of the fertilized egg to provide new set of desired genes to the fetus which could remain stable during meiosis, thus leading to germ-line genetic changes. Thus the ability to construct artificial chromosomes could become a tool for genetic engineering of animals including humans. This methodology has a great prospect, particularly for the enhancement gene therapy in humans (Mishra, 2001) and certainly that of domesticated animals.

Such construction of artificial chromosomes can be used for the simultaneous introduction of multiple genes controlling the structure of several proteins involved in the biosynthesis of antibiotics, hormones, and other products of pharmaceutical importance, whose expression can be triggered and maximized in response to environmental factors or metabolic need. All these manipulations require the use of nucleases, particularly restriction nucleases without which these constructs were not possible.

IX. NEW METHOD FOR MAPPING EUKARYOTIC CHROMOSOMES

A number of genes (without known mutations or functions) can now be mapped based on the use of restriction fragment length polymorphisms (RFLPs) as genetic markers. This method developed by Botstein et al. (1980) has been used for the mapping of many eukaryotic chromosomes. Such methods of molecular cloning have been used to provide information regarding the chromosomal location of genes, their order, and nucleotide sequences.

A. Chromosome Walking (Overlap Hybridization)

An array of bacterial clones harboring chimeric plasmids containing overlapping eukaryotic DNA segments as inserts is used to generate a genomic library. Genes containing overlapping DNA inserts are identified by their cross-hybridization to a common probe on a Southern blot (Southern, 1975). In several rounds of such hybridizations, one can virtually "walk" on a chromosome arm in a particular direction from a fixed point (e.g., a centromere or a known gene) and reconstruct a stretch of DNA of 30 kbp or longer. The physical maps of a number of chromosomes from a number of organisms are available using the method of chromosome walk. The physical linkage of these DNA segments on a particular chromosome can be checked by reference to established linkage of gene via classical methods of Mendelian genetics. It has been estimated that 1 centimorgan (1 cM) encompasses a distance of 1 Mbp in humans.

B. Role of Nucleases in Transposon Mobility

Transposons that cause gene instability and chromosomal rearrangement were described long before their molecular nature was elucidated using the methods of recombinant DNA technology. Transposons usually encode one or more enzymes responsible for the transposition (see Chapter 7). Eukaryotes usually contain dispersed repeated gene families. At least two groups of repeat sequences have been identified; these include LINES (long interspersed repeat sequences) and SINES (small interspersed repeat sequences) at least. Some of these dispersed repeated DNA sequences (LINES and SINES) are now known to be indeed transposons. Certain retroviruses possess structures similar to the transposons (Shiba and Saigo, 1983). The yeast *ty-1* elements and other eukaryotic transposons have a significant effect on gene expression and may facilitate adaptation of a particular cell in a new environment. Transposition seems to occur widely in both prokaryotes and eukaryotes facilitated by transposon-encoded recombinases. A detailed analysis of transposable elements has provided excellent means for development of new vectors for the transfer of genes to a particular target in eukaryotes.

X. USE OF NUCLEASE IN THE PHYSICAL MAPPING OF A MUTATIONAL SITE

The S_1 nuclease specific for single strand DNA has been used to locate the position of mismatch in a DNA due to a mutation caused by deletion, insertion or substitution of base. S_1 nuclease can make a cut in the mismatch region of DNA duplex which can then be analyzed in a denaturing gel or in an alkaline sucrose gradient to reveal the change in the size of duplex fragment due to a nick introduced by the enzyme. Shenk and others (1975) reported that S_1 can cleave within the heteroduplex region formed between the complementary DNA strands obtained from the temperature-sensitive mutant and wild-type SV40. S_1 nuclease can also be used to map a non-base-paired region in a supercoiled DNA. The negative superhelicity of DNA can cause localized unwinding of the helical base pair. S_1 nuclease can make cleavage in such regions. The S_1 cleavage map of SV40 and polyoma show seven or eight such sites at comparable distances in both viruses. The S_1 cleavage site has been found to coincide with the RNA polymerase binding site in plasmid pNS1 of *Bacillus subtilis*.

XI. BIOLOGICAL ACTIVITY OF A DNA SEGMENT

A restriction fragment may contain an entire gene including controlling elements. Several methods are available to test the biological activity of a restriction fragment; these include (a) assay for activity in a transformation experiment, (b) marker rescue, and (c) assay for activity after deletion of restriction fragment. These methods are described below.

A. Use of Nucleases in the Identification of the Function of a DNA Segment via Transformation Experiments

Mammalian cells are capable of taking up naked DNA molecules that may be expressed upon integration into recipient genome. This process is called *genetic transformation*; by this method, recipient cells acquire the phenotype of the alien DNA. A specific restriction fragment of genome (such as that of SV40) or of adenovirus genomes controlling a specific function (such as malignant transformation of mammalian cells or function of HSV thymidine kinase gene) has been identified in this manner. The identification of the physiological function of a DNA segment generated by digestion with restriction fragment after introduction into a suitable host cell via transfection is a routine procedure now.

B. Use of Nuclease in the Deletion Mapping of Biological Activity

The deletion mapping involves the demonstration of a change in the expression of a DNA segment after removal of a nucleotide sequence from a restriction fragment whose expression is examined by transformation. It has been shown that a nucleotide sequence 113–155 base pairs upstream from the transcription start point is necessary for the wild-type expression of the yeast his-3 gene and that the presence of the canonical TATA sequence alone is not enough for the wild-type expression of this gene. This has elucidated the role of different parts of the promoter region in the mRNA production and the expression of the his-3 gene by deletion mapping. This approach included the removal of a specific nucleotide sequence from the his-3 gene with either ExoIII or Bal31 nuclease treatment and construction of an appropriate chimeric plasmid with deleted his-3 gene which was used to transform auxotrophic (his) yeast cells. Their data showed that deletion of sequence 155–200 base pairs upstream from mRNA coding sequence has no effect on the overall expression of the deleted his-3 gene, and transformants capable of growth on minimal medium were obtained. However, deletion of nucleotides encompassing 113–155 base pairs upstream from the transcription start point of the gene was found to cause a drastic effect on the expression of his-3 gene, and consequently no transformants capable of growth on minimal medium were obtained after transformation of his auxotrophs with plasmid containing deleted his-3 gene. Such deletion mapping using nucleases Bal31 or ExoIII is routinely used to identify the role of a gene or its part in its expression.

C. Use of Nuclease in Identification of the Function of DNA Segment via Marker Rescue Method

This method is based on the rescue of the plaque-forming ability of an individual mutant when infected in combination with a particular wild-type DNA fragment obtained after endonuclease treatment of the wild-type viral DNA (Hutchinson and Edgell, 1971; Lai and Nathans, 1974). This method implies cellular repair or the recombination of mutant DNA strand in the partial heteroduplex formed during the mixed infection. Using this method, Lai and Nathans (1974) have demonstrated the ability of five separate HindIII fragments of the wild-type SV40 DNA to rescue five distinct ts SV40 mutants. Such analysis not only identified the function of each restriction fragments, but was successful in locating each ts mutant to a particular position on the physical map of SV40 genome (Lai and Nathans, 1975).

XII. ORGANIZATION OF EUKARYOTIC CHROMOSOMES

Analysis of chromatin after nuclease digestion supports the view that a eukaryotic chromosome is made up of repeating units called *nucleosomes* (Hewish and Burgoyne, 1973). The organization of chromatin into a series of nucleosomes is supported by the results of electron microscopic studies. A nucleosome is usually comprised of a DNA segment about 200 bp in length which is wrapped around a histone octamer consisting of two molecules each of slightly lysine-rich histones H_2A and H_2B and the arginine rich histones H_3 and H_4. The histone H_1 is not a part of the nucleosomal core but is associated with a stretch of linker DNA which connects the adjacent nucleosomes. The linker DNA is most sensitive to attack by staphylococcal nuclease. A limited digestion of chromatin by staphylococcal enzyme yielded chromatin fragments that consisted of a mixture of mono-, di-, and oligonucleosomes. The nature of the product of nuclease digestion of chromatin was established by sedimentation analysis or by agarose gel electrophoresis. The mononucleosomes appeared as 11S particles upon sedimentation analysis and consisted of approximately 200 bp in length. The oligonucleosomes appeared as integral multiples of mononucleosomes.

Upon agarose gel electrophoresis, a chromatin digested by staphylococcus nuclease appeared as discrete DNA bands that were exact 200-bp multiples of the length of the smallest subunit (nucleosome). Further digestion of nucleosomes by pancreatic DNase (DNaseI) and by spleen DNase (DNaseII) revealed the location of the nuclease-sensitive sites within a nucleosome. DNaseII made cleavage halfway within a nucleosome. Thus DNaseII digestion of chromatin yielded DNA bands as multiples of 100 bp (instead of 200 bp as obtained after digestion with staphylococcal enzyme). DNaseI digestion of chromatin revealed the presence of a cleavage site every 10 nucleotides within the nucleosome. Sites located at 20, 40, 50, 100, 120, and 130 nucleotides from the 5 end in a nucleosome were the most sensitive to cleavage by DNaseI; sites located at 30 and 100 were the least sensitive, while the site located at 80 nucleotides was virtually never cleaved. The differences in susceptibility to DNaseI may reflect the protection afforded by interaction of histone within the nucleosomes. The nucleosomal organization of eukaryotic chromosome as revealed by nuclease digestion has been previously discussed (Felsenfeld, 1978). In addition to DNaseI and micrococcal nucleases, a number of chemical nucleases (chemzymes) have been used to probe into the nature of the chromatin structure (Cartwright and Elgin, 1982; Cartwright et al., 1983; Francois et al., 1988). The characteristics of nuclease-hypersensitive sites in relation to gene action have been reviewed (Elgin, 1984).

XIII. DISTINCTION BETWEEN ACTIVE AND INACTIVE GENES: THE RELATION BETWEEN ACTIVITY OF A GENE AND A NUCLEASE-SENSITIVE SITE

It has been shown that actively transcribing genes are more susceptible to digestion by the enzyme DNaseI than is the inactive gene (Weintraub and Groudine, 1976). The fact that the expression of certain genes is tissue-specific was utilized to demonstrate the DNaseI sensitivity of an active gene. It is seen that globin genes are readily digested by DNaseI in erythrocyte nucleic but not in fibroblast nuclei. Results of such experiment are schematically presented in Table 13.1. The DNaseI sensitivity of a gene was measured by determining the ability of a particular nucleus to hybridize with a specific probe (e.g., globin mRNA or cDNA for globin). In such studies, ^{32}P-labeled cDNA for globin gene was found to show a low level of hybridization (e.g., 20%) with DNaseI-treated erythrocyte nuclei as compared to the level of hybridization (e.g., 100%) of the same probe to undigested erythrocyte nuclei or to DNaseI-treated fibroblast or oviduct nuclei. These results were explained on the differential digestion of globin gene by DNaseI in active cells (erythrocyte) as compared to its lack of digestion in an inactive cell (fibroblast). Likewise, it was demonstrated that ovalbumin gene is more readily digested by DNaseI in oviduct nuclei than in fibroblast nuclei. Furthermore, they showed that in erythrocyte nuclei where globin gene is active but ovalbumin gene is not active, DNaseI treatment destroyed the ability of erythrocyte DNA to hybridize with globin cDNA but not with ovalbumin cDNA. Besides their susceptibility to DNaseI, actively transcribing genes

Table 13.1 Differential hybridization of specific polynucleosomes digested with nucleases

DNA Source	Gene Contained in DNA Segment before Enzyme Treatment	DNase Treatment	Status of DNA Segment after Enzyme Treatment	Hybridization with ^{32}P cDNA for		Comment
				Globin	Ovalbumin	
1. Erythrocyte	Globin	No	Intact	100%	100%	
2. Oviduct	Ovalbumin	No	Intact	100%	100%	
3. Fibroblast	Globin Ovalbumin	Yes	Intact Intact	100%	100%	Both genes inactive in fibroblasts
4. Erythrocyte	Globin Ovalbumin	Yes	Hydrolyzed Intact	Drastically reduced (20%)	100%	Globin gene active but ovalbumin gene inactive in erythrocytes
5. Oviduct	Globin Ovalbumin	Yes	Intact Hydrolyzed	100%	Drastically reduced (20%)	Globin gene inactive but ovalbumin gene active in oviduct

were found hypersensitive to digestion by a number of other enzymes such as S_1 nuclease, DNaseII, and restriction endonuclease. The nuclease-hypersensitive sites have been located to the 5′ end of gene. It has been shown that such sites, once formed during the initiation of gene activity, can be propagated indefinitely. The role of certain DNA-binding proteins in the propagation of a DNase-sensitive site has been invoked. It is possible that binding of the upstream DNA sequence of a transcription unit with certain transcription initiation factors may cause the formation of DNase-sensitive site in an active gene.

XIV. DNase FOOTPRINTING

A protein that specifically binds with a DNA segment can protect it from nuclease digestion. Based on this principle, it is possible to generate a footprint of the DNA segment protected from DNase digestion by a binding protein. A DNA segment labeled with ^{32}P at the 5′ end usually yields a ladder of DNA bands upon electrophoresis and autoradiography of the products of DNaseI digestion in the absence of any binding protein. However, the products of such DNA segments digested in the presence of a binding protein usually show certain bands missing from the autoradiograph. The missing bands correspond to the DNA region protected by the binding protein and can be easily identified by examination of the bands on autoradiograph. The exact sequence of base pairs in the DNA protected by the binding protein can be easily determined by running the base-specific reaction products of the DNA segment obtained by the Maxam and Gilbert (1977) sequencing method alongside the DNase digestion product of the DNA in the presence and absence of the binding protein. This method of DNase footprint was first developed by Galas and Schmidtz (1978). These workers visualized the protective footprint of the lac repressor protein on the DNA sequence of the lac operator region of *Escherichia coli* by this method. The base sequence protected by lac repressor was exactly as determined by another method; this fact confirmed the validity of the usefulness of this method developed by Galas and Schmidtz. This method has now been widely applied in determining the interactions between specific proteins and DNA segments, which is the hallmark of the gene regulation (Saluz and Jost, 1993). A number of protein nucleases (such as DNaseI) and chemzymes have been used in such footprinting studies (VanDyke and Dervan, 1983; Saluz and Jost, 1993). The method of footprinting has been used to map the DNA regions that interact with initiation factors and with RNA polymerase during transcription.

XV. CONSTRUCTION OF MUTANTS: SITE-SPECIFIC MUTATION AND PROTEIN ENGINEERING

In vitro mutagenesis of cloned DNA has been the most direct way to establish correlations between DNA and protein sequences and to analyze their respective functions. This method has been particularly useful in analyzing the function of a particular domain of a protein. Several methods of *in vitro* mutagenesis have been described (Shortle et al., 1981; Smith, 1985). These methods requiring DNA sequencing or oligonucleotide synthesis are quite labor-intensive. However, an efficient method utilizing a hexameric linker has been recently developed (Barany, 1985). The hexameric linkers coding for two amino acids contain palindromic sequences complementary to either a two- or a four-base restriction site overhang. The insertion of the hexameric linker into plasmid creates a new restriction site and adds two amino acids into the encoded protein with disruption of the reading frame of the gene. The addition of two amino acids causes a minor perturbation in the protein structure generating temperature-sensitive mutant proteins or enzymes with altered substrate specificity (Barany, 1985). The gentle perturbation in protein structure created by insertion of hexameric linker can map a protein domain. Using this method, Barany (1985) has shown that the tetracycline gene is made up of two domains connected by a central stretch of amino acids and that mutation in the β-lactamase gene causes the production of enzyme with altered sensitivity to temperature or altered affinity to different β-lactamase inhibitors. Such manipulation of genes would not have been possible without the knowledge of restriction endonucleases, their active site, and role in molecular cloning.

XVI. NUCLEASES IN DIRECTED MUTAGENESIS

Nucleases have been employed to create mutant form of the cloned gene. In such mutagenesis, exonuclease III and Bal 31 nuclease have been extensively used (Sambrook et al., 1989). By the use of these nucleases, mutant forms of particular genes have been generated which have been used in the direct analysis of the function of the gene products such as their enzymatic properties and their physiological role.

XVII. NICK TRANSLATION AND LABELING OF DNA WITH HIGH-SPECIFICITY RADIOACTIVITY

The $5' \rightarrow 3'$ exonucleolytic activity of several DNA polymerases in conjunction with their DNA polymerizing activity has been used in nick translation of DNA molecules. During such nick translation, use of radioactively labeled dNTP has yielded DNA molecules with high radioactivity. Radioactive DNA as high as 10^8 CMP/μg has been obtained via this method. Such DNA has been used in various methodology of molecular biology including Southern and Northern hybridizations (Sambrook et al., 1989).

XVIII. ROLE OF NUCLEASES IN PCR

A. Proofreading by Nuclease During PCR Amplification

The polymerase chain reaction (Mullis and Faloona, 1987) provides methodology for the severalfold amplification of a particular DNA segment over a short period of time. However, during this process, one of the important roles that nuclease is required to play is the proficient proofreading activity of the DNA polymerase such as Pfu. In the absence of such vigorous proofreading activity of the editing nuclease, the amplified DNA sequences would not be a faithful copy of the template DNA segment.

B. Application of 5′ Nuclease in PCR Assay for Rapid Detection of a Known Gene in a DNA Sample(s)

The 5′–3′-exonuclease activity associated with Taq DNA polymerase has been exploited to generate a specific signal via cleavage of a reporter oligonucleotide placed upstream of the DNA strand being synthesized by the Taq DNA polymerase. The reporter oligonucleotide that is complementary to the upstream sequence hybridizes the template strand in front of the growing DNA strand. The probe is blocked at the 3′ end to prevent its extension during PCR amplification and also carries a radioactive label at the 5′ end which is released by the 5′–3′ nuclease activity of the Taq DNA polymerase as it approaches the oligoprobe in front of the growing DNA strand. The radioactive label can be identified and become the measure of the amplification of a particular DNA segment. This assay developed by Holland et al. (1991) provides rapid detection of a gene in question and represents a marked improvement.

The efficiency of this method of detection of a DNA segment by 5′–3′ nuclease activity of the Taq DNA polymerase has been further enhanced by the use of an oligonucleotide (Taq-Man) containing a fluorescent reporter dye placed at the 5′ end and a quenching dye at the nonextendable 3′ end in a manner that the fluorescent reporter dye and the quencher situated in close proximity remain nonfluorescent but become fluorescent as soon as the reporter dye is released by the 5′-nuclease activity of Taq DNA polymerase during the amplification process in a PCR assay (see Figure 13.2). This type of detection device using 5′-nuclease activity of Taq DNA polymerase in a PCR assay of a known DNA segment has

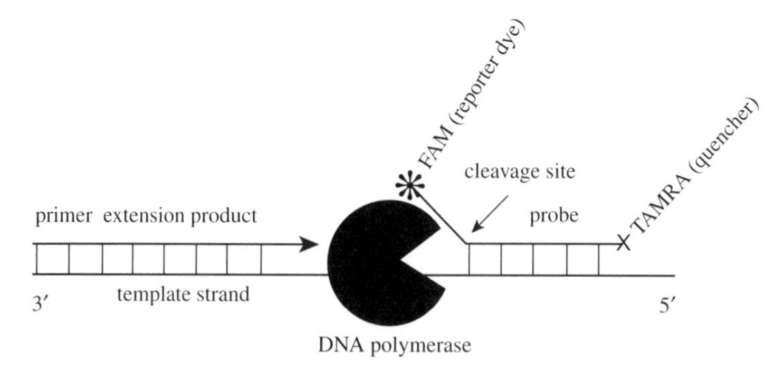

Figure 13.2. Diagram of 5′ → 3′ exonuclease cleavage of 5′-labeled oligonucleotide probe to generate labeled fragements in a PCR. *, The 5′ label on the probe; ×, a 3′-phosphate block on the probe. (Modified from *Proc. Natl. Acad. Sci.*, 1991, **88**: 7276–280, reproduced with permission.)

been automated for a large-scale high-throughput analysis of microbial pathogen (see Chapter 14). This method has also been used efficiently to discriminate between individuals carrying alleles of a known gene. Todd et al. (1995) first used the fluorescent probe to distinguish among the alleles associated with human type 1 diabetes. This methodology can be readily used for epidemiological studies and in pharmacogenetics to identify individuals who are oversensitive to a drug or are more responsive to the drug. This methodology could be used in conjunction to SNP (single nucleotide polymorphism) of a gene as being revealed by genomics.

C. Application of Nucleases in SNP-Genotyping and Pharmacogenetics

A particular gene may possess a single nucleotide difference among different human individuals and for that matter in members of any organism. Such differences, called SNP (single nucleotide polymorphism), are of medical value with respect to tailored therapy of human patients based on specific genetic profile. Such genetic differences may provide the basis for pharmocogenetics by defining which individuals will have an adverse effect from a prescribed medication while others in a human population will respond favorably to the same drug. An SNP map of human genes is being generated by the comparison of the genome sequencing of different individuals. The SNP can also be generated by the use of restriction nucleases in conjunction with amplification of a gene carrying a particular restriction endonuclease site in a PCR assay on a solid-phase matrix followed by fluorescent minisequencing (Shapero et al., 2001). Such methodology is equally accurate and less expensive than the comparison of genomics to generate SNP genotyping by a different SNP consortium. SNP genotyping will be very crucial for the tailored medical therapy of human patients in the next 10 years (Goodman, 2001; Chanock, 2001; Roses, 2001).

XIX. GENE KNOCKOUT

This methodology is used to analyze the functional role of a gene in the lifetime of an organism. A mutant form of a cloned gene generated by directed mutagenesis is introduced into embryonic stem cells of mice via transfection; this type of mutant gene can replace the resident wild-type gene in the mice by homologous recombination. Such procedures that lead into disruption of the resident normal gene are called *gene knockouts.* After the process of gene knockout (Capecchi, 1989), such mice are analyzed for the physiological role of the gene (Silva et al., 1992a,b; Snider, 1994). Such gene knockout experiments have shown that a number of genes are essential for a particular stage in the life of a mouse while others are not. The fact that certain genes are not crucial for the life of mice may be explained in a variety of ways. The gene knockout experiments would not be possible without the use of nucleases. It is plausible that in the future, tissue-specific gene knockout can be performed with site-specific recombinase in order to assign a specific role of an enzyme in the physiology of an organism.

XX. RNase PROTECTION ASSAY

This methodology has been used to identify and characterize a particular RNA in a mixture of total RNA preparation. A specific RNA is hybridized with complementary RNA, and then the mixture is subjected to digestion by single-strand-specific RNA which destroys all RNA except for the specific RNA in the double-stranded form; the latter is then separated by electrophoresis and isolated for further characterization. A number of RNAs have been identified and characterized in this manner (Sambrook et al., 1989).

XXI. USE OF NUCLEASES IN FORENSIC SCIENCE

RFLP of DNA segments from an individual suspected and/or accused of committing a crime can be used to identify the person who committed the crime. In such analysis, the DNA of the victim, DNA from the tissue or cells left at the scene of the crime, and DNA from the individuals accused of it and from suspected criminals are treated with a number of restriction nucleases to generate RFLP of DNA segments, and their comparison can identify the criminal as the person whose RFLP profile matches with RFLP of DNA from cells left at the scene of the crime. Such comparison of RFLP from different individuals can also decide disputed paternity cases.

XXII. HUMAN AND GENOME PROJECTS

The human genome projects aimed at sequencing over 3 billion base pairs contained in the 24 human chromosomes (22 pairs of autosomes, along with X and Y chromosomes) by the year 2005. However, this project was accomplished much ahead of schedule in May 2000. Even before the completion of the human genome project, the entire genome of a number of other organisms including several bacteria, yeast, worm, and fruit fly was aleady accomplished. This objective to generate the entire DNA sequence of various organisms including humans was feasible only because of the ability to generate a restriction map of the different chromosomes and construct a clone of these DNA segments. In this project, use of restriction endonucleases is critical. It is believed that knowing this entire nucleotide sequence of human is essential to any understanding of why we are different from other animals. The problems of the human genome project and the role of nucleases in this project have been reviewed (Olson, 1994). In the spring of 2000, the entire human genome was sequenced. The entire DNA sequence of over 30 other organisms have been made possible in which the use of nucleases was very central. The outcome of the genome projects have been full of surprises. First and foremost, there are more DNA sequences than the number of genes that could be assigned to them. Second, in most organisms, more than 40% of the genes could not be assigned a function. Third, about 30% of genes in any organism are unique to that particular organism. Finally, the understanding of the genome sequence had led to the realization that a cell can survive with less than 500 genes. Above all, it has shattered the previous counts of human gene numbers to be about 100,000. Humans may possess less than 40,000 genes, which is just twice that of a fruit fly or a worm, suggesting the possibility that alternate splicing of exons may account for the larger number of proteins than the number of genes in humans. In reality, humans may have only about 5% of DNA sequences that encode for proteins. Also, humans may differ from chimps by only a few genes. All these data remain to be explored and interpreted to understand the nature of living systems, including that of humans. None of these were possible without the application of nucleases—particularly restriction nucleases, which are central to the development of the recombinant DNA technology.

14

APPLICATION OF NUCLEASES IN BIOTECHNOLOGY, MEDICINE, INDUSTRY, AND ENVIRONMENTS

Application of nucleases in medicine and industry is based on the fact that nucleases are used for the construction of recombinant DNA used for the cloning of genes. Cloned genes are used for the purpose of making drugs such as human insulin, human growth factors, and many more as seen in Table 14.1. Thus cloned gene has become the starting point in the biotechnology of today (Barnum, 1998). Cloned genes are used for mass production of relevant human proteins for therapeutic purposes such as human insulin (humulin) or for the purpose of production of vaccine or for gene therapy or gene replacement therapy (Anderson, 1992, 1995; Lever and Goodfellow, 1995; Blaese, 1997). In addition, RNases, DNases, ribozymes (and possibly DNAzymes) are used as new generation of therapeutics (Barinaga, 1993; Burke, 1997). Nucleases are used to generate different RFLP which are used in the positional cloning of the genes controlling a particular disease leading to possible diagnostic, preventive, or curative treatment of the disease including the development of drug for a particular disease. The applications of nuclease are based on the development of certain food additives and as indicators of infectants in the food industry or as indicators of pollutants in our environment.

I. CONSTRUCTION OF RECOMBINANT DNA AND MOLECULAR CLONING OF GENES

Discovery and application of the class II restriction endonuclease was crucial in developing the recombinant DNA technology. For the construction of recombinant DNA technology, the discovery of certain self-replicating plasmids and the understanding of their biology and that of their bacterial hosts as well as the discovery and characterization of DNA ligases were equally important. These

Table 14.1 List of certain products of medical and other values produced by recombinant DNA-based biotechnology

Gene-Product	Medical or Other Applications
Adenosine deaminase (ADA)	Treatment/gene therapy of SCID patients
α_1-Antitrypsin patients	Treatment of emphysema
Angiogenin	Possible cancer treatment
Atrial natriuvetic factor	Hypertension and heart failure
Blood clotting factor	Treatment of hemophilia
Colony stimulating factor	Leukemia treatment
DNaseI	Improve lung function in cystic fibrosis patients
Epidermal growth factor	Treatment of burn patients
Erythropoietin	Treatment of anemia patients
Hepatatis B vaccine	Prevention against hepititis B virus
Humulin	Treatment of diabetes
Human growth hormone	Treatment of certain form of genetic dwarfism
Interferons (α,γ)	Treatment of certain cancer and viral infections
Interleukin 2	Cancer treatment
Onconase (RNase)	Cancer treatment
Superoxide dismutase	Tissue transplantation
Tissue plasminogen activator	Treatment of heart attacks
Cellulase	Cellulose degradation in animal feed
Renin	Cheese production
Somatotropin (bovine growth hormone)	Increase dairy milk production

formed the important hardwares and tools for the construction and expression of recombinant plasmids carrying a DNA segment, other than its own or that of its host leading to gene cloning.

Thus the plasmid DNA became the vehicle or vector for carrying the foreign DNA segment. The plasmid DNA could be cut open by the action of the restriction endonuclease, and at the same time the DNA segment to be cloned could be generated by the same restriction enzyme such that both the vector DNA and the foreign DNA possessed the complementary sticky ends because of the staggered cuts made by digestion with the same enzyme. The vector DNA and the foreign DNA segments, thus generated by the action of a particular restriction endonuclease when mixed to gether *in vitro*, could join hands via sticky ends with complementary single-stranded hangovers with nicks in between the adjacent nucleotides in each DNA strand of the recombinant DNA molecule. These nicks could be sealed afterwards *in vitro* by the action of the enzyme called ligase. Thus it is obvious that the construction of recombinant DNA molecule which became the routine method for the cloning of a gene was made possible because of the decades of developments in the understanding of the genetics, molecular biology, and biochemistry of the bacterial host and that of plasmids and viral DNAs that were developed to act as vectors in addition to the discovery and characterization of nucleases and ligases. However, among all these factors involved in the construction of recombinant DNA as a method for the cloning of a gene, the role of the restriction nuclease proved to be very crucial in the versatility of this methodology because of the fact that a multitude of different sticky ends could be generated by the use of a large number of restriction enzymes. As a matter of fact, a number of vectors containing a series of unique restriction sites, situated together in one particular region of the vector, were designed and became commercially available. Development of such vectors with the so-called multiple cloning sites (MCS) made it possible to clone any gene in a particular vector. This particular feature (i.e., the restriction site) could itself be designed or manipulated, and the abundance of restriction enzymes could recognize and make cuts within these artificially engineered sites in a vector and make the role of nucleases of such profound significance. That is the distinction of the role of restriction endonucleases in the development of recombinant DNA technology as the methodology for gene cloning. The major steps involved in the construction of recombinant DNA is schematically presented in Figure 14.1.

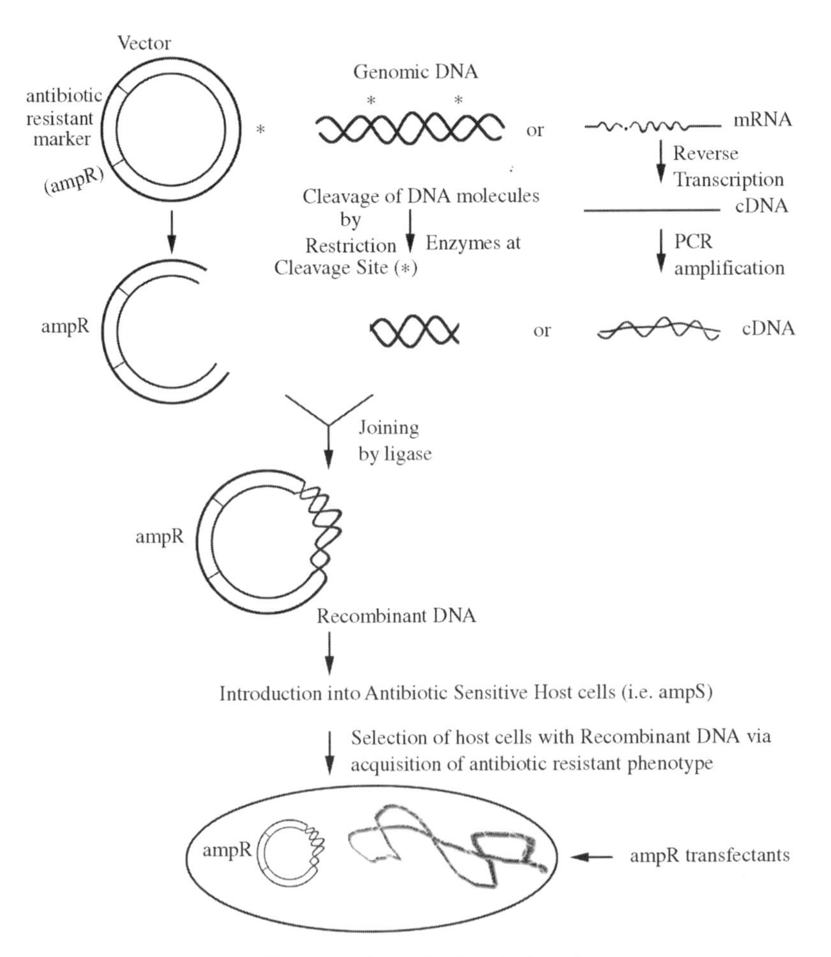

Figure 14.1. General scheme for the cloning of a gene.

II. BIOTECHNOLOGY OF MICROORGANISMS, PLANTS, ANIMALS, AND MARINE ORGANISMS BASED ON RECOMBINANT DNA TECHNOLOGY

The application of nucleases in biotechnology rests on the fact that nucleases, particularly the restriction endonucleases and at times topoisomerases, are used for the cloning of gene via recombinant DNA technology. Once a gene of choice has been cloned, then the cloned gene can be introduced into microorganisms, plants, animals, and marine organisms to confer a new function or phenotype of commercial significance. Such organisms harboring alien gene(s) are called *transgenic* (Palmiter and Brinster, 1986; Church, 1990). Our ability to clone genes has been very useful in the creation of transgenics for biotechnological purposes. However, it should be noted that some of the transgenic animals such as pigs harboring human growth hormone (HGH) gene have several health problems. A brief account of the biotechnology of organisms based on the manipulation of genes via recombinant DNA technology involving nucleases is presented in the following paragraphs.

1. Microorganisms. Microorganisms including bacteria, yeasts, and fungi have been utilized in the industrial production of certain chemicals including organic compounds, pharmaceuticals, and, most importantly, several antibiotics. The use of these microorganisms in this industry is primarily based on our knowledge of their culture conditions, physiology, and genetics. Initially, these microorganisms were subjected to genetic manipulation and selection with respect to the production of a substance. Recently, their biotechnological application has been aided rapidly by our ability of gene cloning and manipulation utilizing recombinant DNA technology.

The ability of microorganisms to convert one substance into another can be speeded up by transfer of a cloned gene of choice to produce pharmaceuticals or chemicals by modifying the existing metabolic pathways or by conferring new enzymatic abilities to microorganisms. For example, the biosynthesis of steroids can be reduced to 11 steps instead of 37 steps required for their chemical synthesis utilizing a precursor obtained from plants. Likewise, genes from different microorganisms involving different steps in bioconversion can be combined into one microorganism in order to eliminate or reduce the steps in chemical synthsis of a substance. The transfer of a gene controlling the first step in the synthesis of vitamin C from *Corynebacterium* sp. into *Erwinia herbicola* lacking the first step but possessing the second step controlling vitamin C synthesis has reduced the number of steps in the chemical synthesis of this substance.

Similarly, recombinant DNA technology has been commissioned to produce microorganisms with the ability to make new antibiotics. This approach to confer microorganisms to produce new secondary metabolites with antibiotic properties or to modify the properties of existing antibiotic is being pursued to generate new antibiotics in order to overcome our difficulity of fighting infectious diseases due to the fact that most disease-causing microorganisms are becoming resistant to currently used antibiotics.

In addition, microorganisms are being manipulated via recombinant DNA technology to produce eco-friendly microbial pesticide and to produce biodegradable plastics and fuel. A good example of biopesticide is the B1 toxin produced in *Bacillus thuringiensis*; the gene for this toxin has been mobilized to crop plants to confer insect resistance.

2. Plants. Use of recombinant DNA technology is the main tool for the production of a genetically modified crop. The "flavor saver" tomato with long self-life represents the first genetically modified crop developed by recombinant DNA technology using antisense strategy. Pectin in the tomato cell wall is readily degraded by the enzyme polygalacturonase; therefore, the expression of pectin-degrading enzyme was blocked by introducing cloned antisense gene for this enzyme. Thus in the genetically modified plant, the sense strand mRNA and the antisense mRNA for genes controlling pectin degradation block each other by complementary hydrogen bonding preventing the translation of the sense strand mRNA into pectin-degrading protein. Transgenic corn plants with bacterial RNase gene barnase are male sterile, whereas other transgenic corn containing genes for barnase and its inhibitor protein barstar are male fertile. The two kinds of transgenic corn plants are used for breeding to produce hybrid corn seeds. In addition, plants possess nucleases that lead to self incompatibility which has potential in plant genetic manipulations.

3. Animals. Like other organisms, animals with desirable characters have been created as transgenic via microinjection of cloned genes into eggs. Such transgenic animals with a much healthier life and desirable products for human consumption are expected to reduce the cost of raising cattle and their products. Herds of transgenic cows, goats, and sheep have been raised to produce a specific human protein used as drugs; such herds are known as "animal pharm." Members of the animal pharm can produce human proteins such as blood clotting factors, growth hormones, and other proteins in the milk, from which they are easily purified and marketed. In addition to the production of transgenic animals with desired characters, the recombinant DNA technology has been used to produce many other useful products; one example is the production of bovine somatotropin (rBST), which is injected into cows to increase the milk production. This practice, even though somewhat controversial, has been approved by the US Federal Drug Adminstration for human consumption without labeling the source of milk. The recombinant DNA technology has been also used in the poultry industry to produce transgenic birds with better health and better products. Recombinant DNA technology is also being used to produce animal with leaner meat and low cholesterol for human consumption.

4. Marine Organisms. Marine organisms constitute a vast gene pool that remains to be harnessed by recombinant DNA technology to provide a greater source of food, drugs, antibiotics, and toxin. Currently, transgenic fish such as rainbow trout and salmon containing the gene for growth harmone have been raised and are being tried in field experiments for sustaining their increased growth size.

Marine organism remain an inexhaustible source of genes for new drugs, antibiotics and toxin to be exploited by recombinant DNA technology for human use.

5. Pharming. As mentioned above, different organisms including microorganisms, plants, and animals have been genetically engineered to produce substances of medical significance and use; this process has been dubbed "pharming." There are two ways for such pharming: In one method the organisms producing medically valuable substance are transgenics harboring a cloned alien gene and thus have undergone permanent genetic changes. whereas in the other method of pharming the plants or animals only transiently harbor a foreign gene and produce medically important substance such as vaccine or monoclonal antibodies (McCormick et al., 1999; Stevenson, 1999). Recently, one of the biotechnology companies has taken the lead in producing proteins of choice by inserting the desired gene in tobacco mosaic virus which are then used to infect tobacco plants to produce the desired pharmaceuticals. This methodology has been used to produce vaccine against non-Hodgkin's lymphoma, a cancer of the lymph system which accounts for 4% of all deaths due to cancer (McCormick et al., 1999).

III. APPLICATION IN MEDICINE

Recombinant DNA technology is expected to be extensively used to maintain healthy humans both crop and ornamental plants, and domestic animals. The recombinant DNA technology is being used for predictive, preventive, and curative aspects of medical practices. Application of nucleases in medicine is based on several facts: Cloned genes could be used for the gene therapy as well as for the production of therapeutic drugs or for the design of new drugs. A number of nucleases involved in human disease, including different kinds of cancer, can be targeted for the treatment of disease; for example, topoisomeraes have been targeted by several inhibitors. Nucleases have been used for diagnostic purposes and for the treatment of human diseases (such as cystic fibrosis) to remove the excessive accumulation of DNA in the septum and to improve the function of lungs under clinical conditions.

A. Role of Nucleases in Predictive, Preventive and Curative Medicine

1. Predictive Medicine. The role of nucleases lies in the prediction of possible occurrence of a disease. Members of the families with possible genetic diseases may be identified by their RFLP phenotype. This methodology, first initiated by Kan and Dozy (1978), has been extensively used for diagnostic purpose leading to the prediction of individuals that are predisposed to a particular disease. Microsatellite instability or RFLP have been used to predict certain kinds of cancer (Ward et al., 2001). RFLP has also been used to clone the gene controlling a particular genetic disease: The gene causing Huntington's chorea was the first one to be cloned and identified to be located on human chromosome IV via this method of positional cloning—that is, cloning of a gene based on its association to a particular RFLP phenotype. Following the identification of Huntington's chorea, a large number of human genes have been identified by positional cloning with reference to a particular RFLP phenotype. The use of RFLP represents the direct application of nucleases not only in predictive medicine and in forensic science but in all aspects of genetics, molecular biology, bioinformatics, and biotechnology involving gene cloning and genome projects. SNP-genotying (see Chapter 13) is expected to influence pharmacogenetics leading to tailored therapy for individual patients.

2. Preventive Medicine. The preventive medicine is largely based on early diagnosis of disease via RFLP. Once it is determined that a particular individual is predisposed to a genetic disease, certain measures can be outlined by the practicing physician to delay the onset of a disease or to ameliorate the adverse effect of the disease or to undertake measures to treat the disease. However, until now not much has been accomplished in the prevention or cure of genetic diseases despite our ability to identify them based on their RFLP phenotype. Nevertheless, the lack of our ability to prevent or cure genetic disease should not be undermined by our ability to establish or eliminate the possibility of the occurrence of a particular disease in a particular individual. It seems that curative medicine will follow the predictive medicine, and perhaps the methodology for preventive medicine for a disease during the lifetime of an individaual would come later. However, diagnosis of certain genetic disease based on a particular RFLP phenotype could be utilized at the prenatal or preimplantation stage to prevent the birth of children with such genetic predisposition.

3. Curative Medicine. The role of nucleases in curative medicine is based on the availability of the wild-type allele (i.e., the normal form) of the gene(s) controlling the disease due to molecular cloning via recombinant DNA technology. The cloned gene could be given to the patient via gene therapy as discussed later.

B. Drug Designs

The role of nucleases in the design of a new drug to treat a human disease is again based on the utilization of nucleases in recombinant DNA technology leading to cloning of a gene and in determining the nucleotide sequences of a gene. Once the DNA sequence is known, the nature of the encoded protein is inferred and its three-dimensional structure can be determined, and then using combinatorial methodology a drug could be designed which can interfere with the function of that protein and thus be used to treat a patient.

C. Antisense Strategy

It is established that the translation of an mRNA could be blocked in the presence of an antisense mRNA because the mRNA and antisense mRNA possessing complementary nucleotide sequence forms double-stranded structure by hydrogen bonds such that mRNA becomes unavailable for translation. Antisense strategy constitutes such blocking of translation of mRNA by complementary antisense mRNA. The antisense mRNA has been used to prevent the expression of a gene transiently by introduction of antisense mRNA or permanently by transfection with the cloned gene encoding antisense-mRNA in the transgenic organisms. Antisense RNA not only blocks the translation of sense RNA via forming a duplex structure, but such a duplex could further be subjected to degradation by RNaseH (Agrawal and Tang, 1992; Stein and Cheng, 1993).

Our ability to clone a gene or a specific DNA sequence via recombinant DNA technology involving the participation of nucleases has been ulilized to introduce a gene into organism to produce antisense mRNA which can prevent the translation of sense mRNA transcribed from the resident gene controlling a human disease. This strategy, called *antisence strategy*, has not yet been effectively used for any known human disease, but this strategy definitely holds lots of promise because such strategy has been very effective in genetic alteration of the tomato and in increasing the self-life of the tomato by blocking the translation of pectinase protein from the sense mRNA by the antisense mRNA transcribed from the introduced gene present in the transgenic plants.

IV. NUCLEASE THERAPEUTICS AND THERAPEUTIC TARGETS

A. DNaseI and DNAzyme-Based Therapeutics

DNaseI has been used as an important clinical agent in the treatment of cystic fibrosis (Ramsey, 1996). Aerosol of DNaseI reduces the viscoelasticity of septum in the lungs and air passage of patients with cystic fibrosis by degrading high-molecular-weight and extensive fibers of DNA into smaller fragments. Such DNase treatment of the cystic fibrosis patients improves the function of lungs. Human DNaseI has been cloned and engineered to degrade the DNA in the cystic fibrosis septum more efficiently by *in vitro* mutagenesis of amino acids in the key position of this enzyme. Such genetically engineered enzymes are 10,000 times more active than the naturally occurring enzyme. In addition to improved catalytic activity of the enzyme, the human DNaseI has also been engineered to increase its resistance to inhibition by G-actin and by salts in the cystic fibrosis patients. DNaseI acts by both degrading the DNA and by depolymerizing F-actin in the septum of the patients suffering from cystic fibrosis (Pan et al., 1998). Cystic fibrosis is controlled by a recessive allele predominantly found in the white population in the United States. The genetically engineered DNaseI, also called Pulmozyme and marketed by a biotechnology company, has the potential of being effective in the treatment of patients suffering from systemic lupus erythematosus (Lachmann, 1996; Macanovic et al., 1996).

Recenly, DNA-based nucleases have been developed by *in vitro* selection; these DNAzymes (Santoro and Joyce, 1997, Sen and Geyer, 1998) possess great promise in being effective in the treatment of several diseases. However, it must be recognized that the application of DNAzyme, like that of antisense RNA or that of ribozyme and other nucleases, is faced with the same problems of delivery, stabilization, and physiological activity as are the therapeutic agents. The use of DNAzyme for several diseases such as aids, hepatitis, atherosclerosis, Cancer, and others are being developed and were discussed earlier (Breaker, 2000; Finkel, 1999; Wu et al., 1999; Yen et al., 1999; Zhang et al., 1999; Sun et al., 1999; Sioud and Leirdal, 2000) because DNAzyme can destroy the mRNAs specific for these diseases.

B. RNaseA and Ribozyme-Based Therapeutics

RNases and ribozymes are being developed as a major therapeutic agent against human diseases and will find application in maintaining the heath of domesticated animals and crops (Singwi and Joshi, 2000). RNaseA and angiogenin are being developed to treat different kinds of cancers (Olson et al., 1994, 1995; Youle et al., 1994). Onconase, a member of the RNaseA family, obtained from frog eggs, is the best-known ribonuclease used for therapeutic treatment of certain cancer in humans. Onconase is under clinical trial stage III. Onconase is a very stable protein and is resistant to proteases. These facts could become a problem during long-term therapeutic use. Attempts are being made to render onconase more suitable for better therapeutic use by eliminating two blocks at the N-terminal cyclized glutamine and the C-terminal disulfide bridge between cysteins residues at the terminal 104 and 87 positions. Such changes in the structure of onconase was found to make it less stable without affecting its biological activity and anticancer activity (Notomista et al., 2001). The therapeutic applications of RNaseA is further improved by producing variants of this enzyme which are less sensitive to the naturally occurring RNase-inhibitor protein (see Chapter 10). It has been shown that RNaseA with amino acid changes in certain key amino acid positions have less catalytic activity like onconase but more cytotoxic effects; for example, RNaseA K42R/G88R is more cytotoxic (Bretscher et al., 2000). In addition, the efficacy of RNase therapeutics can further be improved by coupling RNase to cell binding ligands in order to convert them into cell-type-specific cytotoxins. RNaseA has been linked to human transferring or to antibodies to the transferring receptor, and such hybrid proteins have found to possess cytotoxic activities (Rybak et al., 1992).

Also, certain lectins with RNase activity are being developed to treat different kinds of cancers. Lectins are part of the natural defense mechanisms in animals (Barondes, 1981). At least two sialic acid-binding lectins (SBLs) from frogs have been found to possess RNase as well as cytotoxic activities against certain tumor cell lines (Nitta et al., 1994) . These lectins with RNase activities, called leczymes, possess significant amino acid homology to the RNaseA family (Nitta et al.,1994; Ardelt et al., 1991). Lectins other than leczymes are distinct in lacking nuclease activity.

Ribozymes, both naturally occurring and artificially synthesized, have the potentiality to treat a number of viral infections including HIV infection causing AIDS in humans. Ribozymes such as hammerhead, group I introns, and RNaseP are being engineered for this purpose. Ribozymes have been found to degrade HIV in tissue cultures. One of the important aspects of ribozyme strategy is the fact that they could cleave the regulatory sequences commonly present in different viruses such that there is very little room for the virus to change one of its proteins and become resistant to our immune system. The fact that ribozymes occur naturally and that they have a folded structure and are therefore physiologically more stable offer advantages over the use of antisense to modify the action of a gene under clinical situations. However, a combination of two—that is, a ribozyme in combination with an antisense sequence—is also being developed; such an approach will have the best of both strategies—that is, ribozyme and antisense strand. Ribozymes can also be used to keep domisticated plants and animal healthy.

C. Gene Therapy and Enhancement Therapy

The fact that genes could be cloned and introduced into an organism and expressed under physiological conditions has presented the opportunity to use cloned genes with or without desired modifications as therapeutic agents. This approach in which a patient suffering from a genetic disease is given a gene that is missing or defective in the patients is called *gene therapy* (Anderson, 1995). A number of human diseases such as cystic fibrosis are being targeted to gene therapy with some success. Gene targeting is extended not only to provide for the defective or missing alleles in the patients but also to nullify the effect of other genes; for example, gene therapy for cancer in humans is being developed following this approach. A large number of human diseases are subject to experimentation via the gene therapy approach.

Gene therapy to alleviate a clinical situation in the lifetime of a human, termed *somatic gene therapy*, would eventually be extended to affect the next generation via what is termed *germ-line gene therapy*. Once the germ-line gene therapy becomes accepted practice, it would be used to add a number of additional genes to provide immunity and many other desirable characteristics to humans, and all these forms of genetic engneering in humans will be perpetuated via what may be called *enhancement gene therapy*. Eventually, our ability to create artificial human chromosomes may be used to introduce an extra pair of human chromosomes containing several desirable genes into humans at the time of *in vitro* fertilization before implantation into the uterus of the prospective mother.

D. Gene Silencing

Some proactive or overactive genes causing gene silencing can be controlled by introduction of new genes that can target the silencing of that gene by degradation of a message as is shown to happen naturally in fungi, plants, and animals with the use of nuclease called *dicer*. Gene silencing could be used to maintain the better health of organisms and their biotechnological manipulation by controlling the expression of certain genes.

E. RnaseL and Interferon-Mediated Control of Viral Infection and Treatment of Cancer

Likewise, the RNaseL system could be used to control the development of cancer in humans. RNaseL could also be exploited to control the diseases that are caused by improper folding of certain proteins as in cystic fibrosis, emphysema, and osteoporosis imperfecta. The role of RNase and interferon is presented in Figure 11.16.

F. Recombinase-Mediated Control of Gene Expression

A certain site-specific recombinase, the kind involved in phase variation, could be used to activate or inactivate the gene that could be temporally controlled. This has potential application in gene therapy of human and domesticated animals to activate or inactivate a gene in response to external stimulus and for the temporal control of a particular gene.

G. Poisioning of Topoisomerase–DNA Intermediates

Inhibitors of human topoisomerase I and topoisomersae II have been used to treat cancer patients based on the fact that these inhibitors can kill the rapidly proliferating cells but not the nondividing normal cells in the cancer patients.

V. APPLICATION IN FORENSICS

The application of restriction enzymes in forensics is based on its ability to generate a distinct profile of DNA fragments in an individual. This difference in DNA profile is based on two facts: First, the DNA sequence site where a particular restriction enzyme produces a cleavage may vary just in one nucleotide due to mutation; thus a restriction site may be lost or acquired due to one nuleotide change in the recognition and cleavage site for a restriction enzyme. Second, a difference in DNA profile can be generated by a difference in the number of VNTR due to differences in the satellite or interspersed repeat sequences of DNA in a chromosome. The fact that only 5–7% of human DNA represents the coding sequences provides an opportunity for a variable number of these repeats in the DNA, making a distinct DNA profile for each individual. However, the DNA profile could be used in forensics only to eliminate the paternity or noninvolvement in criminal activity of an individual. It can never be used to establish the paternity or involvement in the criminality of an individual with 100% certaintity. Nevertheless, DNA profiling, which is a cornerstone of modern forensics, could not have been possible without the application of restriction nucleases (Jeffreys, et al., 1985).

VI. APPLICATION IN INDUSTRY: PRODUCTION OF FLAVOR ENHANCER OF FOOD AND BEVERAGE

In addition to the large-scale production of proteins or other substances for medical purposes via recombinant DNA technology, a number of food additives and/or flavor enhancers are industrially produced. Industrial production of mononucleotides such as 5′-GMP and 5′-IMP is achieved by P1 nuclease. As described in Chapter 8, certain fruit plants use nucleases to produce 5′-GMP to add sweetness and flavor to fruits. Also, staphylococcal nuclease is utilized to remove the nucleic acid during the preparation of a single cell protein from yeast used as feed for domesticated animals. In the industrial settings these nucleases are immobilized to a solid matrix for continuous participation of enzymatic activity by these nucleases. The production of immobilized nucleases is achieved by coupling of nucleases to cyanogens bromide (CNBr)-activated cellulose or other matrix (Rokugawa et al.,1979). The industrial production of certain flavor enhancers also possesses anticancer activity (Schaeffer, 1983).

VII. APPLICATION IN ENVIRONMENTAL PROBLEMS

A. Bioremediation

Most industrial productions leading to hazardous waste removal by degradation by metabolic activity of transgenic organisms are now possible. Plants that can remove lethal metals or plastic or many other xenobiotic chemicals have been created and are being utilized to clean our environment (Dietz and Schnoor, 2001). Some of these transgenic microorganisms are being utilized for getting rid of oil spill or for biomining. The field of bioremediation or phytoremediation is just emerging and have great potential in the cleanup of our environment involving the application of nucleases. A DNAzyme that requires Pb^{2+} ions as a cofactor for its catalytic activity has been used to detect the presence of lead, a highly toxic metal in our environment (Geyer and Sen, 1998; Li and Lu, 2000).

B. Detection of Microbial Pathogens to Prevent Bioterrorism by 5′ Nuclease in PCR Assay

Nuclease can be used to detect the presence of a specific bacterial pathogen that may occur in nature under certain adverse environmental conditions or by an act of bioterrorism. The method utilizes the 5′–3′-exonuclease activity of the thermostable DNA polymerase, Taq, during amplification of a pathogen-specific gene or DNA segment in a PCR assay. In such a PCR assay an oligonucleotide probe (called Taq-Man) containing a reporter fluorescent dye FAM at the 5′end and a quencher, TAMRA, at the 3′ blocked end is used in a PCR assay as shown in Figure 13.2 . Such a Taq-Man probe does not show fluorescence because the reporter dye FAM remained quenched by the quencher, TAMRA, because these two components of the Taq-Man probe are situated in close proximity to each other. However, when the reporter dye is released from the probe by the activity of a nuclease, it causes an increase in the fluorescence which can be measured readily. The Taq-Man probe contains nucleotides complementary to the template DNA segment upstream of the DNA strand being synthesized by the Taq DNA polymerase during the process of amplification in a PCR assay. In such an assay the Taq DNA polymerase extending

Table 14.2 List of microorganisms with potential use for bioterrorism[a]

Pathogen	Nature of Diseases Caused
Bacillus anthracis	Anthrax
Brucella	Brucellosis, undulant fever
Francisella tularensis	Tularemia
Yersinia pestis	Plague
Giardia[b]	Human diarrhea
Cryptosporidium[c]	Gasteroenteritis
Salmonella	Food poisoning
Listeria monocytogenes[d]	Listeriosis/meningitis, septicemia; abortion and still birth in expecting human females
Escherichia coli O157:H7	Diarrhea
Escherichia spp.	Kidney infection/meningitis

[a] Most of these microorganisms are also in the list of DoD and CDC bioterrorism preparedness program.
[b] A water pathogen that infects 250 million individuals worldwide.
[c] Found in Milwaukee, Wisconsin municipal water supply system in 1993, causing more than 100 deaths.
[d] Commonly found in dairy and meat products.

the newly synthesized DNA strand encounters the fluorescent Taq-Man probe and cleaves the nucleotide containing the reporter dye by the associated 5′-3′-exonuclease activity of the Taq DNA polymerase. The released reporter dye causes an increase in the fluorescence of the assay mixture which can be measured accurately and rapidly for a large number of samples indicating the presence of the pathogen-specific gene. Such detection of bacterial pathogens as described for *Yersinia pestis* causing plague (Higgins et al., 1998) is being developed for a large number of pathogens that could be used for an act of bioterrorism (Burke, 2001). An abbreviated list of microorganisms that could cause such an epidemic due to environmental problems and due to bioterrorism is presented in Table 14.2.

Such detection of microorganisms involved in bioterrorism and epidemic by the application of 5′-nuclease activity of Taq DNA polymerase is very rapid. However, it requires the knowledge of a marker gene or DNA segment that is specific for the organism to be detected. This method developed for *Yersinia pestis* causing plague utilizes the plasminogen activator gene [pla of *Yersinia pestis* (Higgins et al., 1998)].

It is not difficult to find such a marker gene for various pathogens, and the genome sequence of many of these pathogens is being determined. Because most organisms possess a number of genes or DNA segments specific for that organism as discussed in Chapter 13, it is plausible to develop such a 5′-nuclease assay for most organisms readily.

15

NUCLEASES AND EVOLUTION

I. RIBOZYME AS EVIDENCE FOR THE EARLY WORLD OF RNA

The molecule of life most likely originated in the form of RNA in the primordial soup or other appropriate medium almost three and one half billion years ago (Crick, 1968; Orgel, 1968). The world of RNA evolved into RNP (RNA and protein) and then to DNA (Darnell and Dolittle, 1986; Gilbert, 1987a,b). Several facts support this possible course of molecular evolution. Some of these facts include the following: (a) The present-day DNA synthesis proceeds with RNA as primers; (b) telomere synthesis requires involvement of RNA; (c) replication of several DNA viruses involves an RNA intermediate; (d) transposition of several DNA segments including LINES and SINES occurs via reverse transcription of an RNA intermediate; (e) deoxynucleotides are also synthesized as derivatives of ribonucleotides; (f) most coenzymes have a ring structure and are derived from nucleotides; (g) the biosynthesis of histidine, the amino acid that mostly exists at the active center of protein enzymes, also involves nucleotide precursors (White, 1976; Kornberg and Baker, 1992; Lamond and Gibson, 1990); (h) the processes of RNA editing, (i) peptidyl transferase activity of ribosomal RNA during translation (Noller et al., 1992); and (j) the involvement of telomerase, a ribozyme in the synthesis of eukaryotic telomere (Weiner, 1988; Blackburn, 1992). These facts provide good suggestive evidence for the idea that life originated in the form of RNA. However, the suggestion that the world of RNA existed before the world of proteins and DNA lacks critical evidence in the absence of documentation of molecules (other than proteins) with ability to catalyze the synthesis of macromolecules. Therefore, the discovery of ribozyme (i.e., RNA enzyme) provides the most direct evidence for the existence of the world of RNA before the evolution of DNA as the chemical basis of hereditary material.

It is plausible that the world of RNA would have evolved in the primordial soup as a result of ongoing chemical reactions first assisted by chemical catalysts or chemzymes and later by the catalytic activity of the RNA molecules (i.e., ribozyme) themselves. The ribozymes were helpful because of their multiple catalytic functions. They acted as polymerase as well as ribonucleases and ligases. The latter two functions might have evolved as recombinases with ability for site-specific cleavage and rejoining of RNA molecules leading to the formation of an array of RNA molecules.

Some of these in conjunction with certain proteins developed into ribosomes and led to the development of the translation system. Later some of the RNAs were presumably recruited to function as RNA genomes with ribozyme as intron. These RNA genomes were reverse transcribed to generate their DNA copies, leading to the arrival of the world of DNA (Wintersberger and Wintersberger, 1987). DNA molecules were chosen over RNA molecules as genomic material because of DNA's greater chemical stability. The DNA genome contained the introns and genes encoding diverse proteins evolved by the shuffling of exons via recombination facilitated by introns (Gilbert, 1987b). Once the DNA genome was stabilized, the cell and organisms were developed and evolved into present-day prokaryotes and eukaryotes. The evolution of eukaryotes also involved the incorporation of mitochondria and chloroplasts. This view of evolution (Darnell and Doolittle, 1986; Gilbert, 1987a) suggests that the eukaryotes preserved the introns while the prokaryotes lost them during their evolution. Prokaryotes lost their introns because of their rapid growth lifestyle (Darnell and Doolittle, 1986) and the fact that prokaryotic chromosome was not organized into a nucleus; the lack of a nuclear membrane was a hindrance in providing opportunity for the processing of transcript into mRNA and its subsequent translation into right proteins (Nowak, 1994). The role and sequence of events that might have occurred during the evolution of the present-day living world from the prebiotic soup via a world of RNA are depicted by Darnell and Doolittle (1986) and presented in Figure 15.1.

II. CHEMZYME, RIBOZYME, AND PROTEINZYME

The study of nucleases clearly indicates the evolution of enzymes or biocatalysts and their roles during the process of evolution. In the prebiotic soup or in other appropriate media, the site-specific cleavage and joining of RNA molecules must have been facilitated by certain chemicals acting as catalysts (Orgel, 1986). Later, certain nucleotide derivatives might have first acted as chemzymes and later been relegated to the function of coenzymes of modern enzymes. Once a substantial number of RNAs with enzymatic function became prevalent in the prebiotic soup, these ribozymes facilitated the evolution of RNA genome by their role as recombinase which generated diverse kinds of RNA molecules by facilitating their recombination; later, their job was taken over by the protein enzymes with the evolution of a translation system (Lamond and Gibson, 1990).

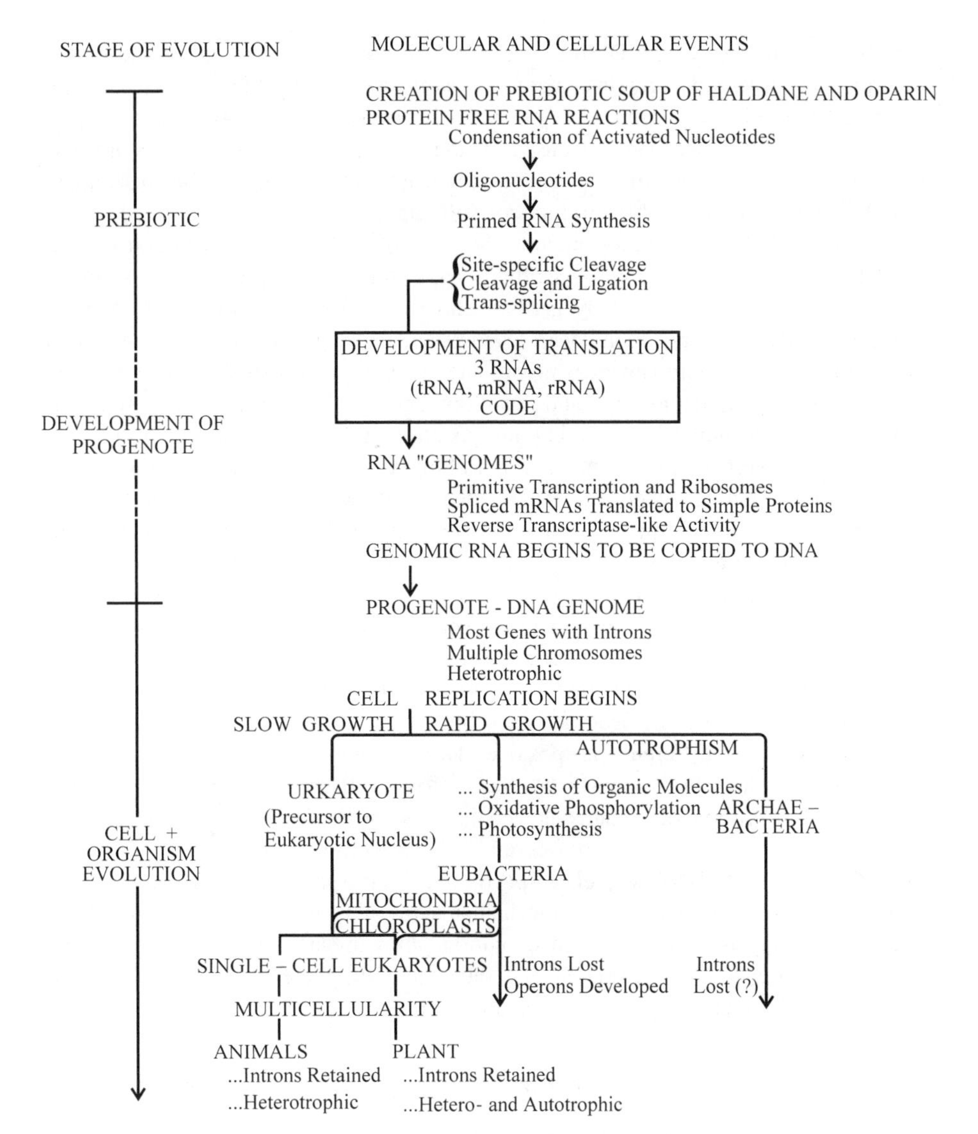

Figure 15.1. Some possible events in evolution. [From Darnell and Doolittle (1986), used with permission.]

III. THE ROLE OF RECOMBINASE IN EVOLUTION

Recombinases must have been an important instrument in the generation of a diverse group of first RNA genomes in the world of RNA and later in the evolution of DNA genes. It has been suggested that DNA genes encoding diverse proteins evolved via the shuffling of exons. It has been estimated that only about 1000–7000 originally occurring exons were shuffled to create the diverse genes encoding all proteins in eukaryotes (Dorit et al., 1990); this view is based on the fact that a number of divergent proteins share the same exons (Südhof et al., 1985a,b; Germaine, 1985). Recombination mediated by introns might have played a part in exon shuffling. This view is based on the fact that even some of the present-day introns act as ribozymes or encode nucleases and maturases that participate in DNA and RNA recombination and/or splicing.

Modern ideas on evolution include the view of a punctuated evolution (Gould and Eldridge, 1977). According to this idea, evolution has progressed by jumps (called *macroevolution*) and by gradual changes (called *microevolution*) leading to major differences between species. However, many groups of organisms (such as humans and chimpanzees) possessing similar genes differ anatomically, physiologically, behaviorally, and developmentally. Wilson has proposed that mutations in regulatory genes, but not in structural genes, are responsible for such vast differences in the humans and chimpanzees (King and Wilson, 1975). In my view, regulatory changes may not be sufficient to account for the vast differences between chimpanzees and humans. Such major changes can be brought about by extensive genome reorganization catalyzed by recombinases in response to environmental factors including SOS response. A comparison of the human and chimpanzee chromosomes does indeed reveal the occurrence of chromosomal re-arrangements leading to fusion of two chimpanzee chromosomes and several in-versions to yield the equivalent of human chromosomes. Such chromosomal reorganization must have been achieved by the action of different recombinases. It is plausible that different transposable elements might have mediated such vast reorganization of chromosomes by the action of their tranposases during the process of evolution of organisms via their versatile nature in such a process. The versatile nature of transposases and other recombinases is indicated by the fact that they have DNA sequence specificity, tissue specificity, and the ability to control their actions by diverse mechanisms (as described in Chapter 7). They must have played a great role in evolution, allowing a number of genes to be expressed, silenced, and expelled and thus facilitating the control of immunity, neuron development, and other major differentiation.

A. Present-Day Selfish DNA-Possible Origin From Transposon

Eukaryotic genomes contain certain DNA segments that have not yet been assigned to control any obvious function. These DNA segments have been erroneously called selfish DNA or junk DNA, implying that they exist only to be perpetuated as molecular parasites or as DNA segments without any function (Doolittle and Spienza, 1980; Cavalier-Smith, 1985; Dawkins, 1989). However, it is becoming obvious that such DNA segments are neither selfish nor junk (Nowak, 1994). These DNA segments are now shown to be involved in several regulatory functions (Nowak, 1994). In addition, in my view at least some of these DNA seg-ments are the residual transposable elements that have lost their mobility due to mutation in transposases genes. For example, the Ds element of corn may appear as selfish DNA, but in fact the Ds element is a deleted form of the full-fledged transposable element called Ac. Thus it is pertinent to think that at least a good part of so-called selfish or junk DNAs might have been, at one time, functional transposons that propelled the evolution by extensive gene rearrangements. It has been shown that in humans, certain repeat sequences called LINES are indeed transposons with intact reverse transcriptase activities responsible for their mobility (Dombroski et al., 1991; Mathias et al., 1991). Likewise, introns might be considered to be residual transposons that facilitated the movement of its own gene or other genes. This is supported by the fact that even some of the present-day introns indeed encode nucleases involved in mobility of genes. The role of present-day selfish genes as ancient transposable elements with transposase genes fulfills the need for the reorganization of chromosomes leading to origin of species. It is possible that mutations in these transposable elements leading to the loss of transposase gene would have fixed the structure of the reorganized genome in humans.

Furthermore, the gene rearrangement might have been facilitated by the nuclease associated with the virus and with the reverse transcriptase of several viruses and retroelements. Certain viral recombinases capable of integration and expulsion of DNA sequences such as human immunodeficiency virus (HIV) integrase might have played significant roles in sequence reorganization during evolution. Herpes simplex virus (HSV) carries nuclease(s) that can cause breakage of the human chromosomes upon infection. It is further seen that HSV infection does not lead to chromosomal damage in the wild-type Chinese hamster ovary (CHO) cells, but causes extensive chromosomal damage in CHO mutant cells defective in DNA replication and/or repair (Nauchtigal and Mishra, unpublished work). Ribozyme with recombinase activity present in the world of RNA was passed on as introns into the world of DNA. These introns participated in the genome reorganization either directly by encoding a nuclease or indirectly as retroelements that encoded for reverse transcriptase and integrase. Thus these introns acted as transposons and bring about genome reorganization, leading to a burst of speciation perhaps in response to certain environmental stimuli. Later, these transposons became inactive by point mutation or deletion and fixed the level of rapid speciation at a certain point in the process of evolution which then progressed gradually by point mutation. These inactive transposons or residual introns with lost recombinase activities represent, at least in part, the so-called selfish or junk DNAs. It has been estimated that among 100,000 or more LINES sequences in human, only a few of them are left active. It is possible that recombinases played a dominant role in short bursts of speciation by chromosomal rearrangements during the long periods of gradual evolution by gene mutation causing macroevolution and microevolution.

IV. NUCLEASES AND CONTROL OF DNA TRANSACTIONS AND THEIR ROLES IN EVOLUTION

Different DNA transactions such as replication, repair, and recombination are interrelated and are mediated by a number of common proteins, including nucleases. It is known that transcription is involved in DNA replication, recombination, and repair (Hanawalt and Mellon, 1993; Wu and Lichten, 1994). Nucleases might have also played a role in evolution by controlling the accuracy of the DNA transactions. The proofreading function of DNA polymerases might have contributed to the fidelity of DNA synthesis during DNA replication, repair, and recombination while maintaining the DNA structure and function with significant constancy over a period of generations of cell growth. However, occasional failures on the part of these editing exonucleases or those nucleases involved in DNA repair might have generated a large number of mutants on which evolution might have worked. The proofreading activity of the RNAse domain present in RNA polymerase might also have contributed in a similar manner, first by maintaining the precision of information transfer during transcription of DNA to RNA and then by generating different kinds of transcript which might have provided different translation products with different evolutionary consequences.

V. ROLE OF NUCLEASES IN DIRECTED MUTAGENESIS: ADAPTIVE MUTATION AN SOS RESPONSE

Almost all mutations are assumed to arise randomly. Their selection is based on their adaptability to prevailing environment during the process of evolution. Such a mechanism of mutation may not provide selectable mutation at a speed needed by the changing environment and will result in a slower rate of evolution. In view of this fact, recent evidence for directed mutation or adaptive mutation (also called Cairnsian mutation by Hall, 1990) arising in response to environmental condition has aroused much interest. Such mutation occurs more frequently under conditions where it has a selective value. For example, lac^- mutants can produce more lac^+ revertants on a medium containing lactose than on a medium lacking lactose. Thus such mutations are nonrandom in nature. A number of such mutations have been now demonstrated to occur in both prokaryotes and eukaryotes (Craine et al., 1980; Hall, 1988, 1990, 1991). The common features of these mutations are that they appear under conditions of starvation. A number of mechanisms have been proposed to explain the basis for the occurrence of these mutations. These include the following: (a) Cells may enter into a hypermutable state (Hall, 1990) that leads to adaptive mutations, (b) cells may undergo a slow mismatch repair of the gene in question and lead to directed mutations (Stahl, 1988), or (c) cells may reverse transcribe and lead to incorporation of new genetic information in the gene and thus result in directed mutagenesis (Davis, 1989). At present, there is no support for any of these views regarding the mechanism of the occurrence of adaptive mutation. However, it is possible that adaptive mutation is an SOS response in which nucleases play an important role. This is based on the following facts:

1. All adaptive mutations occur under conditions of stress (Cairns et al., 1988; Hall, 1988; Symonds, 1989; Shapiro and Higgins, 1989; Mittler and Lenski, 1990; Zambrano et al., 1993) so it is plausible that they are SOS responses.

2. A number of adaptive mutations require precise excision of a particular transposon present within a gene. This can only be facilitated by the participation of the nuclease called transposase. The frequency of transposition is usually increased under conditions of stress (Hall, 1988; Shapiro and Higgins, 1989; Mittler and Lenski, 1990), and that transposition is induced by environmental factors (Symonds, 1989; Morishima et al., 1990; Ratner et al., 1991.
3. It is known that a damage in an actively transcribing gene is repaired preferentially over the damage in a nontranscribing gene (Hanawalt and Mellon, 1993). Nucleases are involved in DNA repair and known to interact with proteins that are transcription factors and act as transcription DNA repair coupling factor (TRCF) as discussed in Chapter 10.

In view of these facts, it is plausible that in lac$^-$ mutants plated on a medium containing lactose, accumulation of lac$^-$ transcript (or its nonutilizable translation product) may be identified by the SOS system and may commission nucleases and other enzymes of DNA repair which may then cause the reversion of Lac$^-$ to lac$^+$ which will readily grow on the selective medium. Thus SOS response may allow the cells to enter what has earlier been called a hypermutable stage (Hall, 1990) and lead to the occurrence of adaptive mutation in a nonrandom fashion via a DNA repair mechanism involving the participation of nucleases. This model for the occurrence of adaptive mutation can be tested in *E. coli* or other organism containing mutation in genes involved in SOS response and/or in DNA repair nucleases (or transposase) as well as in proteins interacting with these systems—that is, genes of SOS response or of DNA repair nuclease (Witkin, 1976; Radman, 1980; Walker, 1985; Kenyon and Walker, 1980; Shinagava et al., 1988; Little and Mount, 1982; Lu et al., 1986; Bonner et al., 1990; Ratner et al., 1992; Roca and Cox, 1990; Naas et al., 1994). The mechanism for the occurrence of adaptive mutation as proposed here is consistent with the fact that nucleases (including topoisomerases, recombinase, and transposases) and a number of interacting proteins and the proteins of SOS response have been known to act in specific manner in response to environmental factors. Such occurrence of adaptive mutation can explain the occurrence of punctuated evolution more readily than the occurrence of random mutation.

VI. NUCLEASES AS MULTIFUNCTIONAL MOLECULES

The world of nuclease seems to be unique in at least two ways; First, most nucleases may possess many functions in addition to nuclease function such as kinase, helicase, topoisomerase, or methylase; second, there is such an abundance of nucleases. Molecules with several functions might have played a role during the process of evolution by participation of the same molecule during diverse metabolic pathways. Several ribozymes, DNAzymes, and protein nucleases have multiple functions. Ribozymes have been shown to possess several functions besides the cleavage of the phosphodiester linkages. Likewise, protein nucleases possess functions other than cleavage of phosphodiester linkages. Certain RNases have been known to possess diverse properties in addition to their nucleolytic functions. One RNase has been found to possess angiogenic properties. As a matter of fact, they were first discovered as molecules with angiogenic properties and only later established as possessing RNA's functions. Likewise, a number of glycoside proteins with a self-incompatibility role in plants have been shown to possess RNase activity both *in vivo* and *in vitro* (Haring et al., 1990). Such RNases controlling the self-incompatibility in plants could play significant roles in the evolution of plants. It is difficult to ascertain whether these different functions in RNases coevolved or were later acquired by mutation. It is known that an intron encoding a site-specific nuclease ("homing enzyme") in yeast mitochondria could acquire "maturase" function by mutation. Likewise, it is now established that SnRNA4 and SnRNA6 can undergo mutation to acquire the properties of a hammerhead ribozyme (Yang et al., 1994). A number of transposases act as repressors of transcription (Reed et al., 1982). Likewise, introns control expression of genes in addition to their role as recombinase or as ribozyme or as gene within gene encoding an endonuclease (Reid et al., 1990). Intron-encoded endonucleases can mediate nonhomologous recombination (Kell and Belcour, 1989). Several AP endonucleases are known to possess functions other than nucleolytic activity; one such activity has been identified as a redox function that reduces the cysteine residues in the DNA-binding domains of several oncogenes (such as fos and jun) that act as transcription factors. It is thus obvious that AP endonuclease can regulate the transcription activity *in vivo* (Babiychuk et al., 1994). Furthermore, the rag-2 gene, which controls the recombinase involved in immunoglobulin gene(s) assembly or a gene involved in *Neurospora* RNA processing, is related to cell cycle events in these organisms (Lin and Desiderio, 1994; Turcq et al., 1994).

VII. DNA SEQUENCE ANALYSIS, CRYSTAL STRUCTURE, AND BIOINFORMATICS

The DNA sequence of a large number of nucleases has become known because of the cloning of these genes followed by the determination of their nucleotide sequences. Additionally, the availability of the entire genomic DNA sequences of over 30 organisms, emerging as a result of the completion of the different genome projects involving bacteria to human, has made it possible to examine the sequence homology or lack of it in a number of nuleases via the application of the methods of bioinformatics. A comparision of the sequences of a number of nuclease genes show general to limited homology of DNA sequences among them. However, a number of the nucleases, catalyzing the same nucleolytic reaction, do not possess nucleotide homology in their structure. However, they show conservation of certain nucleotide (and hence amino acid) sequences at key sites that endow these nucleases either (a) the property of a site-specific activity such as the restriction enzymes or (b) homing enzymes or certain recombinases or the structure selectivity of their substrates such as many DNA repair nucleases; the classical examples of this group of nucleases are FEN-1 enzymes. The basis for the selectivity of the structure-specific nucleases are revealed by the crystal structure of these enzymes (Suck, 1998). The knowledge of the nucleotide sequences and the three-dimensional structure of certain nucleases have been crucial in understanding (a) the evolution of these enzymes, (b) the basis for their multienzyme activities, and (c) engineering of these nucleases for therapeutic and biotechnological applications. For example, the study of the structure-specific nucleases have shown that how the nuclease activity was associated with DNA polymerse activity in the same protein in *E. coli*, but later these functions were transferred to separate proteins in later bacteria and in eukaryotes. Also, the nucleases with more than one enzymatic activities do possess different motifs in their N- and C-terminal segments corresponding to different activities. Such structural organization of nucleases provide strong support for the evolution of genes and proteins by exon reshuffling.

VIII. POSSIBLE HORIZONTAL TRANSMISSION OF NUCLEASE GENE AND INTRON

Certain introns encode site-specific endonucleases required for the transfer of intron from an intron plus organism to an intron minus organism during a genetic cross involving intron plus and intron minus organisms (Jacquiver and Dijon, 1985). There is a great similarity in the homing site (integration site in the exons out flanking the introns) and in the intron sequences and in the protein encoded by the ORF of these introns. The data suggest the horizontal transmission of these mobile introns in fungi and bacteriophages (Syvanen, 1986; Michel and Dujon, 1986; Koll et al., 1987). Further support for horizontal transmission of recombinase genes in phages and fungi comes from the fact that T4 endonuclease VII and yeast X-solvase-3 involved in resolution of Holliday intermediate during the process of general recombination are immunologically related proteins. These data explain the role of introns and intron-coded proteins in recombination and the significance of other recombination endonucleases in the process of evolution.

IX. CONCLUSIONS AND OUR FUTURE IN THE WORLD OF NUCLEASES

The fact that nucleases participate in different biological functions that are tissue-specific and time-specific and act in response to environmental factors suggests that a better understanding of their nature and role can provide a greater insight into the workings of several biological processes and that of the process of evolution itself. Further study and use of nucleases will lead to a better understanding of several human diseases and their alleviation. In addition, nucleases will find their application in the improvement of agricultural crops and domestic animals via their use in the methods of recombinant DNA technology. It is plausible that some of the future nucleases will be computer-designed with specific properties for specific purposes (DiLisi, 1988).

It is known that gene duplication (Ohno, 1970), shuffling of exons (Gilbert, 1987), and intron capture (Golding et al., 1994) are the major processes in evolution of genes and that nucleases have played important roles in all these processes.

Moreover, the structure of several nucleases support the theory of exon-shuffling by the fact many of them possess more than one enzyme activity—for example, nuclease and ligase or nuclease and helicase activity.

Events leading to major speciation (i.e., macroevolution) seem to result from the SOS response of the organism to drastically changing environments. In view of this possibility it is of considerable interest to recognize the fact that most recombinases and many other nucleases involved in DNA transaction are sensitive to environmental stimuli. It is important to note that a number of DNA repair enzymes are under SOS control. A yeast recombinase (SceI) has a heat shock protein (HSP 70) as the major component. The P-element transposases of *Drosophila* are germ-cell-specific. The recombinases involved in assembly of mammalian antibody genes act in a very programmed manner. The excision of other transposable elements is also controlled by environmental factors such as temperature (Carpenter et al., 1987). All these features of recombinases suggest their greater role in the reorganization of genomes, leading to major evolution of species such as human in response to major SOS factors in a drastically changing environment.

One of the most important questions pertaining the world of nucleases is why so many of them. Perhaps the simplest answer is that nucleases play a role in all aspects of nucleic acid metabolism and encounter nucleic acids as substrates in different situations that require action or cleavage of the same phosphodiester linkages in an array of metabolic events and physiological conditions. The final answer to this question will also provide an insight into the question of the abundace of nucleotide sequences present in the genome of higher organisms. The second most important thing is the development of nuclease-based technology or utilization of nucleases in medicine and biotechnology; for example, a method could be devised to disable the parasites like trypanosomes to elaborate its 5′-NT/NU nuclease, which is the mechanism by which these parasites acquire purine from the host cells to establish themselves. This approach would be much more feasible than to interfere with the immune system of these parasites, which have a repetitor of antigens. Such an approach by developing an inhibitor of 5′-NT/NU can prevent these parasites from parasitizing human hosts and help in controlling the devastating diseases of blindness and kalazar. The opportunities in the world of nucleases seem to be limitless, providing the engineering of nucleases for use in angioplasty to xeroderma pigmentosum!

REFERENCES

Aaij, C., and Borst, P. 1972. The gel electrophoresis of DNA. *Biochim. Biophys. Acta* **269**: 192.

Abelson, J. 1979. RNA processing and the intervening sequence problem. *Annu. Rev. Biochem.* **48**: 1035.

Abremski, K., and Hoess, R. 1984. Bacteriophage P1 site-specific recombination. *J. Biol. Chem.* **259**: 1509.

Agrawal S., and Tang, J. Y. 1992. GEM 91—an antisense oligonucleotide phosphorothioate as a therapeutic agent for AIDS. *Antisense Res. Dev.* **2**: 261.

Ahmed, A., Holloman, W. K., and Holliday, R. 1975. Nuclease that preferentially inactivates DNA containing mismatched bases. *Nature* **258**: 54.

Ailliton, B. J. 1985. Gene therapy: Research in public. *Science* **227**: 493.

Akaboshi, E., Guerrier-Takada, C., and Altman, S. 1980. Veal heart ribonuclease-P has an essential RNA component. *Biochem. Biophys. Res. Commun.* **96**: 861.

Alani, E., Padmore, R., and Kleckner, N. 1990. Analysis of wild-type and *rad* 50 mutants of yeast suggests an intimate relationship between meiotic chromosome synapsis and recombination. *Cell* **61**: 419.

Alleva, J. L., and Doetsch, P. W. 2000. The nature of the 5′-terminus is a major determinant for DNA processing by *Schizosaccharomyces pombe* Rad2p, a FEN-1 family nuclease. *Nucleic Acid Res.* **28**: 2893–2901.

Almasan, A., and Mishra, N. C. 1988. Molecular characterization of the mitochondrial DNA of a new stopper mutant ER-3 of *Neurospora crassa. Genetics* **120**: 935.

Almasan, A., and Mishra, N. C. 1991. Recombination by sequences repeats with formation of suppressive or residual mitochondrial DNA in *Neurospora. Proc. Natl. Acad. Sci. U.S.A.* **88**: 7684.

Altuvia, S., Lockes-Giladi, H., Kobry, S., Ben-Nun, D., and Oppenhem, A. B. 1987. RNase III stimulates the translation of the CIII gene of bacteriophage λ. *Proc. Natl. Acad. Sci. U.S.A.* **84**: 6511.

Anderson, W. F. 1992. Human gene therapy. *Science* **256**: 808.

Anderson, W. F. 1995. Gene therapy. *Sci. Am.* **273**: 124–128.

Ando, S. 1966. A nuclease specific for heat-denatured DNA isolated from a product of *Aspergillus oryzae. Biochim. Biophys. Acta* **114**: 158.

Andoh, T., Ishii, K., Suzuki, Y., Ikegami, Y., Kusunoki, V., Takemoto, Y., and Okado, K. 1987. Characterization of a mammalian mutant with a campothecin-resistant DNA topoisomerase I. *Proc. Natl. Acad. Sci. U.S.A.* **84**: 5565.

Andrews, B. J., Proteau, G. A., Beatty, L. G., and Sadowski, P. D. 1985. The FLP recombinase of the 2μ circle DNA of yeast. Interaction with its target sequences. *Cell* **40**: 795.

Andrews, B. J., Beatty, L. G., and Sadowski, P. D. 1987. Isolation of intermediates in the binding of the FLP recombinase of the yeast plasmid 2-micron circle to its target sequence. *J. Mol. Biol.* **193**: 345.

Anfinsen, C. B. 1963. *The Molecular Basis of Evolution.* John Wiley & Sons, New York.

Anfinsen, C. B. 1964. On the possibility of predicting tertiary structure from primary sequence. In Sela, M. (ed.), *New Perspectives in Biology.* American Elsevier, New York.

Anfinsen, C. B. 1973. Principles that govern the folding of protein chains. *Science* **181**: 223.

Anfinsen, C. B., Cuatrecases, P., and Taniuchi, H. 1971. Staphylococcal nuclease, chemical properties and catalysis. In *The Enzymes.* Vol. 4. Academic Press, New York.

Aoyagi, S., Sugiyama, M., and Fukuda, H. 1998. BEN1 and ZEN1 cDNAs encoding S1-type DNases that are associated with programmed cell death in plants. *FEBS Lett.* **429**: 134.

Apirion, D., and Lassar, A. B. 1978. A conditional lethal mutant of *Escherichia coli* which affects the processing of ribosomal RNA. *J. Biol. Chem.* **253**: 1738.

Aplan, P. D., Lombardi, D. P., Ginsberg, A. M., Cossman, J., Bertness, V. L., and Kirsch, I. R. 1990. Disruption of the human SCL locus by "illegitimate V(D)J recombinase activity. *Science* **250**: 1426.

Araki, T. 1903. Enzymatic decomposition of nucleic acids. *Z. Physiol. Chem.* **38**: 84.

Araujo, S. J., Nigg, E. A., and Wood, R. D. 2001. Strong functional interactions of TFIIH with XPC and XPG in human DNA nucleotide excision repair, without a preassembled repairsome. *Mol. Cell. Biol.* **7**: 2281–2291.

Araya, A., Blanc, V., Begu, D., Crabier, F., Mouras, A., and Litvak, S. 1995. RNA editing in wheat mitochondria. *Biochimie.* **77**: 87–91.

Arber, W., and Dussoix, D. 1962. Host specificity of DNA produced by *Escherichia coli.* I: Host controlled modification of bacteriophage λ. *J. Mol. Biol.* **5**: 18.

Arber, W., and Linn, S. 1969. DNA modification and restriction. *Annu. Rev. Biochem.* **38**: 467.

Arber, W., and Wauter-Willems, D. 1970. Host specificity of DNA produced by *Escherichia coli* the two restriction and modification systems of strain 15T$^-$. *Mol. Gen. Genet.* **108**: 203.

Argos, P., Landy, A., Abremski, K., Egan, J. B., Haggard-Ljungquist, E., Hacas, R. H., Kahn, M. L., Kalionis, B., Naryanna, S. V. L., Pierson, L. S., III, Sternberg, N., and Leong, J. 1986. The integrase family of site-specific recombinases: Regional similarities and global diversity. *EMBO J.* **5**: 433.

Armel, P. R., and Wallace, S. S. 1978. Apurinic endonucleases from *Saccharomyces cerevisiae. Nucleic Acid Res.* **5**: 3347.

Ardelt, W., Mikulski, S. M., and Shogen, K. 1991. Amino acid sequence of an anti-tumor protein from *Rana pipens* oocytes and early embryos. Homology to pancreatic ribonucleases. *J. Biol. Chem.* **266**: 245–251.

Ashahara, H., Wistort, P. M., Bank, J. F., Bakerian, R. H., and Cunningham, R. P. 1989. Purification and characterization of *Escherichia coli* endonuclease III from the cloned *n*th gene. *Biochemistry* **28**: 4444.

Atkinson, N. S., Dunst, R. W., and Hopper, A. K. 1985. Characterization of an essential *Saccharomyces cerevisiae* gene related to RNA processing: Cloning of RNA1 and generation of a new allele with a novel phenotype. *Mol. Cell. Biol.* **5**: 907.

Austin, S., Ziese, M., and Sternberg, N. 1981. A novel role for site-specific recombination in maintenance of bacterial replicons. *Cell* **25**: 729.

Avery, O. T., Macleod, C. M., and McCarty, M. 1944. Studies on the chemical nature of the substance inducing transformation of pneumococcal types. I. Induction of transformation by a deoxyribonucleic acid fraction isolated from *Pneumococcus* type III. *J. Exp. Med.* **79**: 137.

Babiychuk, E., Kushnir, S., Montague, M. V., and Inze, D. 1994. The *Arabidopsis thaliana* apurinic endonuclease AP reduces transcription factors Fos and Jun. *Proc. Natl. Acad. Sci. U.S.A.* **91**: 3299.

Bachmann, B. J., and Low, K. B. 1980. Linkage map of *Escherichia coli* K12. Edition 6. *Microbiol. Rev.* **44**: 1.

Badman, R. 1972. Deoxyribonuclease mutants of *Utilago maydis* with altered recombination frequencies. *Genet. Res.* **20**: 213.

Bae, S.-H., and Seo, Y.-S. 2000. Characterization of the enzymatic properties of the yeast Dna2 helicase/nuclease suggests a new model of Okazaki fragment processing. *J. Biol. Chem.* **275**: 38022–38031.

Baer, M., and Altman, S. 1985. A catalytic RNA and its gene from *Salmonella typhimurium. Science* **228**: 999.

Baker, B. S. 1989. Sex in flies: The splice of life. *Nature* **340**: 521.

Baker, T. A., and Luo, L. 1994. Identification of residues in the Mu transposase essential for catalysis. *Proc. Natl. Acad. Sci. U.S.A.* **91**: 6654.

Baldi, M. I., Mattoccia, E., Ciafre, S., Attardi, D. G., and Tocchini-Valentini, G. P. 1986. Binding and cleavage of pre-tRNA by the *Xenopus* splicing endonuclease: Two separable steps of the intron excision reaction. *Cell* **47**: 965.

Baldi, M. I., Benedetti, P., Mattoccia, E., and Tocchini-Valentini, G. P. 1988. *In vitro* catenation and decatenation of DNA and a novel eucaryotic. ATP-dependent topoisomerase. *Cell* **20**: 461.

Balganesh, T. S., and Lacks, S. A. 1985. Heteroduplex DNA mismatch repair system of *Streptococcus pneumoniae*: Cloning and expression of the hexA gene. *J. Bacteriol.* **162**: 979.

Banroques, J., Perea, J., and Jocq, C. 1987. Efficient splicing of two yeast mitochondrial introns controlled by a nuclear encoded maturase. *EMBO J.* **6**: 1085.

Barany, F. 1985. Single stranded hexameric linkers: A system for phase insertion mutagenesis and protein engineering. *Gene* **37**: 111.

Barinaga, M. 1993. Ribozyme: Killing the messenger. *Science* **262**: 1512–1514.

Barker, R. B., Gleich, G. J., and Pease, L. R. 1988. Acidic precursor revealed in human eosinophil granular major basic protein cDNA. *J. Exp. Med.* **169**: 1493.

Barnard, E. A. 1969. Ribonucleases. *Annu. Rev. Biochem.* **38**: 677.

Barnes, R. 1985. Regulated expression of endonuclease EcoRI in *Saccharomyces cervisiae*: Nuclear entry and biological consequences. *Proc. Natl. Acad. Sci. U.S.A.* **82**: 1354.

Barnum, S. 1998. *Biotechnology—An Introduction.* Wadsworth Press, Belmont, CA, p. 225.

Barondes, S. H. 1981. Lectins: Their multiple endogenous cellular functions. *Annu. Rev. Biochem.* **50**: 207–231.

Barth, K. A., Powell, D., Trupin, M., and Mosig, G. 1988. Regulation of two nested proteins from gene 49 (recombination endonuclease VII) and of a λ rexA-like protein of bacteriophage T4. *Genetics* **120**: 329.

Barton, J. K. 1986. Metals and DNA: Molecular left handed complements. *Science* **233**: 727.

Barton, J. K., and Raphael, A. L. 1985. Site-specific cleavage of left-handed DNA in pBR322 by Λ-tris(diphenylphenanthroline) cobalt (III). *Proc. Natl. Acad. Sci. U.S.A.* **82**: 6460.

Barton, J. K., Basile, L. A., Danishefsky, A., and Alexandrescu, A. 1984. Chiral probes for the handedness of DNA helices: Enantiomers of tris(4,7-diphenylphenanthroline) ruthenium (II). *Proc. Natl. Acad. Sci. U.S.A.* **81**: 926.

Bass, B. L. 2000. Double-stranded RNA as a template for gene silencing. *Cell* **101**: 235–238.

Battula, N., and Loeb, L. A. 1974. The infidelity of avian myeloblastosis virus deoxyribonucleic acid polymerase in polynucleotide replication. *J. Biol. Chem.* **249**: 4086.

Beadle, G. W., and Tatum, E. L. 1941. Genetic control of biochemical reactions in *Neurospora. Proc. Natl. Acad. Sci. U.S.A.* **27**: 499.

Beard, P., Morrow, J. F., and Berg, P. 1973. Cleavage of circular, superhelical simian virus 40 DNA to a linear duplex by S1 nuclease. *J. Virol.* **12**: 1303.

Becker, A., and Gold, M. 1978. Enzymatic breakage of the cohesive end site of phage λDNA: Terminase (ter) reaction. *Biochemistry* **75**: 4199.

Becker, M. M., Lesser, D., Kurpicwski, M., Baranger, A., and Jen-Jacobsen, L. 1988. "Ultraviolet footprinting" accurately maps sequence-specific contacts and DNA kinking in the EcoRI endonuclease–DNA complex. *Proc. Natl. Acad. Sci. U.S.A.* **85**: 6247.

Been, M. D., and Champoux, J. J. 1981. DNA breakage and closure by rat liver topoisomerase: Separation of the half-reactions by using a single-stranded DNA substrate. *Proc. Nat. Acad. Sci. U.S.A.* **78**: 2883.

Been, M. D., Barfod, E. T., Burke, J. M., Price, J. V., Tanner, N. K., Zaug, A. J., and Cech, T. R. 1987. Structures involved in *Tetrahymena* rRNA self-splicing and RNA enzyme activity. *Cold Spring Harbor Symposia on Quantitative Biology* **52**: 147.

Behmooras, T., Toulmé, J.-J., and Helene, C. 1981. Specific recognition of apurinic sites in DNA by a tryptophan-containing peptide. *Proc. Natl. Acad. Sci. U.S.A.* **78**: 926.

Beintema, J. J., Wieetzes, P., Weickman, J. L., and Glitz, D. G. 1984. The amino acid sequence of human pancreatic ribonuclease. *Anal. Biochem.* **136**: 48.

Beintema, J. J., Hofsteenge, J., Iwama, M., Morita, T., Ohgi, K., Irie, M., Sugiyama, R. H., Schiven, G. L., Dekker, C. A., and Glitz, D. G. 1988. Amino acid sequence of the non-secretory ribonuclease of human urine. *Biochemistry* **27**: 4530.

Belfort, M., Pedersen-Lane, J., Ehrenman, K., Chu, F., Maley, G. F., Maley, F., McPheeters, D. S., and Gold, L. 1986. RNA splicing and *in vivo* expression of the intron-containing td gene of bacteriophage T4. *Gene* **41**: 93.

Bell-Pedersen, D., Quirk, S. M., Aubrey, M., and Belfort, M. 1989. A site-specific endonuclease and co-conversion of flanking exons associated with the mobile td intron of phage T4. *Gene* **82**: 119.

Bell-Pedersen, D., Quirk, S. M., Bryk, M., and Belfort, M. 1991. I-Tev-I, the endonuclease encoded by the mobile td intron, recognizes binding and cleavage domains on its DNA target. *Proc. Natl. Acad. Sci. U.S.A.* **88**: 7719.

Bender, J., and Kleckner, N. 1992. Transposase mutations that specifically alter target site recognition. *EMBO J.* **11**: 741.

Benedik, M. J., and Strych, U. 1998. Serratia marcescens and its extracellular nuclease. *FEMS Microbiol. Lett.* **165**: 461–468.

Benjamin, H. W., Matzuk, M. M., Kransnow, M. A., and Cozzarelli, N. R. 1985. Recombination site selection by Tn resolvase topological tests of a tracking mechanism. *Cell* **40**: 147.

Benne, R. 1988. Aminoacyl-tRNA synthetases are involved in RNA splicing in fungal mitochondria. *Trends Genet.* **4**: 181.

Benyajati, C., and Worcel, A. 1976. Isolation, characterization and structure of the folded interphase genome of *Drosophila melanogaster. Cell* **9**: 393.

Berg, D. E., and Howe, M. M. 1989. *Mobile DNA*. American Society for Microbiology, Washington, D.C.

Bernard, E. A. 1969. Ribonucleases. *Annu. Rev. Biochem.* **38**: 677.

Bernstein, C. 1981. Deoxyribonucleic acid repair in bacteriophage. *Microbiol. Rev.* **45**: 72.

Bernstein, E., Caudy, A. M., Hammond, S. C., and Hannon, G. J. 2001. Role for a bidentate ribonuclease in the initiation step of RNA interference. *Nature* **409**: 295–256.

Berrios, M., Osherobb, N., and Fisher, P. A. 1985. *In situ* localization of DNA topoisomerase II, a major polypeptide component of the *Drosophila* nuclear matrix fraction. *Proc. Natl. Acad. Sci. U.S.A.* **82**: 4142.

Bertani, G., and Weigle, J. J. 1953. Host controlled variation in bacterial viruses. *J. Bacteriol.* **65**: 113.

Bickle, T. A. 1982. ATP dependent restriction endonucleases. In Linn, S., and R. J. Roberts (eds.). *Nucleases*. Cold Spring Harbor Press, New York.

Biggins, J. B., Prudent, J. R., Marshall, D. J., Ruppen, M., and Thorson, J. S. 2000. A continuous assay for DNA cleavage: The application of "break light" to enediynes, iron-dependent agents, and nucleases. *Proc. Natl. Acad. Sci. U.S.A.* **97**: 13537–13542.

Bisbal, C., Silhol, M., Lubenthal, H., Kaluza, T., Carnac, G., Milligan, L., LeRoy, F., and Salehzada, T. 2000. The 2′–5′ oligoadenylate/RnaseL/RnaseLinhibitor pathway regulates both MyoD mRNA stability and muscle cell differentiation. *Mol. Cell. Biol.* **20**: 4959–4769.

Bishop, J. O. 1979. A DNA sequence cleaved by restriction endonuclease *Eco*RI in only one strand. *J. Mol. Biol.* **128**: 545.

Bishop, D. K., and Kolodner, R. D. 1986. Repair of heteroduplex plasmid DNA after transformation into *Saccharomyces cerevisiae. Mol. Cell. Biol.* **6**: 3401.

Bishop, D. K., Anderson, J., and Kolodner, R. 1989. Specificity of mismatch repair following transformation of *Saccharomyces cerevisiae* with heteroduplex plasmid DNA. *Proc. Natl. Acad. Sci. U.S.A.* **86**: 3713.

Bishop, D. K., Williamson, M. S., Fogel, S., and Kolodner, R. D. 1987. The role of heteroduplex correction in gene conversion in *Saccharomyces cerevisiae. Nature* **328**: 362.

Blackburn, E. H. 1992. Telomerases. *Annu. Rev. Biochem.* **61**: 113.

Blackwell, T. K., and Alt, F. W. 1989. Mechanism and developmental program of immunological gene management in mammals. *Annu. Rev. Genet.* **23**: 605.

Blackburn, P., and Moore S. 1982. Pancreatic ribonuclease. In: *The Enzymes*, Vol. XV. Academic Press, New York.

Blaese, R. M. 1997. Gene therapy for cancer. *Sci. Am.* **276**: 111–115.

Blaustein, A. R., Hoffman, P. D., Hokit, D. G., Kiesecker, J. M., Walls, S. C., and Hays, J. B. 1994. UV repair and resistance to solar UV-B in amphibian eggs: A link to population decline. *Proc. Natl. Acad. Sci. U.S.A.* **91**: 1791.

Bodley, A. L., and Liu, L. F. 1988. Topoisomerases as novel targets for cancer chemotherapy. *Biotechniques*, 1315.

Boeke, J. D., and Corces, V. G. 1989. Transcription and reverse transcription of retro-transposons. *Annu. Rev. Microbiol.* **43**: 403.

Boeke, J. D., Garfinkel, D. J., Styles, C. A., and Fink, G. R. 1985. Ty elements transpose through RNA intermediate. *Cell* **40**: 491.

Bojanowski, K., Lelievre, S., Markovits, J., Couprie, J., Jaquemin-Sablon, A., and Larsen, A. K. 1992. Suramin is an inhibitor of DNA topoisomerase II in *in vitro* and in Chinese hamster fibrosarcoma cells. *Proc. Natl. Acad. Sci.* **89**: 3025.

Bonner, C. A., Hays, S., McEntee, K., and Goodman, M. F. 1990. DNA Polymerase II is encoded by the DNA damage inducible dinA gene of *Escherichia coli*. *Proc. Natl. Acad. Sci. U.S.A.* **87**: 7663.

Borst, P., and Greaves, D. R. 1987. Programmed gene rearrangements altering gene expression. *Science* **235**: 658.

Borst, P., and Cross, G. A. M. 1982. Molecular basis for trypanosome antigenic variation. *Cell* **29**: 291.

Botstein D., and Shortle, D. 1985. Strategies and application of *in vitro* mutagenesis. *Science* **229**: 1193.

Botstein, D., White, R. L., Skolnick, M., and Davis, R. W. 1980. Construction of a genetic linkage map in man using restriction fragment length polymorphism. *Am. J. Hum. Genet.* **32**: 314.

Boyd, J. B., Mason, J. M., Yamamoto, A. H., Brodberg, R. K., Banga, S., and Sakaguchi, K. 1987. A genetic and molecular analysis of DNA repair in *Drosophila*. *J. Cell Sci. Suppl.* **6**: 39.

Boyer, P. D. (ed.). 1981. *The Enzymes*, 3rd ed., Vol. 14. Academic Press, New York.

Boyer, P. D. (ed.). 1982. *The Enzymes*, 3rd ed., Vol. 15. Academic Press, New York.

Branch, A. D., and Robertson, H. D. 1991. Efficient trans cleavage and a common structural motif for the ribozymes of the human hepatitis δ agent. *Proc. Natl. Acad. Sci. U.S.A.* **88**: 10163.

Breaker, R. R. 2000. Making catalytic DNAs. *Science*. **290**: 2095–2096.

Breaker, R. R., and Joyce, G. F. 1994. A DNA enzyme that cleaves RNA. *Chem. Biol.* **1**: 223–229.

Breaker. R. R., and Joyce, G. F. 1995. A DNA enzyme with Mg(2+)-dependent RNA phosphoesterase activity. *Chem. Biol.* **2**: 655–660.

Breitbart, R. E., Andreadis, A., and Nadal-Ginard, B. 1987. Alternate splicing: A ubiquitous mechanism for the generation of multiple protein isoforms from single genes. *Ann. Rev. Biochem.* **56**: 467.

Bretscher, L. E., Abel, R. L., and Raines, R. T. 2000. A ribonuclease A variant with low catalytic activity but high cytotoxicity. *J. Biol. Chem.* **275**: 9893–6.

Brill S. J., and Sternglanz, R. 1988. Transcription-dependent DNA supercoiling in yeast DNA topoisomerase mutants. *Cell* **54**: 403.

Brill, S. J., DiNardo, S., Voelkel-Meiman, K., and Sternglanz, S. R. 1987. Need for DNA topoisomerase activity: A swivel for DNA replication for transcription of ribosomal RNA. *Nature* **326**: 414.

Broach, J. R., Guarascio, V. R., and Jayaram, M. 1982. Recombination within yeast plasmid 2μ circle is site-specific. *Cell* **29**: 227.

Brosh, R., Von Kobbe, M. C., Sommers, J. A., Karakkar, P., Opresko, P. L., Piotroski, J., Dianova, I., Dianov, G. L., and Bohr, V. A. 2001. Werner syndrome protein interacts with human Flap endonuclease 1 and stimulates its cleavage activity. *EMBO J.* **20**: 5791–5801.

Broker, T. R., and Lehman, I. R. 1971. Branched DNA molecules: Intermediates in T4 recombination. *J. Mol. Biol.* **60**: 131.

Brown, D. M., and Todd, A. R. 1955. Evidence on the nature of the chemical bonds in nucleic acids. In Chargaff, E., and J. N. Davidson (eds.), *The Nucleic Acids*, Vol. 1. Academic Press, New York.

Brown, P. O., and Cozzarelli, N. R. 1981. Catenation and knotting of duples DNA by type 1 topoisomerases. A mechanistic parallel with type 2 topoisomerases. *Proc. Natl. Acad. Sci. U.S.A.* **78**: 843.

Bruist, M. F., Horvath, S. J., Hood, L. E., Steitz, T. A., and Simon, M. I. 1987. Synthesis of a site-specific DNA binding peptide. *Science* **235**: 777.

Brun, G., Vannier, P., Scovassi, I., and Callen, J. C. 1981. DNA topoisomerase I from mitochondria of *Xenopus laevis* oocytes. *Eur. J. Biochem.* **118**: 407.

Bucala, R., Model, P., and Cerami, A. 1984. Modification of DNA by reducing sugars: a possible mechanism for nucleic acid aging and age-related dysfunction in gene expression. *Proc. Natl. Acad. Sci. U.S.A.* **81**: 105.

Budd, M. E., and Campbell, J. L. 1995. A yeast gene required for DNA replication encodes a protein with homology to DNA helicases. *Proc. Natl. Acad. Sci. U.S.A.* **92**: 7642–7646.

Budd, M. E., and Campbell, J. L. 1997. A yeast replicative helicase, Dna2 helicase, interacts with yeast FEN-1 Nuclease in carrying out its essential function. *Mol. Cell. Biol.* **17**: 2136–2142.

Bugg, B. Y., Danks, M. K., Beck, W. T., and Suttle, D. P. 1991. Expression of mutant DNA topoisomerase II in CCRF-CEM human leukemic cells selected for resistance to teniposide. *Proc. Natl. Acad. Sci. U.S.A.* **88**: 7654.

Bullas, L. R., Colson, C., and Van Pel, A. 1975. DNA restriction and modification systems in *Salmonella*. SQ, a new system derived by recombination between the SB system of *Salmonella typhimurium* and the SP system of *Salmonella potsdam. J. Gen. Microbiol.* **95**: 166.

Bullas, L. R., Colson, C., and Neufeld, B. 1980. Deoxyribonucleic acid restriction and modifications systems in *Salmonella*: Chromosomally located systems of different serotypes. *J. Bacteriol.* **141**: 275.

Bullock, P., Champoux, J. J., and Botchan, M. 1985. Association of crossover points with topoisomerase I cleavage sites: A model for nonhomologous recombination. *Science* **230**: 954.

Burckhardt, S. E., Woodgate, R., Scheuermann, R. H., and Echols, H. 1988. UmuD mutagenesis protein of *Escherichia coli*: overproduction, purification and cleavage by *RecA*. *Proc. Natl. Acad. Sci. U.S.A.* **85**: 1811.

Burke, A. J. 2001. Genomics on high alert. *Genome Tech.* **1**: 33–35.

Burke, J. M. 1988. Molecular genetics of group I introns: RNA structures and protein factors required for splicing—a review. *Gene* **73**: 273.

Burke, J. M. 1997. Clearing the way for ribozymes. *Nature Biotechnol.* **15**: 414–415.

Burton, W. G., Roberto, R. J., Myers, P. A., and Sager, R. 1977. A site-specific single-strand endonuclease from the eukaryote *Chlamydomonas*. *Proc. Natl. Acad. Sci. U.S.A.* **74**: 2681.

Bushman, F. D., Fujiwara, T., and Craigie, R. 1990. Retroviral DNA integration directed by HIV integration protein *in vitro*. *Science* **249**: 1555.

Buxton, J. 1992. Detection of an unstable fragment of DNA specific to individuals with myotonic dystrophy. *Nature* **355**: 547.

Buzayan, J. M., Gerlach, W. L., and Breuning, G. 1986. Satellite tobacco ringspot virus RNA: A subset of the RNA sequence is sufficient for autolytic processing. *Proc. Natl. Acad. Sci. U.S.A.* **83**: 8859.

Cairns, J. 1980. Efficiency of the adaptive response of *Escherichia coli* to alkylating agents. *Nature* **286**: 176.

Cairns, J., Overbaugh, J., and Miller, S. 1988. The origin of mutants. *Nature* **335**: 142.

Cals, S., Tan, K. L., Megregor, A., and Connolly, B. A. 1998. Conversion of bovine pancreatic DNaseI to repair endonuclease with a high selectivity for abasic sites. *EMBO J.* **17**: 128–138.

Cameron, F. H., and Jennings, P. A. 1989. Specific gene suppression by engineered ribozymes in monkey cells. *Proc. Natl. Acad. Sci. U.S.A.* **86**: 9139.

Campbell, A. 1962. Episomes. *Adv. Genet.* **11**: 101.

Canals, A., Ribo, M., Benito, A., Bosch, M., Mombelli, E., and Vilanova, M. 1999. Production of engineered human pancreatic ribonuclease, solving expression and purification problems and enhancing thermostability. *Protein Expr. Purif.* **17**: 169–181.

Capecchi, M. R. 1989. Altering the genome by homologous recombination. *Science* **244**: 1288.

Carignani, G., Groundinsky, P., Trezza, D., Schigvon, E., Bergantino, E., and Slonimski, P. P. 1983. An mRNA maturase is encoded by the first intron of the mitochondrial gene for the subunit 1 of cytochrome oxidase in *S. cerevisiae*. *Cell* **35**: 733.

Carmi, N., and Breaker, R. R. 2001. Characterization of a DNA-cleaving deoxyribozyme. *Bioorg Med. Chem.* **9**: 2589.

Carpenter, R., Martin, C., and Coen, E. S. 1987. Comparison of genetic behaviour of the transposable element Tam-3 at two unlinked pigment loci *Antirrhinum majus*. *Mol. Gen. Genet.* **207**: 82.

Carpousis, A. J., Van Houwe, G., Ehretsamm, C., and Krisch, H. M. 1994. Copurification of *E. coli* RNase E and pNase: Evidence for a specific association between two enzymes. *Cell* **76**: 889.

Carson, P., Kushner, S. R., and Grossman, L. 1985. Involvement of helicase II (*uvrD* gene product) and DNA polymerase I in excision mediated by the *uvrABC* protein complex. *Proc. Natl. Acad. Sci. U.S.A.* **82**: 4925.

Cartwright, I. L., and Elgin, S. C. R. 1982. Analysis of chromatin structure and DNA sequence organization: use of the 1,10-phenanthroline-cuprous complex. *Nucleic Acids Res.* **10**: 5835.

Cartwright, I. L., Hertzberg, R. P., Dervan, P. B., and Elgin, S. C. R. 1983. Cleavage of chromatin with methidiumpropyl-EDTA-iron (II). *Proc. Natl. Acad. Sci. U.S.A.* **80**: 3213.

Caskey, C. T. 1987. Disease diagnosis by recombinant DNA methods. *Science* **236**: 1223.

Castora, F. J., and Simpson, M. V. 1979. Search for a DNA gyrase in mammalian mitochondria. *J. Biol. Chem.* **254**: 11193.

Catcheside, D. G. 1977. *The Genetics of Recombination*. Arnold, London.

Cavalier-Smith, T. 1985. Selfish DNA and the origin of introns. *Nature* **315**: 283.

Cech, T. R. 1983. RNA splicing: Three themes with variations. *Cell* **34**: 713.

Cech, T. R. 1986. The generality of self-splicing RNA relationship to nuclear mRNA splicing. *Cell* **44**: 207.

Cech, T. R. 1988. Conserved sequences and structures of group I introns: Building an active site for RNA catalysis—a review. *Gene* **73**: 259.

Cech, T. R. 1990. Self-splicing of group I introns. *Annu. Rev. Biochem.* **59**: 543.

Cech, T. R., and Bass, B. L. 1986. Biological catalysis by RNA. *Annu. Rev. Biochem.* **55**: 599.

Cech, T. R., Zaug, A. J., and Grabowski, P. J. 1981. *In vitro* splicing of the ribosomal RNA precursor of tetrahymena: Involvement of a guanosine nucleotide in the excision of the intervening sequence. *Cell* **27**: 487.

Celander, D. W., and Cech, T. R. 1991. Visualizing the higher order folding of a catalytic RNA molecule. *Science* **251**: 401.

Ceska, T. A., and Sayers, J. R. 1998. Structure-specific DNA cleavage by 5′ nucleases. *Trends Biochem. Sci.* **23**: 331–336.

Chambers, R. W., Sledziewska-Gojska, E., Hojatti, S. H., and Borowski, H. B. 1985. *UvrA* and *RecA* mutations inhibit a site-specific transition produced by a single O^4-methyl-guanine in gene O of bacteriophage φX174. *Proc. Natl. Acad. Sci. U.S.A.* **82**: 7173.

Champoux, J. J., and Dulbecco, R. 1972. An activity from mammalian cells that untwists superhelical DNA-A possible swivel for DNA replication. *Proc. Natl. Acad. Sci. U.S.A.* **69**: 143.

Champoux, J. J., and McConaughy, B. L. 1976. Purification and characterization of the DNA untwisting enzyme from rat liver. *Biochemistry* **15**: 4638.

Champoux, J. J. 1977a. Strand breakage by the DNA untwisting enzyme results in covalent attachment of the enzyme to DNA. *Proc. Natl. Acad. Sci. U.S.A.* **74**: 3488.

Champoux, J. J. 1977b. Renaturation of complementary single-stranded DNA circles: Complete rewinding facilitated by the DNA untwisting enzyme. *Proc. Natl. Acad. Sci. U.S.A.* **74**: 5328.

Chang, D. D., and Clayton, D. A. 1987. A mammalian mitochondrial RNA processing activity contains nucleus encoded RNA. **235**: 1178.

Chanock, S. 2001. Candidate genes and SNPs in the study of human diseases. *Dis. Markers* **17**: 89–98.

Chase, J. W., and Masker, W. E. 1977. Deoxyribonucleic acid repair in *Escherichia coli* mutants deficient in the 53 exonuclease activity of deoxyribonucleic acid polymerase I and exonuclease VII. *J. Bacteriol.* **130**: 667.

Chase, J. W., and Richardson, C. C. 1974. Exonuclease VII of *Escherichia coli. J. Biol. Chem.* **249**: 4553.

Chase, J. W., and Richardson, C. C. 1977. *Escherichia coli* mutants deficient in exonuclease VII. *J. Bacteriol.* **129**: 934.

Chase, J. W., Masker, W. E., and Murphy, J. B. 1979. Pyrimidine dimer excision in *Escherichia coli.* Strains deficient in exonucleases V and VII and in the 53 exonuclease of DNA polymerase I. *J. Bacteriol.* **137**: 234.

Chen, T. L., and Manuelidis, L. 1989. SINEs and LINEs cluster in distinct DNA fragments of giemsa band size. *Chromosoma* (Berl) **98**: 309.

Chen, C-H. B., and Sigman, D. S. 1986. Nuclease activity of 1,10-phenanthroline copper: Sequence specific targeting. *Proc. Natl. Acad. Sci. U.S.A.* **83**: 7147.

Cherniak, A. D., Garriga, G., Kittle, J. D., Akins, R. A., and Lambowitz, A. M. 1990. Function of *Neurospora* mitochondrial tyrosyl-tRNA synthetase in RNA splicing requires an idiosyncratic domain not found in other synthetases. *Cell* **62**: 745.

Chow, T. Y.-K., and Fraser, M. J. 1979. The major intracellular alkaline deoxyribonuclease activities expressed in wild type and *Rec*-like mutants of *Neurospora crassa*. *Can. J. Biochem.* **57**: 889.

Chow, T. Y.-K., and Fraser, M. J. 1983. Purification and properties of single stranded DNA binding endo–exonuclease of *Neurospora crassa*. *J. Biol. Chem.* **258**: 12012.

Chow, T. Y.-K., and Resnick, M. A. 1988. An endo–exonuclease activity of yeast that requires a functional RAD 52 gene. *Mol. Gen. Genet.* **211**: 41.

Chu, B. C. F., and Orgel, L. E. 1985. Nonenzymatic sequence-specific cleavage of single-stranded DNA. *Proc. Natl. Acad. Sci. U.S.A.* **82**: 963.

Chu, F. K., Maley, G. F., Maley, F., and Belfort, M. 1984. Intervening sequence in the thymidylate synthase gene of bacteriophage T4. *Proc. Natl. Acad. Sci. U.S.A.* **81**: 3049.

Chu, F. K., Maley, G. F., Belfort, M., and Maley, F. 1985. *In vitro* expression of the intron-containing gene for T4 phage thymidylate synthase. *J. Biol. Chem.* **260**: 10680.

Chu, F. K., Maley, G. F., West, D. K., Belfort, M., and Maley, F. 1986. Characterization of the intron in the phage T4 thymidylate synthase gene and evidence for its self-excision from the primary transcript. *Cell* **45**: 157.

Chu, F. K., Maley, G. F., and Maley, F. 1987a. Mechanism and requirements of *in vitro* RNA splicing of the primary transcript from the T4 bacteriophage thymidylate synthase gene. *Biochemistry* **26**: 3050.

Chu, F. K., Maley, F., Martinez, J., and Maley, G. F. 1987b. Interrupted thymidylate synthase gene of bacteriophages T2 and T6 and other potential self-splicing introns in the T-even bacteriophages. *J. Bacteriol.* **169**: 4368.

Chu, F. K., Maley, G. F., and Maley, F. 1988. RNA splicing in the T-even bacteriophage. *FASEB J.* **2**: 216.

Chu, F. K., Maley, G., Pedersen-Lane, J., Wang, A.-M., and Maley, F. 1990. Characterization of the restriction site of a prokaryotic intron-encoded endonuclease. *Proc. Natl. Acad. Sci. U.S.A.* **87**: 3574.

Chu, G., and Chang, E. 1988. *Xeroderma pigmentosum* group E cells lacks a nuclear factor that binds to damaged DNA. *Science* **242**: 564.

Chun, J. J. M., Schatz, D. G., Oettinger, M. A., Jaenisch, R., and Baltimore, D. 1991. The recombination activating gene-1 (RAG-1) transcript is present in the marine central nervous system. *Cell* **64**: 189.

Church, R. B. (ed.). 1990. *Transgenic Animals*. Butterworth-Heinemann, Boston.

Ciarrocchi, G., and Linn, S. 1978. A cell-free assay measuring repair DNA synthesis in human fibroblasts. *Proc. Natl. Acad. Sci. U.S.A.* **75**: 1887.

Clark, A. J. 1971. Towards a metabolic interpretation of genetic recombination of *E. coli* and its phages. *Annu. Rev. Microbiol.* **25**: 437.

Clark, A. J. 1973. Recombination deficient mutants of *E. coli* and other bacteria. *Annu. Rev. Genet.* **7**: 67.

Clark, A. J. 1991. rec genes and homologous recombination proteins in *Escherichia coli*. *Biochimie*. **73**: 523.

Clarke, N. D., Lien, D. C., and Schimmel, P. 1988. Evidence from cassette mutagenesis for a structure–function motif in a protein of unknown structure. *Science* **240**: 521.

Claverys, J. P., and Lacks, S. A. 1986. Heteroduplex deoxyribonucleic acid based mismatch repair in bacteria. *Micro. Rev.* **50**: 133.

Claverys, J.-P., Mejean, V., Gasc, A., and Sicard, A. M. 1983. Mismatch repair in *Streptococcus pneumoniae*: Relationship between base mismatches and transformation efficiencies. *Proc. Natl. Acad. Sci. U.S.A.* **80**: 5956.

Cleaver, J. E. 1994. It was a very good year for DNA repair. *Cell* **76**: 1.

Clements, J., Rogers, S. G., and Weiss, B. 1978. A DNase for apurinic/apyrimidinic sites associated with exonuclease III of *Hemophilus influenzae*. *J. Biol. Chem.* **253**: 2990.

Colleaux, L. 1986. Universal code equivalent of a yeast mitochondrial intron reading frame is expressed into *E. coli* as a specific double strand endonuclease. *Cell* **44**: 521.

Colleaux, L., D'Auriol, L., Galibert, F., and Dujon, B. 1988. Recognition and cleavage site of the intron-encoded omega transposase. *Proc. Natl. Acad. Sci. U.S.A.* **85**: 6022.

Collins, A., Johnson, R. T., and Boyle, J. M. 1987. Molecular biology of DNA repair. *J. Cell Sci. Suppl.*, 1–353.

Collins, R. A., and Lambowitz, A. M. 1985. RNA splicing in *Neurospora* mitochondria defective splicing of mitochondrial mRNA precursors in the nuclear mutant cyt 18-1. *J. Mol. Biol.* **184**: 413.

Colwill, R. W., and Sheinin, R. 1983. *ts* A1S9 locus in mouse L cells may encode a novobiocin binding protein that is required for DNA topoisomerase II activity. *Proc. Natl. Acad. Sci. U.S.A.* **80**: 4644.

Connolly, B., and West, S. C. 1990. Genetic recombination in *E. coli* Holliday junction made by *RecA* protein are resolved by frationated cell force extracts. *Proc. Natl. Acad. Sci. U.S.A.* **87**: 8476.

Connolly, B., Parsons, C. A., Benson, F. E., Dunderdale, H. J., Sharples, G. A., Lloyd, R. G., and West, S. C. 1991. Resolution of Holliday junction *in vitro* requires the *Escherichia coli* ruvC gene product. *Proc. Natl. Acad. Sci. U.S.A.* **88**: 6063.

Constantinou, A., Gunz, D., Evans, E., Lalle, P., Bater, P. A., and Wood, R. D. 1999. Conserved residues of human XPG protein important for nuclease activity and function in nucleotide excision repair. *J. Biol. Chem.* **274**: 5637–5648.

Cook, P. R., and Brazell, I. A. 1975. Supercoils in human DNA. *J. Cell. Sci.* **19**: 261.

Cooley, L., Kelley, R., and Spradling, A. 1988. Insertional mutagenesis of the *Drosophila* genome with single P elements. *Science* **239**: 1121.

Cooper, E. J., Trautmann, M. L., and Laskowski, M., Sr. 1950. Occurrence and distribution of an inhibitor for desoxyribonuclease in animal tissues. *Proc. Soc. Exp. Biol. Med.* **73**: 219.

Corey, D. R., and Schultz, P. G. 1987. Generation of a hybrid sequence-specific single-stranded deoxyribonuclease. *Science* **238**: 1401.

Corey, E. J., and Reichard, G. A. 1989. Enantioselective and practical synthesis of R and S fluoxetines. *Tetrahedron Lett.* **30**: 5207.

Cotten, M., Gick, O., Vasserot, A., Schaffner, G., and Birnstiel, M. L. 1988. Specific contacts between mammalian U7 snRNA and histone precursor RNA are indispensable for the *in vitro* 3RNA processing reaction. *EMBO* **7**: 801.

Cotton, F. A., Hazen, E. E., and Legg, M. J. 1979. Staphylococcal nuclease: Proposed mechanism of action based on structure of enzyme thymidine 3,5 biophosphate–calcium ion complex at 1–5 Å resolution. *Proc. Natl. Acad. Sci. U.S.A.* **76**: 2551.

Covello, P. S., and Gray, M. W. 1989. RNA editing in plant mitochondria. *Nature* **341**: 662.

Cox, M. M. 1983. The FLP protein of the yeast 2μm plasmid: Expression of a eukaryotic genetic recombination system in *Escherichia coli*. *Proc. Natl. Acad. Sci. U.S.A.* **80**: 4223.

Cozzarelli, N. R. 1980. DNA gyrase and the supercoiling of DNA. *Science* **207**: 953.

Craig, N. L. 1985. Site-specific inversions: enhancers, recombination proteins, and mechanism. *Cell* **41**: 649.

Craig, N. L., and Nash, H. A. 1983. The mechanism of phage λ site-specific recombination: Site-specific breakage of DNA by Int. topoisomerase. *Cell* **35**: 795.

Craigie, R., and Mizzuchi, K. 1986. Role of DNA topology in mu transposition mechanism, assessing the relative orientation of two DNA segments. *Cell* **45**: 793.

Crick, F. H. C. 1968. Origin of genetic code. *J. Mol. Biol.* **38**: 367.

Croce, C. M. 1987. Role of chromosome translocation in human neoplasia. *Cell* **49**: 155.

Crouch, R. J., and Dirksen, M. L. 1982. In Linn, S., and R. J. Roberts (eds.), *Ribonuclease H in Nucleases*. Cold Spring Harbor Press, New York, 211–247.

Cudny, H., Roz, P., and Deutscher, M. P. 1981. Alteration of *Escherichia coli* RNase D by infection with bacteriophage T_4. *Biochem. Biophys. Res. Commun.* **98**: 337.

Cuenoud, B., and Szostak, J. W. 1995. A DNA metalloenzyme with DNA ligase activity. *Nature* **375**: 611.

Cunningham, L., Catlin, B. W., and Privat de Garilhe, M. 1956. A deoxyribonuclease of *Micrococcus pyogenes*. *Am. Chem. Soc.* **78**: 4642.

Cunningham, R. P., Wu, A. M., Shibata, T., DasGupta, C., and Radding, C. M. 1981. Homologous pairing and topological linkage of DNA molecules by combined activation of *E. coli RecA* protein and topoisomerase I. *Cell* **24**: 213.

Dake, E., Hofman, J. J., McIntire, S., Hudson, A., and ZassenHaus, H. P. 1988. Purification and properties of the major nuclease from mitochondria of *Saccharomyces cerevisiae*. *J. Biol. Chem.* **263**: 7691.

Dani, Z. M., and Zakian, V. A. 1983. Mitotic and meiotic stability of linear plasmids in yeast. *Proc. Natl. Acad. Sci. U.S.A.* **80**: 3406.

Danna, K. J., and Nathans, D. 1971. Sequence specific cleavage of simian virus 40 DNA by restriction endonucleases of *Haemophilus influenzae*. *Proc. Natl. Acad. Sci. U.S.A.* **68**: 2913.

Darnell, J. E., and Doolittle, W. F. 1986. Speculations on the early course of evolution. *Proc. Natl. Acad. Sci. U.S.A.* **83**: 1271.

Dash, P., Knapp, I. L. M., Kandel, E. R., and Goelet, P. 1987. Selective elimination of mRNA's *in vivo:* Complementary oligodeoxynucleotides promote RNA degradation by an RNase H-like activity. *Proc. Natl. Acad. Sci. U.S.A.* **84**: 7896.

Davidson, J. N. 1972. Nucleases and related enzymes. In the biochemistry of the nucleic acids. 183–214.

Davies, J. F., Hostomska, Z., Hostomsky, Z., Jordan, S. R., and Matthews, D. A. 1991. Crystal structure of the ribonuclease H domain of HIV-1 reverse transcriptase. *Science* **252**: 88.

Davies, K. E. 1991. The application of DNA recombinant technology to the analysis of the human genome and genetic disease. *Hum. Genet.* **58**: 35.

Davies, R. W. 1980. DNA sequence of the int-xis-P_I region of the bacteriophage λ; overlap of the int and xis genes. *Nucleic Acids Res.* **8**: 1765.

Davies, R. W., Waring, R. B., Ray, J. A., Brown, T. A., and Scazzocchio, C. 1982. Making ends meet: A model for RNA splicing in fungal mitochondria. *Nature* **300**: 719.

Davis, B. D. 1989. Transcription bias: A non-lamarckian mechanism for substrate-induced mutations. *Proc. Nat. Acad. Sci. U.S.A.* **86**: 5005.

Davis, M., Kim, K., and Hood, L. 1980. Developmentally regulated DNA rearrangements during differentiation immunoglobulin class switching. *Cell* **22**: 1.

Dawkins, R. 1989. *The Selfish DNA*. Oxford University Press, New York, pp. 1–350.

de Massy, B. D., and Weisberg, R. A. 1987. Gene endonuclease of bacteriophage T7 resolved conformationally branched structure in double-stranded DNA. *J. Mol. Biol.* **193**: 359.

De Robertis, E. M., Black, P., and Nishikura, K. 1981. Intranuclear location of the tRNA splicing enzymes. *Cell* **23**: 89.

Debenham, P. G., Webb, M. B. T., Jones, N. J., and Cox, R. 1987. Molecular studies on the nature of the repair defect in *Ataxia telangiectasia* and their implication for cellular radiobiology. *J. Cell. Sci. Suppl.* **6**: 177.

Debrabant, A., Gottleiband, M., and Dwyer, D. M. 1995. Isolation and characterization of the gene encoding the surface membrane 3′-nucleotidase/nuclease of *Leishmania donovani. Mol. Biochem. Parasitol.* **71**: 51–63.

Deininger, P. L. 1989. SINEs: Short interspersed repeated DNA elements in higher eucaryotes. In Berg, D. E., and M. M. Howe (eds.), *Mobile DNA*. American Society of Microbiology, Washington, D.C.

Delahodde, A. 1989. Site-specific DNA endonuclease and RNA maturase activities of two homologous intron-encoded proteins from yeast mitochondria. *Cell* **56**: 431.

Delange, A. M., and Mishra, N. C. 1981. The isolation of MMS and histidine sensitive mutants of *Neurospora crassa. Genetics* **97**: 247.

Delange, A. M., and Mishra, N. C. 1982. Characterization of mutagen sensitive mutants of *Neurospora. Mutat. Res.* **24**: 1.

Demidov, V., Frank-Kamanetskii, M. D., Egholm, M., Buchardt, O., and Nielson, P. E. 1993. Sequence selective double strand DNA cleavage by peptide nucleic acid (PNA) targeting using nuclease S1. *Nucl. Acid Res.* **21**: 2103–2107.

Demple, B., and Linn, S. 1980. DNA *N*-glycosylases and UV repair. *Nature* **287**: 203.

Demple, B., and Linn, S. 1982. On the recognition and cleavage mechanism of *Escherichia coli* endonuclease, V. A possible DNA repair enzyme. *J. Biol. Chem.* **257**: 2848.

Depew, R. E., Liu, L. F., and Wang, J. C. 1978. Interaction between DNA and *Escherichia coli* protein ω formation of a complex between single-stranded DNA and protein *w. J. Biol. Chem.* **253**: 511.

Derbyshire, V., Freemont, P-S., Sanderson, M. R., Beese, L., Friedman, J. M., Joyce, C. M., and Steitz, T. A. 1988. Genetic and crystallographic studies of the 35 exonucleolytic site of DNA polymerase I. *Science* **199**: 239.

Dervan, P. B. 1986. Design of sequence specific DNA binding molecules. *Science* **232**: 464.

Desiderio, S., and Baltimore, D. 1984. Double stranded cleavage by cell extracts near recombinational signal sequences of immunoglobulin gene. *Nature* **308**: 860.

Deutscher, M. P. 1984. Processing of tRNA in prokaryotes and eukaryotes. *CRC Crit. Rev. Biochem.* **17**: 45.

Deutscher, M. P., and Kornberg, A. 1969. Enzymatic synthesis of deoxyribonucleic acid XXIX hydrolysis of deoxyribonucleic acid from the 5 terminus by an exonuclease function of deoxyribonucleic acid polymerase. *J. Biol. Chem.* **244**: 3029.

Dewey, R. E., Levings, C. S., III, and Timothy, D. H. 1986. Novel recombinations in the maize mitochondrial genome produce a unique transcriptional unit in the Texas male-sterile cytoplasm. *Cell* **44**: 439.

Dickie, P., McFadden, G., and Morgan, A. R. 1987. The site-specific cleavage of synthetic Holliday junction analogs and related branched DNA structures by bacteriophage T7 endonuclease I. *J. Biol. Chem* **262**: 14826.

Dietz, A. C., and Schnoor, J. L. 2001. Advances in phytoremediation. *Environ. Health Perspect.* **1**: 163–168.

Di Gioacchino, M., Cavallucci, E., Di Stefano, F., Verna, N., Ramondo, S., Ciuffreda, S., Roccioni, G., and Boscolo, P. 2000. Influence of total IgE and seasonal increase of eosinophil cationic protein on bronchial hyperactivity in asthmatic grass-sensitized farmers. *Allergy* **55**: 1030–1034.

DiLisi, C. 1988. Computers in molecular biology: Current applications and emerging trends. *Science* **240**: 47.

DiNardo, S., Voekel, K. A., and Sternglanz, R. 1984. DNA topoisomerase II mutant of *Saccharomyces cerevisiae:* Topoisomerase II is required for segregation of daughter molecules at the termination of DNA replication. *Proc. Natl. Acad. Sci. U.S.A.* **81**: 2616.

Domachowske, J. B., Dryer, K. D., Bonneville, C. A., and Rosenberg, H. F. 1998. Recombinant human Eosinophil-derived neurotoxin/RNase2 functions as an effective antiviral agent against respiratory syncytial virus. *J. Infect. Dis.* **177**: 1458–1464.

Dombroski, B. A., Mathais, S. L., Nathakumar, E., Scott, A. F., and Kazazian, H. H., Jr. 1991. Isolation of an active human transposable element. *Science* **254**: 1805.

Dong, B., and Silverman, R. H. 1997. Abipartite model of 2–5-dependent Rnase L. *J. Biol. Chem.* **272**: 22236–22242.

Doniger, J., and Grossman, L. 1976. Human correxonuclease. Purification and properties of DNA repair exonuclease from placenta. *J. Biol. Chem.* **251**: 4579.

Donis-Keller, H., Green, P., Helms, C., Cartinhour, S., Weiffenbach, B., Stephens, K., Keith, T. P., Bowden, D. W., Smith, D. R., Lander, E. S., Botstein, D., Akots, G., Rediker, K. S., Gravius, T., Brown, V. A., Rising, M. B., Parker, C., Powers, J. A., Watt, D. E., Kauffman, E. R., Bricker, A., Phipps, P., Muller-Kahle, H., Fulton, T. R., Ng, S., Schumm, J. W., Braman, J. C., Knowlton, R., Barker, D. F., Crooks, S. M., Lincoln, S. E., Daly, M. J., and Abrahamson, J. 1987. A genetic linkage map of the human genome. *Cell* **51**: 319.

Doolittle, W. F., and Spienza, C. 1980. Selfish genes, the phenotype paradigm and genome evolution. *Nature* **284**: 601.

Döring, H.-P., and Starlinger, P. 1984. Barbara McClintock's controlling elements: Now at the DNA level. *Cell* **39**: 253.

DIring, H.-P., and Starlinger, P. 1986. Molecular genetics of transposable elements in plants. *Annu. Rev. Genet.* **20**: 175.

Dorit, R. L., Schoenbach, L., and Gilbert, W. 1990. How big is the universe of exons? *Science* **250**: 1377.

Drake, F. H., Hofmann, G. A., Bartus, H. F., Mattern, M. R. R., Croke, S. T., and Mirabelli, C. K. 1989. Biochemical and pharmalogical properties of p170 and p180 forms of topoisomerase II. *Biochemistry* **28**: 8154.

Drapkin, R., Sancor, A., and Reinberg, D. 1994. Where transcription meets repair. *Cell* **77**: 9.

Drlica, K. 1984. Biology of bacterial deoxyribonuclic acid topoisomerases. *Microbiol. Rev.* **48**: 273.

Dryer, G. B., and Dervan, P. B. 1985. Sequence-specific cleavage of single stranded DNA: Oligodeoxynucleotide-EDTA-Fe(II). *Proc. Natl. Acad. Sci. U.S.A.* **82**: 963.

Dubnair, E., Lenny, A. B., and Margolin, P. 1973. Nonsense mutation of the supX focus: Further characterization of the SupX mutant phenotype. *Mol. Gen. Genet.* **126**: 191.

Dubnau, E., and Margolin, P. 1972. Suppression of promoter mutations by the phenotropic supX mutations. *Mol. Gen. Genet.* **117**: 91.

Duesberg, P. H. 1987. Cancer genes: rare recombinant instead of activated oncogenes. *Proc. Natl. Acad. Sci. U.S.A.* **84**: 2117.

Dujon, B. 1989. Group I introns as mobile genetic elements: facts and mechanistic speculations—A review. *Gene* **82**: 91.

Dunn, J. J., and Studier, F. W. 1975. Effect of RNase III cleavage on translation of bacteriophage T_7 messenger RNAs. *J. Mol. Biol.* **99**: 487.

Dwyer, M. A., Huang, A. J., Pan, C. Q., and Lazarus, R. A. 1999. Expression and Characterization of a DnaseI-Fc fusion enzyme *J. Biol. Chem.* **274**: 9738–9743.

Ebright, R. H., Ebright, Y. W., Pendergast, P. S., and Gunasekera, A. 1990. Conversion of a helix-turn-helix motif sequence-specific cDNA binding protein into a site-specific DNA cleavage agent. *Proc. Natl. Acad. Sci. U.S.A.* **87**: 2882.

Echols, H., and Goodman, M. F. 1991. Fidelity mechanisms in DNA replication. *Annu. Rev. Biochem.* **60**: 477.

Eddy, S. R., and Gold, L. 1994. Artificial mobile DNA element constructed from the EcoRI gene. *Proc. Natl. Acad. Sci. U.S.A.* **91**: 1544.

Eder, P. S., and Walder, J. A. 1991. RibonucleaseH from K562 human erythroleukemia cells. Purification, characterization and substrate specificity. *J. Biol. Chem.* **266**: 6472–6479.

Egami, F., and Nakamura, K. 1969. *Microbial Ribonucleases.* Springer, New York.

Eichinger, D. J., and Boeke, J. D. 1988. The DNA intermediate in yeast TY1 element transposition copurifies with virus-like particles: Cell-free Ty1 transposition. *Cell* **54**: 955.

Eisen, H. 1988. RNA editing: Who's on first. *Cell* **53**: 331.

Eisenstein, B. 1995. Enterobacteriaceae. In Mandell, G. L., et al. (eds.), *Principle and Practice of Infectious Diseases*, 4th ed. Churchill Livingstone, New York, 1964–1980.

Elborough, K. M., and West, S. C. 1990. Resolution of synthetic Holliday junction in DNA by an endonuclease activity from calf thymus. *EMBO J.* **9**: 2938.

Elgin, S. C. R. 1984. Anatomy of hypersensitive sites. *Nature* **309**: 213.

Emmert, S., Schneider, T. D., Khan, S. G., and Kraemer, K. H. 2001. The human XPG gene: gene architecture, alternate splicing and single nuleotise polymorphism. *Nucleic Acid Res.* **29**: 1443–1452.

Endlich, B., and Linn, S. 1981. Type I restriction enzymes. In Boyer, P. D. (ed.), *The Enzymes*, Vol. 14, part A, 3rd ed. Academic Press, New York.

Engelke, D. R., Gegenheimer, P., and Abelson, J. 1985. Nucleolytic processing of a $tRNA^{Arg}$–$tRNA^{Asp}$ dimeric precursor by a homologous component from *Saccharomyces cerevisiae. J. Biol. Chem.* **260**: 1271.

Epstein, L. M., and Gall, J. G. 1987. Self-cleaving transcripts of satellite DNA from the newt. *Cell* **48**: 535.

Eron, L. J., and McAuslan, B. R. 1966. Inhibition of deoxyribonuclease by actinomycin D and ethedium bromide. *Biochem. Biophys. Acta* **114**: 663.

Esposito, D., and Scocca, J. J. 1995. The family of Tyrosine recombinases: evolution of a conserved active site domain. *Nucleic Acid Res.* **25**: 3605–3610.

Evans, D. H., and Kolodner, R. 1987. Construction of a synthetic Holliday junction analog and characterization of its interaction with a *Saccharomyces cerevisiae* endonuclease that cleaves Holliday junctions. *J. Biol. Chem.* **262**: 9160.

Evans, D. H., and Kolodner, R. 1988. Effect of DNA structure and nucleotide sequence on Holliday junction resolution by a *Saccharomyces cerevisiae* endonuclease. *J. Mol. Biol.* **20**: 69.

Fabrizio, P., and Abelson, J. 1990. Two domains of yeast U6 small nuclear RNA required for both steps of nuclear precursor messenger RNA splicing. *Science* **250**: 404.

Fairfield, F. R., Bauer, W. R., and Simpson, M. V. 1979. Mitochondria contain a distinct DNA topoisomerase. *J. Biol. Chem.* **254**: 9352.

Fairweather, N. F., Orr, E., and Holland, I. B. 1980. Inhibition of deoxyribonucleic acid gyrase: Effects on nucleic acid synthesis and cell division in *Escherichia coli* K12. *J. Bacteriol.* **142**: 153.

Fanning, T. G., and Singer, M. F. 1988. LINE-1: A mammalian transposable element. *Biochem. Biophys. Acta* **910**: 203.

Fedor, M. J., and Uhlenbec, O. C. 1990. Substrate sequence effects on "hammerhead" RNA catalytic efficiency. *Proc. Natl. Acad. Sci. U.S.A.* **87**: 1668.

Federoff, N. 1989. About maize transposable elements and developments. *Cell* **56**: 181.

Feher, Z., and Mishra, N. C. 1994. Aphidicolin resistant Chinese hamster ovary cells possess altered DNA polymerases of the α-family. *Biochim. Biophys. Acta* **1218**: 35.

Feher, Z., and Mishra, N. C. 1995. An aphidicolin mutant of Chinese hamster ovary cell with altered DNA polymerase and 3′ exonuclease activities. *Biochem Biophys. Acta* **1263**: 141.

Feiss, M. 1986. Terminase and the recognition, cutting, and packaging of λ chromosomes. *Trends Genet.* **2**: 100.

Felix, C. A. 2001. Leukemias related to treatment with DNA topoisomerase II inhibitors. *Med. Pediatr. Oncol.* **36**: 525–535.

Felsenfeld, G. 1978. Chromatin. *Nature* **271**: 115.

Feng, J. A., Johnson, R. C., and Dickerson, R. E. 1994. Hin recombinase bound to DNA: The origin of specificity in major and minor groove interactions. *Science* **263**: 348.

Fett, J. W., Strydom, D. J., Lobb, R. R., Alderman, E. M., Bethune, J. L., Riordan, J. F., and Vallee, B. L. 1985. Amino acid sequences of human tumor derived angiogenin. *Biochemistry* **24**: 5480.

Finkel, E. 1999. DNA cuts its teeth—as an enzyme. *Science* **286**: 2441–2442.

Fishel, R. A., James, A. A., and Kolodner, R. 1981. *RecA*-independent general genetic recombination of plasmids. *Nature* **294**: 184.

Fishel, R. A., Siegel, E. C., and Kolodner, R. 1986. Gene conversion in *Escherichia coli* resolution of heteroallelic mismatched nucleotides by co-repair. *J. Mol. Biol.* **188**: 147.

Fisher, M. 1981. DNA supercoiling by DNA gyrase. *Nature* **294**: 607.

Fisher, M. 1984. DNA supercoiling and gene expression. *Genetics* **307**: 686.

Fleischmann, G., Pelugfelder, G., Steiner, E. K., Javaherian, K., Howard, G. C., Wang, J. C., and Elgin, S. C. 1984. *Drosophila* DNA topoisomerase I is associated with transcriptionally active regions of the genome. *Proc. Natl. Acad. Sci. U.S.A.* **81**: 6958.

Floyd-Smith, G., Yoshie, O., and Lengyel, P. 1982. Interferon action. *J. Biol. Chem.* **257**: 8584.

Forsthoefel, A. M., and Mishra N. C.1983. Biochemical genetics of *Neurospora nucleases I. Genet. Res.* **41**: 271.

Francois, J. C., Saison-Behmoaras, T., Chassiquol, M., Thuong, N. T., Jian-Sheng-Sun, and Helene, C. 1988. Periodic cleavage of Poly (dA) by oligothymidylates covalently linked to the 1,10-phenanthroline–copper complex. *Biochemistry* **27**: 2272.

Frank, G., J. Qiu, Zheng, L., and Shen, B. 2001. Stimulation of Flap endonuclease-1 activities by PCNA is independent of its in vitro interaction via a consensus PCNA binding region. *J. Biol. Chem.* **276**: 36295–36302.

Frankel, G. D., and Richardson, C. C. 1971. The deoxyribonuclease induced after infection of *Escherichia coli* by bacteriophage T5. I. Characterization of the enzyme as a 5-exonuclease. *J. Biol. Chem.* **256**: 4839.

Frankel, G. D., Randles, K., and Burns, H. 1981. Purification and properties of a new DNase activity from KB cells. *Nucleic Acids Res.* **9**: 6635–6644.

Franklin, N. C. 1971. In the *Bacteriophage Lambda*. (ed: A. D. Hershey) p. 175–210. Cold Spring Harbor Laboratory Press, Cold Spring Harbor, NY.

Fraser, M. J. 1979. Alkaline deoxyribonucleases released from *Neurospora crassa* mycelia: Two activities not released by mutants with multiple sensitivities to mutagens. *Nucleic Acids Res.* **6**: 231.

Fraser, M. J., Tjeerde, R., and Matsumoto, K. 1976. A second form of the single strand specific endonuclease of *Neurospora crassa* which is associated with a double-strand exonuclease. *J. Can. Biochem.* **54**: 971.

Fraser, M. J., Chow, T. Y.-K., and Käfer, E. 1980. Nucleases and their control in wild type and *nuh* mutants of *Neurospora*. In Generoso, W. M., M. D. Shelby, and F. J. de Serres (eds.), *DNA Repair and Mutagenesis in Eukaryotes*. Plenum Press, New York, Chapter 5.

Fraser, M. J., Koa, H., and Chow T. Y.-K. 1990. Neurospora endo-exonuclease is immunologically related to the *RecC* gene product of *E. coli*. *J. Bacteriol.* **172**: 507.

Freudenreich, C. H., Kantrow, S. H., and Zakin, V. A. 1998. Expansion and length-dependent fragility of CTG repeats in yeast. *Science* **279**: 853–856.

Friedberg, E. C. 1985. *DNA Repair*. W. H. Freeman, San Francisco.

Friedberg, E. C., and Goldthwait, D. A. 1969. Endonuclease II of *Escherichia coli*. I. Isolation and purification. *Proc. Natl. Acad. Sci. U.S.A.* **62**: 934.

Friedman, E. A., and Smith, H. O. 1972. An adenosine triphosphate-dependent deoxyribonuclease from *Hemophilus influenzae* Rd. I. Purification and properties of the enzyme. *J. Biol. Chem.* **247**: 2846.

Froster, A. C., Davies, C., Sheldon, C. C., Jeffries, A. C., and Symons, R. H. 1988. Self-cleaving viroid and new tRNAs may only be active as dimers. *Nature* **334**: 265.

Froster, A. C., and Symons, R. H. 1987. Self-cleavage of virusoid RNA is performed by the proposed 5S-nucleotide active site. *Cell* **50**: 9.

Fujimoto, M., Kuninaka, A., and Yoshino, H. 1974. Purification of a nuclease from *Penicillium citrinum*. *Agr. Biol. Chem.* **38**: 777.

Galas, D. J., and Schmidtz, A. 1978. DNA foot printing: A simple method for the deletion of protein–DNA binding specificity. *Nucleic Acid Res.* **5**: 3157.

Gafurov, Iu M., Terent'ev, L. L., and Rasskazov, V. A. 1979. Isolation and some properties of ATP-dependent DNAse from sea urchin (Strongylocentrotus intermedius) embryo. *Biokhimiia*. **44**: 996–1004.

Garber, R. L., and Altman, S. 1979. *In vitro* processing of *B. mori* transfer RNA precursor molecules. *Cell* **17**: 389.

Gariga, G., and Lambowitz, A. M. 1986. Protein-dependent splicing of a group I intron in ribonucleoprotein particles and soluble fractions. *Cell* **46**: 669.

Gates, C. A., and Cox, M. M. 1988. FLP recombinase is an enzyme. *Proc. Natl. Acad. Sci. U.S.A.* **85**: 4628.

Gates, F. T., and Linn, S. 1977a. Endonuclease V of *Escherichia coli*. *J. Biol. Chem.* **252**: 1647.

Gates, F. T., and Linn, S. 1977b. Endonuclease from *Escherichia coli* that acts specifically upon duplex DNA damaged by ultraviolet light, osmium tetraoxide, acid, or x-rays. *J. Biol. Chem.* **252**: 2802.

Gegenheimer, P., and Apirion, D. 1981. Processing of procaryotic ribonucleic acid. *Microbiol. Rev.* **45**: 502.

Geiduschek, E. P., and Daniels, A. 1965. A simple assay for DNA endonucleases. *Anal. Biochem.* **11**: 133.

Gellert, M. 1981. DNA topoisomerases. *Annu. Rev. Biochem.* **60**: 879.

Gellert, M., Mizuuchi, K. O'Dea, M. H., and Nash, H. A. 1976a. DNA gyrase: An enzyme that introduces superhelical turns into DNA. *Proc. Natl. Acad. Sci. U.S.A.* **73**: 3872.

Gellert, M., O'Dea, M. H., Itoh, T., and Tomizawa, J. 1976b. Novobiocin and coumermycin inhibit DNA supercoiling catalysed by DNA gyrase. *Proc. Natl. Acad. Sci. U.S.A.* **73**: 4474.

Gellert, M., Fisher, L. M., Ohmori, H. O'Dea, M. H., and Mizuuchi, K. 1980. DNA gyrase: Site-specific interactions and transient double-strand breakage of DNA. *Cold Spring Harbor Symp. Quant. Biol.* **45**: 391.

Gellert, M., Mizuuchi, K. O'Dea, M. H., Itoh, T., and Tomizawa, J. 1977. Nalidixic acid resistance: A second genetic character involved in DNA gyrase activity. *Proc. Natl. Acad. Sci. U.S.A.* **74**: 4772.

George, J., Blakesley, R. W., and Chirikjian, J. G. 1980. Sequence-specific endonuclease *Bam*HI: Effect of hydrophobic reagents on sequence recognition and catalysis. *J. Biol. Chem.* **255**: 6521.

Gerlt, J. A. 1992. Phosphate ester hydrolysis. *Enzymes* **20**: 95.

Gerlt, J. A. 1993. Mechanistic principles of enzyme-catalyzed cleavage of phosphodiester bonds. In Linn, S. M., R. S. Lloyd, and R. J. Roberts (eds.), *Nucleases*, 2nd ed., Cold Spring Harbor Laboratory Press, Cold Spring Harbor, NY.

Germaine, R. N. 1985. "Exon-shuffling" maps control of antibody-and T-cell-recognition sites to the NH_z-terminal domain for the class II major histocompatibility polypeptide A_B. *Proc. Natl. Acad. Sci. U.S.A.* **82**: 2940.

Germond, J. E., Hirt, B., Oudet, P., Gross-Belard, M., and Chambon, P. 1975. Folding of the DNA double helix in chromatin-like structures from simian virus 40. *Proc. Natl. Acad. Sci. U.S.A.* **72**: 1843.

Geyer, C. R., and Sen, D. 1998. Lanthanide probes for a phosphodiester-cleaving, lead-dependent, DNAzyme. *J. Mol. Biol.* **23**: 483.

Ghora, B. K., and Apirion, D. 1978. Structural analysis and *in vitro* processing a RNA molecule isolated from a τne mutant of *E. coli*. *Cell* **15**: 1055.

Ghora, B. K., and Apirion, D. 1979. Identification of a novel RNA molecule in a new RNA processing mutant of *Escherichia coli* which contains 5sτRNA sequence. *J. Biol. Chem.* **254**: 1951.

Ghosh, R. K., and Deutscher, M. P. 1978. Purification of potential 3 processing nucleases using synthetic tRNA precursors. *Nucleic Acid Res.* **5**: 3831.

Gierl, A., Saedler, H., and Peterson, P. A. 1989. Maize transposable elements. *Annu. Rev. Genet.* **23**: 71.

Gilbert, W. 1987a. The RNA world. *Cold Spring Harbor Symp. Quant. Biol.* **52**: 901.

Gilbert, W. 1987b. The exon theory of genes. *Cold Spring Harbor Symp. Quant. Biol.* **52**: 901.

Gilbert, W., and Dressler, D. 1969. DNA replication: The rolling circle model. *Cold Spring Harbor Symp. Quant. Biol.* **33**: 473.

Gilmour, D. S., Pflugfelder, G., Wang, J. C., and Lis, J. T. 1986. Topoisomerase I interacts with transcribed regions in *Drosophila* cells. *Cells* **44**: 401.

Ginsburg, D., and Steitz, J. A. 1975. The 30S ribosomal precursor RNA from *Escherichia coli*. A primary transcript containing 23s, 16s, and 5s sequences. *J. Biol. Chem.* **250**: 5657.

Giphart-Gassler, M., Plasterk, R. H. A., and van de Putte, P. 1982. G inversion in bacteriophage Mu: A novel way of gene splicing. *Nature* **297**: 339.

Glasgow, A. C., Hughes, K. T., and Simon, M. I. 1989. Bacterial DNA inversion systems. In Berg D. E., and M. M. Howe (eds.), *Mobile DNA*. American Society for Microbiology, Washington, D.C.

Gleich, G. J., Loegering, D. A., Bell, M. P., Checkelm, J. L., Ackerman, S. J., and McKean, D. J. 1986. Biochemical and functional similarities between human eosinophil-derived neurotoxin and eosinophil cationic protein: Homology with ribonuclease. *Proc. Natl. Acad. Sci. U.S.A.* **83**: 3146.

Goedel, W., and Helinski, D. R. 1970. Nicking activity of an endonuclease I-transfer ribonucleic acid complex of *Escherichia coli. Biochemistry* **9**: 4793.

Gold, H. A., and Altman, S. 1986. Reconstitution of RNAse P activity using inactive subunits from *E. coli* and HeLa cells. *Cell* **44**: 243.

Gold, H. A., Craft, J., Hardin, J. A., Bartkiewicz, M., and Altman, S. 1988. Antibodies in human serum that precipitate ribonuclease P. *Proc. Natl. Acad. Sci. U.S.A.* **85**: 5483.

Gold, H. A., Topper, J. N., Clayton, D. A., and Craft, J. 1989. The RNA processing enzyme RNase MRP is identical to the RNP and related to RNase P. *Science* **245**: 1377.

Goldfarb, A., and Daniel, V. 1980. An *E. coli* endonuclease responsible for primary cleavage of *in vitro* transcripts of bacteriophage T4 tRNA gene cluster. *Nucleic Acids Res.* **8**: 4501.

Golding, G., Tsao, N., and Pearlman, R. E. 1994. Evidence for intron capture: An unusual path for evolution of proteins. *Proc. Natl. Acad. Sci. U.S.A.* **93**: 7506.

Goodman, H. M. 1975. Restriction and modification of a self-complementary octanucleotide containing the *Eco*RI site. *J. Mol. Biol.* **99**: 237.

Goodman, N. 2001. Cruising in the single polynucleotide polymorphism scene. *Genome Tech.* 11.01: 48–51.

Gomes, X. V., and Burgers, P. M. 2000. Two modes of FEN1 binding to PCNA regulated by DNA. *EMBO J.* **17**: 3811–3821.

Goto, T., and Wang, J. C. 1982. Yeast DNA topoisomerase II. *J. Biol. Chem.* **257**: 5866.

Goto. T., and Wang, J. C. 1984. Yeast DNA topoisomerase II is encoded by a single-copy, essential gene. *Cell* **36**: 1073.

Goto, T., and Wang, J. C. 1985. Cloning of yeast TopI, the gene encoding DNA topoisomerase I, and construction of mutants defective in both DNA topoisomerase I and DNA topoisomerase II. *Proc. Natl. Acad. Sci. U.S.A.* **82**: 7178.

Gould, S. J., and Eldridge, N. 1977. Punctuated equilibria—The tempo and mode of evolution reconsidered. *Paleobiology* **3**: 115.

Govind, N. S., and Jayaram, M. 1987. Rapid localization and characterization of random mutations within the 2μ circle site-specific recombinase: A general strategy for analysis of protein function. *Gene* **51**: 31.

Grafstrom, R. H., and Hoess, R. H. 1983. Cloning of mutH and identification of the gene product. *Gene* **22**: 245.

Grafstrom, R. H., and Hoess, R. H. 1987. Nucleotide sequence of the *Escherichia coli mutH* gene. *Nucleic Acids Res.* **15**: 3073.

Graw, J., Schlager, E.-J., and Knippers, R. 1981. A lymphocyte ATP-dependent deoxyribonuclease. *J. Biol. Chem.* **256**: 13207.

Gray, H. B., Ostrander, D. A., Hodnett, J. L., Legerski, R. J., and Robberson, D. L. 1975. Extracellular nucleases of *Pseudomonas* BAL 31. I. Characterization of the single-strand-specific deoxyriboendonuclease and double-strand deoxyriboexonuclease. *Nucleic Acids Res.* **2**: 1459.

Gray, H. B., Ostrander, D. A., Hodnett, J. L., Legerski, R. J., Nees, D. W., Wel, C.-F., and Robberson, D. L. 1981. The extracellular nuclease from *Alteromonas espejiana:* An enzyme highly specific nonduplex structure in nominally duplex DNAs. In *Gene Amplification and Analysis*, Vol. 2: *Structural Analysis of Nucleic Acid,* Elsevier North-Holland, New York.

Green, P. 1994. The ribonucleases of higher plants. *Annu. Rev. Plant Physiol. Plant Mol. Biol.* **45**: 421.

Greene, A. L., Snipe, J. R., Gordenin, D. A., and Resnick, M. A. 1999. Functional Analysis of human FEN-1 in Saccharomyces cerevisiae and its role in genome stability. *Hum. Mol. Genet.* **8**: 2263–2273.

Greene, P. J., Heyneker, H. L., and Bolivar, F. 1978. A general method for the purification of restriction enzymes. *Nucleic Acid. Res.* **5**: 2373.

Greene, P. J., Gupta, M., Boyer, H. W., Brown, W. E., and Rosenberg, J. M. 1981. Sequence analysis of the DNA encoding *Eco*RI endonuclease and methylase. *J. Biol. Chem.* **256**: 2143.

Greer, C. L., Soll, D., and Willis, I. 1987. Substrate recognition and identification of splice sites by the tRNA-splicing endonuclease and ligase from *Saccharomyces cerevisiae. Mol. Cell. Biol.* **7**: 76.

Greer, C. L. 1986. Assembly of a tRNA splicing complex: Evidence for concerted excision and joining steps in splicing *in vitro. Mol. Cell. Biol.* **6**: 635.

Greer, C. L., Peebles, C. L., Gegenheimer, P., and Abelson, J. 1983. Mechanism of action of a yeast RNA ligase in tRNA splicing. *Cell* **32**: 537.

Grindley, N. D. F. 1983. Transposition of Tn 3 and related transposons. *Cell* **32**: 3.

Grindley, N. D. F., and Reed, R. R. 1985. Transpositional recombination in prokaryotes. *Annu. Rev. Biochem.* **54**: 863.

Gromkova, R., and Goodgal, S. H. 1976. Biological properties of a Haemophilus influenzae restriction enzyme, Hind I. *J. Bacteriol.* **127**: 848.

Gromkova, R., Bendler, J., and Goodgal, S. H. 1973. Restriction and modification of bacteriophage S2 in *Haemophilus influenzae. J. Bacteriol.* **114**: 1151.

Gronostajski, R. M., and Sadowski, P. D. 1985. The FLP recombinase of the *Saccharomyces cerevisiae* 2 μm plasmid attaches covalently to DNA via a phosphotyrosyl linkage. *Mol. Cell. Biol.* **5**: 3274.

Gross, W. A., and Cook, T. M. 1975. Nalidixic acid—Mode of action. In Corcoran, J. W., and F. E. Hahn (eds.), *Antibiotics*, Vol. III., Springer-Verlag, New York.

Gross, W. A., Deitz, W. H., and Cook, T. M. 1965. Mechanism of action of nalidixic acid on *Escherichia coli*. II. Inhibition of deoxyribonucleic acid synthesis. *J. Bacteriol.* **89**: 1068.

Guerrier-Takada, C., Gardiner, K., Marsh, T., Pace, N., and Altman, S. 1983. The RNA moiety of ribonuclease P is the catalytic subunit of the enzyme. *Cell* **35**: 849.

Guerrier-Takada C., and Altman, S. 1984. Catalytic activity of an RNA prepared by transcription *in vitro*. *Science* **223**: 285.

Gullberg, U., Widegren, B., Arnason, U., Egeston, A., and Olsson, I. 1986. The cytotoxic eosinophil cationic protein (ECP) has ribonuclease activity. *Biochem. Biophys. Res. Commun.* **139**: 1239.

Gunther, J. K., and Goodgal, S. H. 1970. An exonuclease specific for double stranded deoxyribonucleic acid. *J. Biol. Chem.* **245**: 5341.

Gusella, J. F. 1986. DNA polymorphism and human disease. *Annu. Rev. Biochem.* **55**: 831.

Guthrie, C. 1991. Messenger-RNA splitting in yeast—Clues to why the splicosome is a ribonucleoprotein. *Science* **253**: 157.

Gutte, B. 1992. The beginning of protein engineering. In Doerfler, W. (ed.), *Molecular Biology of the Cell*. VCH Publishers, New York.

Gutte, B., and Merrifield, R. B. 1971. Synthesis of ribonuclease-A. *J. Biol. Chem.* **246**: 1922.

Haber, L. T., Pang, P. P., Sobell, D. I., Mankovich, J. A., and Walker, G. C. 1988. Nucleotide sequence of the *Salmonella typhimurium* mutS gene required for mismatch repair: Homology of mutS and hexA of *Streptococcus pneumoniae*. *J. Bacteriol.* **170**: 197.

Habraken, Y., Sung, P., Prakash, L., and Prakash, S. 1993. Yeast excision repair gene RAD2 encodes a single-stranded DNA endonucleas. *Nature* **366**: 365–368.

Hadi, S-M., and Goldthwait, D. A. 1971. Endonuclease II of *Escherichia coli* degradation of partially depurinated deoxyribonucleic acid. *Biochemistry* **10**: 4986.

Halford, S. E. 1983. How does *Eco*RI cleave its recognition site on DNA? *Trends Biochem. Sci.* **8**: 455.

Halford, S. E., Johnson, N. P., and Grinstead, J. 1979. The reactions of *Eco*RI and other restriction endonucleases. *Biochem. J.* **179**: 353.

Hall, B. G. 1988. Adaptive evolution that requires multiple spontaneous mutations I mutation involving an insertion sequence. *Genetics* **120**: 887.

Hall, B. G. 1990. Spontaneous point mutations that occur more often when advantageous than when neutral. *Genetics* **129**: 5.

Hall, B. G. 1991. Adaptive evolution that requires multiple spontaneous mutations; mutations involving base substitutions. *Proc. Natl. Acad. Sci. U.S.A.* **88**: 5882.

Hall, J. D., and Mount, D. W. 1981. Mechanisms of DNA replication and mutagenesis in ultraviolet-irradiated bacteria and mammalian cells. *Prog. Nucleic Acids Res. Mol. Biol.* **25**: 53.

Halligan, B. D., Davis, J. L., Edwards, K. A., and Liu, L. F. 1982. Intra-and inter-molecular strand transfer by Hela DNA topoisomerase I. *J. Biol. Chem.* **257**: 3995.

Han, S., Udverdy, A., and Schedl, P. 1985. Novobiocin blocks the *Drosophila* heat shock response. *J. Mol. Biol.* **183**: 13–29.

Hanawalt, P., and Mellon, I. 1993. Stranded in an active gene. *Current Biology* **3**: 67.

Hanawalt, P. C., Friedberg, E. C., and Fox, C. F. (eds.). 1978. *DNA Repair Mechanisms*. Academic Press, New York.

Hanawalt, P. C., Cooper, P. K., Ganesan, A. K., and Smith, C. A. 1979. DNA repair in bacteria and mammalian cells. *Annu. Rev. Biochem.* **48**: 783.

Hane, M. W., and Wood, T. H. 1969. *Escherichia coli* K-12 mutants resistant to nalidixic acid: Genetic mapping and dominance studies. *J. Bacteriol.* **99**: 238.

Haniford, D. B., and Pulleyblank, D. E. 1983. Facile transition of poly [d(TG)-d(CA)] into a left-handed helix in physiological conditions. *Nature* **302**: 632.

Haniford, D. B., Chelouche, A. R., and Kleckner, N. 1989. A specific class of Is10 transposase mutants are blocked for target site interaction and promote formation of an excised transposon fragment. *Cell* **59**: 385.

Haring, V., Gray, J. E., McClure, B. A., Anderson, M. A., and Clarke, A. E. 1990. Self-incompatibility: A self-recognition system in plants. *Science* **250**: 937.

Harper, J. W., and Vallee, B. L. 1988. Mutagenesis of aspartic acid-116 enhances the ribonucleolytic activity and angiogenic potency of angiogenin. *Proc. Natl. Acad. Sci. U.S.A.* **85**: 7139.

Harrington, J. J., and Lieber, M. R. 1994a. The characterization of a mammalian DNA structure-specific endonuclease. *EMBO J.* **13**: 1235–1246.

Harrington, J. J., and Lieber, M. 1994b. Functional domains within FEN-1 and RAD2 define a family of structure-specific endonucleases: Implications for nucleotide excision repair. *Genes Dev.* **8**: 1344–1355.

Haseloff, J., and Gerlach, W. L. 1988. Simple RNA enzymes with new and highly specific endoribonuclease activities. *Nature* **334**: 265.

Hasunuma, K., and Ishikawa, T. 1967. Biochemical and genetical studies on regulatory mechanisms of nucleases. *Jpn. J. Genet.* **42**: 410.

Hasunuma, K., and Ishikawa, T. 1972. Properties of two nuclease genes in *Neurospora crassa. Genetics* **70**: 31.

Heffron, F., McCarthy, B. J., Ohtsubo, H., and Ohtsubo, E. 1979. DNA sequence analysis of the transposon Tn3: Three genes and three sites involved in transposition of Tn3. *Cell* **18**: 1153.

Heichman, K. A., and Johnson, R. C. 1990. The Hin invertasome: Protein mediated joining of distant recombination sites at the enhancer. *Science* **249**: 511.

Heininger, K., Horz, W., and Zachau, H. G. 1977. Specificity of cleavage by a restriction nuclease from *Bacillus subtilis. Gene* **1**: 289.

Hendrickson, E. A., Qin, X-Q, Bump, E. A., Schata, D. G., Oettinger, M., and Weaver, D. T. 1991. A link between double strand break related reapair and V(D)J recombination; The SCID mutation. *Proc. Natl. Acad. Sci. U.S.A.* **88**: 4061.

Henning, K. A., Novotny, E. A., and Compton, S. T. 1999. Human artificial chromosome generated by modification of a yeast artificial chromosome. *Proc. Natl. Acad. Sci. U.S.A.* **96**: 592–597.

Herbert, C. J., Labouesse, M., Dujardin, G., Slonimski, P. P. 1988. The NAM2 proteins from cerevisiae S. and S. douglasii are mitochondrial leucyl-tRNA synthetases, and are involved in mRNA splicing. *EMBO J.* **7**: 473.

Herr, W. 1985. Diethyl pyrocarbonate: A chemical probe for secondary structure in negatively supercoiled DNA. *Proc. Natl. Acad. Sci. U.S.A.* **82**: 8009.

Hertzberg, R. P., and Dervan, P. B. 1982. Cleavage of double helical DNA by (methidiumpropyl-EDTA) iron (II). *J. Am. Chem. Soc.* **104**: 313.

Hertzberg, R. P., and Dervan, P. B. 1984. Cleavage of DNA with methidiumpropyl-EDTA-Iron (II): Reaction conditions and product analyses. *Biochemistry* **23**: 3934.

Hevroni, D., and Livneh, Z. 1988. Bypass and termination at apurinic sites during replication of single-stranded DNA *in vitro*: A model for apurinic site mutagenesis. *Proc. Natl. Acad. Sci. U.S.A.* **85**: 5046.

Hewish D. R., and Burgoyne, L. A. 1973. The calcium dependent endonuclease activity of isolated nuclear preparations. Relationships between its occurrence and the occurrence of other classes of enzymes found in nuclear preparations. *Biochem. Biophys. Res. Commun.* **52**: 475.

Hickey, D. A., Bunkelard, B. F., and Abukashawa, S. M. 1989. A general model for the evolution of nuclear premRNA introns. *J. Theor. Biol.* **137**: 41.

Hiestand-Nauer, R., and Iida, S. 1983. Sequence of the site-specific recombinase gene cin and of its substrates serving in the inversion of the C segment of bacteriophage PI. The *EMBO J.* **2**: 1733.

Higgins, J. A., Ezzell, J., Hinnebusch, B. J., Shipley, M., Henchal, E. A., and Ibrahim, M. S. 1998. 5′-Nuclease PCR assay to detect *Yersinia pestis. J., Clin, . Microbiol.* **36**: 2284–2288.

Higgins, N. P., Peebles, C. L., Sugino, A., and Cozzarelli, N. R. 1978. Purification of subunits of *Escherichia coli* DNA gyrase and reconstitution of enzymatic activity. *Proc. Natl. Acad. Sci. U.S.A.* **75**: 1773.

Hill, C., Dodson, G. G., Heinemann, U., Saenger, W., Mitsui, Y., Nakamura, K., Bonsov, V. V., Tischenko, G. N., Polyakov, K. M., and Pavlovsky, A. G. 1983. The structural and sequence homology of a family of microbial ribonucleases. *Trends Biochem. Sci.* **8**: 364.

Hilmoe, R. J., Heppel, L. A., Springhorn, S. S., and Koshland, D. E. 1961. Cleavage point during hydrolysis catalyzed by ribonuclease and phosphodiesterase. *Biochem. Biophys. Acta* **53**: 214.

Hime, G., Prior, L., and Saint, R. 1995. The *Drosophila melanogaster* contains a member of the Rh/T2/S glycoprotein family of ribonuclease-encoding gene. *Gene* **158**: 203–207.

Hinsch, B., Mayer, H., and Kula, M.-R. 1980. Binding of nonsubstrate nucleotides to a restriction endonuclease: A model for the interaction of *Bam*HI with its recognition sequence. *Nucleic Acids Res.* **9**: 3159.

Hiraoka, L. A., Harrington, J. J., Gerhard, D. S., Lieber, M. R., and Hsieh, C.-L. 1995. Sequence of human FEN-1, a structure-specific endonuclease, and chromosomal localization in mouse and human. *Genomics* **25**: 220–225.

Hizi, A., Hughes, S. H., and Shaharabany, M. 1990. Mutational analysis of ribonuclease H activity of human immuno deficiency virus I reverse transcriptase. *Virology* **175**: 575.

Hoeijmakers, J. H. J. 1993a. Nucleotide excision repair I from *E. coli* to yeast. *Trends Cell. Biol.* **9**: 173.

Hoeijmakers, J. H. J. 1993b. Nucleotide excision repair II from yeast to mammals. *Trends Cell Biol.* **9**: 21.

Hoess, R. H., and Abremski, K. 1984. Interaction of the bacteriophage P1 recombinase Cre with the recombining site LoxP. *Proc. Natl. Acad. Sci. U.S.A.* **81**: 1026.

Hoess, R. H., and Abremski, K. 1985. Mechanism of strand cleavage and exchange in the Cre-Lox site-specific recombination system. *J. Mol. Biol.* **181**: 351.

Hoess, R. H., and Landy, A. 1978. Structure of the λ att sites generated by independent deletions. *Proc. Natl. Acad. Sci. U.S.A.* **75**: 5437.

Hoess, R. H., Foeller, C., Bidwell, K., and Landy, A. 1980. Site-specific recombination functions of bacteriophage λ: DNA sequence of regulatory regions and overlapping structural genes for Int and Xis. *Proc. Natl. Acad. Sci. U.S.A.* **77**: 2482.

Holland, P. M., Abramson, R. D., Watson, R., and Gelfand, D. H. 1991. Detection of a specific polymerase chain reaction product by utilizing the 5′-3′ exonuclease activity of Thermus aquaticus DNA polymerase. *Proc. Natl. Acad. Sci. U.S.A.* **88**: 7276–7280.

Holley, R. W. 1965. The nucleotide sequence of a nucleic acid. *Sci. Am.* **214**: 30.

Holliday, R. 1964. A mechanism for gene conversion in fungi. *Genet. Res.* **5**: 282.

Holliday, R., Holloman, W. K., Banks, G. R., Unrau, P., and Pugh, J. E. 1974. Genetic and biochemcial studies of recombination in *Ustilago maydis*. In Grell, R. F. (ed.), *Mechanisms in Recombination*. Plenum Press, New York.

Hollis, G. F., and Grossman, L. 1981. Purification and characterization of DNAse VII, a 3′→5′-directed exonuclease from human placenta. *J. Biol. Chem.* **256**: 8074.

Holloman, W. K., and Holliday, R. 1973. Studies on a nuclease from *Ustilago maydis*. I., Purification, properties, and implication in recombination of the enzyme. *J. Biol. Chem.* **248**: 8107.

Holloman, W. K. 1973. Studies on a nuclease from *Ustilago maydis*. II. Substrate specificity and mode of action of the enzyme. *J. Biol. Chem.* **248**: 8114.

Holloman, W. K., Wiegand, R., Hoessli, C., and Radding, C. M. 1975. Uptake of homologous single-stranded fragments by superhelical DNA: A possible mechanism for initiation of genetic recombination. *Proc. Natl. Acad. Sci. U.S.A.* **72**: 2394.

Holm, C., Goto, T., Wang, J. C., and Botstein, D. 1985. DNA topoisomerase II is required at the time of mitosis in yeast. *Cell* **41**: 553.

Holmes, A. M., and Haber, J. E. 1999. Double strand break repair in yeast requires both leading and lagging strand DNA polymerase. *Cell* **96**: 415–424.

Holmes, J., Clark, S., and Modrich, P. 1990. Strand specific mismatch correction in nuclear extracts of human and *Drosophila melanogaster* cell lines. *Proc. Natl. Acad. Sci. U.S.A.* **87**: 5837.

Honigberg, S. M., and Radding, C. M. 1988. The mechanics of winding and unwinding helices in recombination: Torsional stress associated with strand transfer promoted by *RecA* protein. *Cell* **54**: 525.

Hopper, A. K., and Schultz, L. D. 1980. Processing of intervening sequences: A new yeast mutant which fails to excise intervening sequences from precursor tRNAs. *Cell* **19**: 741.

Hopper, A. K., Banks, F., and Evangelidis, V. 1978. A yeast mutant which accumulates precursor tRNAs. *Cell* **14**: 211.

Horz, M., and Altenberger, W. 1981. Sequence specific cleavage of DNA by micrococcal nuclease. *Nucleic Acids Res.* **9**: 2643.

Hosfield, D. J., Mol, C. D., Shen, B., and Trainer, J. A. 1998. Structure of the DNA repair and replication endonuclease and exonuclease NEN-1: Coupling DNA and PCNA binding to FEN-1 activity. *Cell* **95**: 135–146.

Hostomsky, Z., Hostomska, Z., Hudson, G. F., Mooman, E. W., and Nodes, B. R. 1991. Reconstitution *in vitro* of RNaseH activity by using purified N-terminal and C-terminal domains of human immunodeficiency virus type 1 reverse transcriptase. *Proc. Natl. Acad. Sci. U.S.A.* **88**: 1148.

Howard-Flanders, P. 1981. Inducible repair of DNA. *Sci. Am.* **245**: 72.

Huai, Q., Colandene, J. D., Chen, Y., Luo, F., Zhao, Y., Topal, M. D., and Ke, H. 2000. Crystal structure of NaeI—An evolutionary bridge between DNA endonuclease and topoisomerase. *EMBO J.* **19**: 3110–3118.

Hsu, P. L., and Landy, A. 1984. Resolution of synthetic Att-site Holliday structures by the integrase protein of bacteriophage λ. *Nature* **311**: 721.

Huang, J. C., Savoda, D. L., Reardon, J. T., and Sancar, A. 1992. Human nucleotide excision nuclease removes thymine dimers from DNA by incising the 22nd phosphodiester bond 5 and the 6th phosphodiester bond 3 to the photodimer. *Proc. Natl. Acad. Sci. U.S.A.* **89**: 3664.

Huang, S., Li, B., Gray, M. D., Oshima, J., Mian, I. S., and Campisi, J. 1998. The premature ageing syndrome protein, WRN is a 3′–5′ exonuclease. *Nat. Genet.* **20**: 114–116.

Huber, H., Iida, S., and Bickle, T. A. 1985. Expression of the bacteriophage P1 Cin recombinase gene from its own and heterologous promoters. *Gene* **34**: 63.

Huess, R., Wierzbicki, A., and Abremski, K. 1987. Isolation and characterization of intermediates in site-specific recombination. *Proc. Natl. Acad. Sci. U.S.A.* **84**: 6840.

Hutchinson, C. A., III, Hardies, S. C., Loeb, D. D., Shehee, W. R., and Edgell, M. H. 1989. LINES and related retroposons: Long interspersed repeated sequences in the eucaryotic genome. In Berg, D. E., and M. M. Howe (eds.), *Mobile DNA*, American Society of Microbiology, Washington, D.C.

Hutchinson, C. A., III, and Edgell, H. 1971. Genetic assay for small fragments of bacteriophage φX174 deoxyribonucleic acid. *J. Virology* **8**: 181.

Hutchinson, C. A., III, Newbold, J. E., Potter, S. S., and Edgell, M. H. 1974. Maternal inheritance of mammalian mitochondrial DNA. *Nature* **251**: 536.

Iida, R., Yasuda, T., Aoyama, M., Tsubota, E., Kobayashi, M., Yuasa, I., Matsuki, T., and Kishi, K. 1997. The fifth allele of the human DNaseI polymorphism. *Electrophoresis* **18**: 1936–1939.

Iida, S., Meyer, J., Kennedy, K. E., and Arber, W. 1982. A site-specific, conservative recombination system carried by bacteriophage PI. Mapping the recombinase gene Cin and the cross-over sites Cix for the inversion of the C segment. *EMBO J.* **1**: 1445.

Iida, S. 1984. Bacteriophage PI carries two related sets of genes determining its host range in the invertible C segment of its genome. *Virology* **134**: 421.

Iino, T. 1984. Role of recombination in genetic regulatory systems. *Proc. Int. Genet. Cong.* (New Delhi, India). **1**: 407.

Ikeda, J.-E., Yudelevich, A., and Hurwitz, J. 1976. Isolation and characterization of the protein coded by gene A of bacteriophage φX174 DNA. *Proc. Natl. Acad. Sci. U.S.A.* **73**: 2669.

Ikehara, M., Ohtsuka, E., Togunaga, T., Nishikawa, S., Uesugi, S., Tanaka, T., Aoyana, Y., Kikyodani, S., Fujimoto, K., Yanase, K., Fuchimura, K., and Morioka, H. 1986. Inquiries into structure and function relationship of ribonuclease T1 using chemically synthesized coding sequences. *Proc. Natl. Acad. Sci. U.S.A.* **83**: 4695.

Iordanov, M. S., Rubinina, O. P., Wong, J., Dinh, T. H., Newton, D. L., Rybak, S. M., and Magun, B. E. 2000. Molecular determinants of apoptosis induced by the cytotoxic ribonuclease onconase: Evidence for cytotoxic mechanisms different from inhibition of protein synthesis. *Cancer Res.* **60**: 1983–1994.

Ishikama, A., Mizumoto, K., Kawakami, K., Kato, A., and Honda, A. 1986. Proofreading function associated with the RNA-dependent RNA polymerase from influenza virus. *J. Biol. Chem.* **261**: 10417.

Ishikawa, T., Toh-e, A., and Hasunuma, K. 1969. Isolation and characterization of nuclease mutants. *Genetics* **63**: 75.

Itaya, M. 1990. Isolation and characterization of a second RNaseH (RNase H-II) of *Eschericia coli K12* encoded by the rnh β gene. *Proc. Natl. Acad. Sci. U.S.A.* **87**: 8587.

Itaya, M., McKelvin, D., Chatterjee, S. K., and Crouch, R. J. 1991. Selective cloning of genes encoding RNaseH from *Salmonella typhimurium, Saccharomyces cerevisiae*, and *Escherichia coli* rnh mutant. *Mol. Gen. Genet.* **227**: 438.

Iwanoff, L. 1903. Fermentative decomposition of thymo-nucleic acid by fungi. *Z. Physiol. Chem.* **39**: 31.

Jackson, D. A., Symons, R. H., and Berg, P. 1972. Biochemical method for inserting new genetic information in DNA of simian virus 40: Circular SV40 DNA molecule containing λ phage genes and the galactose opening of *E. coli. Proc. Natl. Acad. Sci. U.S.A.* **69**: 3370.

Jacquiver, A., and Dijon, B. 1985. An intron-encoded protein is active in a gene conversion process that spreads an intron into a mitochondrial gene. *Cell* **41**: 383.

Jakubsczak, J. L., Burke, W. D., and Eichbush, T. H. 1991. Retrotransposable element R1 and R2 interrupt the rRNA genes of most insects. *Proc. Natl. Acad. Sci. U.S.A.* **88**: 3295.

Jayaram, M. 1985. Two-micrometer circle site-specific recombination: The minimal substrate and the possible role of flanking sequences. *Proc. Natl. Acad. Sci. U.S.A.* **82**: 5875.

Jayaram, M. 1988. The *Int* family of site-specific recombinase: Some thoughts on a general reaction mechanism. *J. Genet.* **67**: 29.

Jayaram, M., Crain, K. L., Parsons, R. L., and Harshey, R. M. 1988. Holliday junctions in FLP recombination: Resolution by step-arrest mutants of FLP protein. *Proc. Natl. Acad. Sci. U.S.A.* **85**: 7902.

Jeffreys, A. J., Wilson, V., and Thein, S. L. 1985. Hypervariable minisatellite regions in human DNA. *Nature* **314**: 67–73.

Jeltsch, A., Kroger, M., and Pingoud, A. 1995. Evidence for an evolutionary relationship among type-II restriction endonucleases. *Gene* **160**: 7–16.

Jeyaseelan, R., and Shanmugam, G. 1988. Human placental endonuclease cleaves Holliday junctions. *Biochem. Biophys. Res. Commun.* **156**: 1054.

Jensch, F., and Kemper, B. 1986. Endonuclease VII resolves Y-junctions in branched DNA *in vitro*. The *EMBO J.* **5**: 181.

Jensch, F., Kosak, H., Seeman, N. C., and Kemper, B. 1989. Cruciform cutting endonucleases from *Saccharomyces cerevisiae* and phage T_4 show conserved reactions with branched DNAs. *EMBO J.* **8**: 4325.

Johnson, R. C., and Simon, M. I. 1985. Hin-mediated site-specific recombination requires two 26 bp recombination sites and a 60 bp recombinational enhancer. *Cell* **41**: 781.

Johnson, R. C., Bruist, M. F., and Simon, M. I. 1986. Host protein requirements for *in vitro* site-specific DNA inversion. *Cell* **46**: 531.

Johnson, R. C., Glasgow, A. C., and Simon, M. I. 1987. Spatial relationship of the Fis binding sites for Hin recombinational enhancer activity. *Nature* **329**: 462.

Johnson, R. C., Ball, C. A., Pfeffer, D., and Simon, M. I. 1988. Isolation of the gene encoding the Hin recombinational enhancer binding protein. *Proc. Natl. Acad. Sci. U.S.A.* **85**: 3484.

Jones, B. K., and Yeung, A. T. 1988. Repair of 4, 5, 8-trimethylpsoralen monoadducts and cross-links by the *Escherichia coli uvrABC* endonuclease. *Proc. Natl. Acad. Sci. U.S.A.* **80**: 8410.

Jones, M., Wagner, R., and Rodman, M. 1987. Repair of a mismatch is influenced by the base composition of the surrounding nucleotide sequence. *Genetics* **115**: 605.

Jorgenson, S. E., and Koerner, J. F. 1966. Separation and characterization of deoxyribonucleases of *Escherichia coli*. B. I. Chromotographic separation and properties of two deoxyribo-oligonucleotidases. *J. Biol. Chem.* 241: 3090.

Joyce, C. M., and Steitz, T. 1987. DNA polymerase I: From crystal structure to function via genetics. *Trends Biochem. Sci.* **12**: 288.

Käfer, E., and Fraser, M. 1979. Isolation and genetic analysis of nuclease halo (*nuh*) mutants of *Neurospora crassa*. *Mol. Gen. Genet.* **169**: 117.

Kahmann, R., Rudt, F., Koch, C., and Mertens, G. 1985. G inversion in bacteriophage Mu DNA is stimulated by a site within the inverse gene and a host factor. *Cell* **41**: 771.

Kamp, D., Kahmann, R., Zipser, D., Broker, T. R., and Chow, L. T. 1978. Inversion of the G1 DNA segment of phage Mu controls phage infectivity. *Nature* **271**: 577–580.

Kamp, D., and Kahmann, R. 1981. The relationship of two invertible segments in bacteriophage Mu and *Salmonella typhimurium* DNA. *Mol. Gen. Genet.* **184**: 564.

Kan, Y. W., and Dozy, A. M. 1978. Polymorphism of DNA sequence adjacent to human β-globin structural gene: Relationship to sickle mutation. *Proc. Natl. Acad. Sci. U.S.A.* **75**: 5631.

Kanaar, R., van de Putte, P., and Cozzarelli, N. R. 1988. Gin-mediated DNA inversion: Product structure and the mechanism of strand exchange. *Proc. Natl. Acad. Sci. U.S.A.* **85**: 752.

Kane, C. M., and Linn, S. 1981. Purification and characterization of an apurinic/apyrimidinic endonuclease from HeLa cells. *J. Biol. Chem.* **256**: 3405.

Kantor, G. J., Barsalou, L. S., and Hanawalt, P. C. 1990. Selective repair of specific chromatin domains in UV irradiated cells from *Xeroderma pigmentosum* complementation group C Mutant Res. *DNA Repair* **235**: 71.

Karanjawala, Z., Shi, E. X., Hsieh, C. L., and Lieber, M. R. 2000. The mammalian FEN-1 locus: Structure and conserved sequence features. *Microb. Comp. Genomics* **5**: 173–177.

Karsen, T., Kuck, U., and Esser, K. 1987. Mitochondrial group I introns from the filamentous fungus *Podospora anserina* code for polypeptides related to maturases. *Nucleic Acids Res.* **15**: 6743.

Kasai, K., and Grunberg-Manago, M. 1967. Sheep kidney nuclease. Hydrolysis of tRNA. *Eur. J. Biochem.* **1**: 152.

Katayanagi, K., Miyagawa, M., Matsushima, M., Ishikawa, M., Kanaya, S., Ikehara, M., Matsuzaki, T., and Marihawa, K. 1990. Three-dimensional structure of ribonuclease H from *E. coli*. *Nature* **347**: 306.

Kato, A. C., Bartik, K., Fraser, M. J., and Denhardt, D. T. 1972. Sensitivity of superhelical DNA to a single-strand specific endonuclease. *Biochim. Biophys. Acta* **308**: 68.

Katz, R. A., Merkel, G., Kulkosky, J., Leis, J., and Skalka, A. M. 1990. The avian retroviral int protein is both necessary and sufficient for integrative recombination *in vitro*. *Cell* **63**: 87.

Katzman, M. J., Mack, P. G., Skalka, A. M., and Leis, J. 1991. A covalent complex between retroviral integrase and nicked substrate DNA. *Proc. Natl. Acad. Sci. U.S.A.*

Kauc, L., and Piekarowicz, A. 1978. Purification and properties of a new restriction endonuclease from *Haemophilus influenzae*. *Eur. J. Biochem.* **92**: 417.

Kawata, Y., Sakiyama, F., and Tamaoki H. 1988. Amino acids sequences of ribonuclease T2 from Aspergillus oryzae. *Eur. J. Biochem.* **176**: 683–697.

Kell, F., and Belcour, L. 1989. Are intron-encoded specific endonucleases responsible for nonhomologous recombination. *Trends Genet.* **5**: 173.

Keller, W., and Crouch, R. 1972. Degradation of DNA–RNA hybrids by ribonuclease H and DNA polymerases of cellular and viral origin. *Proc. Natl. Acad. Sci. U.S.A.* **69**: 3360.

Keller, W. 1975a. Characterization of purified DNA-relaxing enzyme from human tissue culture cells. *Proc. Nat. Acad. Sci. U.S.A.* **72**:

Keller, W. 1975b. Determination of the number of superhelical turns in simian virus 40 DNA by gel electrophoresis. *Proc. Natl. Acad. Sci. U.S.A.* **72**: 4876.

Kelly, T. J., and Smith, H. O. 1970. A restriction enzyme from *Haemophilus influenzae*. II. Base sequence of the recognition site. *J. Mol. Biol.* **51**: 393.

Kemey, S., Eker, A. P. M., Brody, T., Vermeulen, W., Bootsma, D., Hoeijmakers, J. J., and Linn, S. 1994. Correction of the DNA repair defect in *Xeroderma pigmentosum* group E by injection of a DNA damage-binding protein. *Proc. Natl. Acad. Sci. U.S.A.* **91**: 4053.

Kemp, L. M., Sedgewick, S. G., and Jeggo, P. A. 1984. X-ray sensitive mutants of Chinese hamster ovary cells defective in double strand breakjoining. *Mutant Res.* **132**: 189.

Kemper, B., and Brown, D. T. 1976. Function of gene 49 of bacteriophage T4. II. Analysis of intracellular development and the structure of very fast-sedimenting DNA. *J. Virol.* **18**: 1000.

Kenyon, C. J., and Walker, G. C. 1980. DNA-damaging agents stimulate gene expression at specific loci in *Escherichia coli*. *Proc. Natl. Acad. Sci. U.S.A.* **77**: 2819.

Kikuchi, Y., and Nash, H. A. 1978. The bacteriophage λ *int* gene product. A filter assay for genetic recombination, purification of *int* and specific binding to DNA. *J. Biol. Chem.* **253**: 7149.

Kikuchi, Y., and Nash, H. A. 1979. Nicking-closing activity associated with bacteriophage λ *int* gene product. *Proc. Natl. Acad. Sci. U.S.A.* **76**: 3760.

Kikuchi, Y., Sasaki, N., and Ando-Yamagami, Y. 1990. Cleavage of tRNA within the mature tRNA sequence by the catalytic RNA of RNase P: Implication for the formation of the primer tRNA fragment for reverse transcription in *copia* retrovirus-like particles. *Proc. Natl. Acad. Sci. U.S.A.* **87**: 8105.

Kim, I. S., Lee, M. Y., Lee, I. H., Shin, S. L., and Lee, S. Y. 2000. Gene Expression of flap endonuclease-1 during cell proliferation and differentiation. *Biochim. Biophys. Acta* **17**: 333–340.

Kim, S. C., Podhajska, A. J., and Szybalski, W. 1988. Cleaving DNA at any predetermined site with adapter-primers and class-IIS restriction enzymes. *Science* **240**: 504.

Kim, Y.-G., and Chandrasegaran, S. 1994. Chimeric restriction endonuclease. *Proc. Natl. Acad. Sci. U.S.A.* **91**: 883.

Kim, Y., Grable, J. C., Love, R., Greene, P. J., and Rosenberg, J. M. 1990. Refinement of *EcoRI* endonuclease crystal structure: A revised protein chain tracing. *Science* **249**: 1307.

Kindler, P., Keil, T. U., and Hofschneider, P. H. 1973. Isolation and characterization of a ribonuclease III deficient mutant of *Escherichia coli*. *Mol. Gen. Genet.* **126**: 53.

King, M. C., and Wilson, A. C. 1975. Evolution at two levels in humans and chimpanzees. *Science* **188**: 107.

King, K., Benkovic, S. J., and Modrich, P. 1989. Glucose III is required for activation of the DNA cleavage center of *EcoRI* endonuclease. *J. Biol. Chem.* **264**: 11807.

Kippen, A. D., Sancho, J., and Fresht, A. R. 1994. Folding of barnase in parts. *Biochemistry* **33**: 3778.

Kirchhausen, T., Wang, J. C., and Harrison, S. C. 1985. DNA gyrase and its complexes with DNA: Direct observation by electron microscopy. *Cell* **41**: 933.

Kirkgaard, K., and Wang, J. C. 1978. *Escherichia coli* DNA topoisomerase I catalyzed linking of single-stranded rings of complementary base sequences. *Nucleic Acids Res.* **5**: 3811.

Kitts, P. A., and Nash, H. A. 1987. Homology-dependent interactions in phage λ site specific recombination. *Nature* **329**: 346.

Kitts, P., Symington, L., Burke, M., Reed, R., and Sherratt, D. 1982. Transposon-specified site specific recombination. *Proc. Natl. Acad. Sci. U.S.A.* **79**: 46.

Kleckner, N. 1981. Transposable elements in prokaryotes. *Annu. Rev. Genet.* **15**: 341.

Klein, H. L. 1988. Different types of recombination events are controlled by the RAD1 and RAD52 genes of *Saccharomyces cerevisia. Genetics* **120**: 367.

Klenow, H., and Henningsen, I. 1970. Selective elimination of the exonuclease activity of the deoxyribonucleic acid polymerase from *Escherichia coli* by limited proteolysis. *Proc. Natl. Acad. Sci. U.S.A.* **65**: 168.

Klimasauskas, S., Kumar, S. J., Roberts, R. J., and Cheng, X. 1994. HhaI methyl transferase flips its target base out of the DNA helix. *Cell* **76**: 357.

Kline, L., Mishikawa, S., and Soll, D. 1981. Partial purification of RNaseP from *Schizo-saccharomyces pombe. J. Biol. Chem.* **256**: 5058.

Klungsland, A., and Lindahl, T. 1997. Second pathway for completion of human DNA base excision repair: Reconstitution with purified proteins and requirement for DnaseIV (FEN1). *EMBO J.* **16**: 3341–3348.

Kmiec, E. B., and Holloman, W. K. 1986. Homologous pairing of DNA molecules by *Ustilago Rec1* protein is promoted by sequences of Z-DNA. *Cell* **44**: 545.

Kmiec, E. B., and Worcel, A. 1985. The positive transcription factor of the 5S RNA gene induces a 5S DNA specific gyration in *Xenopus* oocyte extracts. *Cell* **41**: 945.

Kmiec, E. B., Kraeger, P. E., Brougham, M. J., and Holloman, W. K. 1986. Topological linkage of circular DNA molecules promoted by *Ustilago Rec1* protein and topoisomerase. *Cell* **34**: 919.

Knapp, G., Beckmann, J. S., Jackson, P. F., Fuhrman, S. A., and John Abelson. 1978. Transcription and processing of intervening sequences in yeast tRNA genes. *Cell* **14**: 221.

Kobayashi, H., Kumagai, F., Itagaki, T., Ninokuchi, Tkoyama, Miwama, Ohgi, K., and Irie, K. 2000. Amino acid sequence of a nuclease (Le1) from *Lentinus edodes. Biosci. Biotechnol. Biochem.* **64**: 948–957.

Kobe, B., and Deisenhofer, J. 1996. Mechanism of ribonuclease inhibition by ribonulease inhibitor protein based on the crystal structure of its complex with ribonuclease A. *J. Mol. Biol.* **264**: 1028–1043.

Koizumi, M., Iwai, S., and Ohtsuka, E. 1988. Construction of a series of several self-cleaving RNA duplexes using synthetic 21-mers. *FEBS.* **228**: 228.

Kole, R., Baer, M. F., Stark, B. C., and Altman, S. 1980. *E. coli* RNaseP has a required RNA component *in vivo. Cell* **19**: 881.

Koll, F., Begel, O., and Belcour, L. 1987. Insertion of short poly(A)d(T) sequences at recombination junctions in mitochondrial DNA of *Podospora. Mol. Gen. Genet.* **209**: 630.

Kolodner, R., Joseph, J., James, A., Dean, F., and Doherty, M. J. 1980. Genetic recombination of bacterial plasmid DNAs *in vivo* and *in vitro*. In Alberts, B. (ed.), *Mechanistic Studies of DNA Replication and Genetic Recombination*. Academic Press, New York.

Kolodner, R. 1980. Genetic recombination of bacterial plasmid DNA: Electron micro-scopic analysis of *in vitro* intramolecular recombination. *Proc. Natl. Acad. Sci. U.S.A.* **77**: 4847.

Kolodner, R., Evans, D. H., and Marrison, P. T. 1987. Purification and characterization of an activity from *Saccharomyces cerevisiae* that catalyzes homologous pairing and strand exchange. *Proc. Natl. Acad. Sci. U.S.A.* **84**: 5560–5564.

Komano, T. 1999. Shufflons: Multiple inversion systems and integrons. *Annu. Rev. Genet.* **33**: 171.

Kondo, S., Katooka, T., Nighe, M., Kodaira, M., Takeda, S., and Honjo, T. 1984. Endonuclease J: A site specific endonuclease cleaving immunoglobulin genes. *Cold Spring Harbor Symp. Quant. Biol.* **49**: 661.

Koob, M., and Szybalski, W. 1990. Cleaving yeast and *Escherichia coli* genomes at a single site. *Science* **233**: 727.

Kooistra, J., and Venema, G. 1970. Fate of donor DNA in some poorly transformable strains of *Hemophilus influenzae. Mutation Res.* **9**: 245.

Kooistra, J., and Venema, G. 1973. Poor transformobility with Novr and cry^r donor DNA of some mitomycin-c-sensitive strains of *Hemophilus influenzae. Mutat. Res.* **20**: 313.

Kooistra, J., and Venema, G. 1974. Fate of donor deoxyribonucleic acid in a highly transformation deficient strain of *Hemophilus influenzae. J. Bacteriol.* **119**: 705.

Kooistra, J., and Venema, G. 1976. Effect of adenosine 5-triphosphate dependant deoxy ribonuclease. Deficiency on properties and transformation of *Hemophilus influenzae* strains. *J. Bacteriol.* **128**: 548.

Korenberg, J. R., and Rykowski, M. C. 1988. Human genome organization: Alu, LINES, and the molecular structure of metaphase chromosome bands. *Cell* **53**: 391.

Kornberg, A. 1961. *Enzymatic Synthesis of DNA*, CIBA Lectures. John Wiley & Sons, New York.

Kornberg, A. 1974. *DNA synthesis.* W. H., Freeman, San Francisco.

Kornberg, A. 1980. *DNA replication.* W. H., Freeman, San Francisco.

Kornberg, A. 1987. Chemistry and biology. *Chem. Eng. News* March 9, Editorial page 3.

Kornberg, A., and Baker, A. 1992. *DNA Replication.* W. H., Freeman, San Francisco.

Kosak, H. G., and Kemper, B. W. 1990. Large-scale preparation of T4 endonuclease VII from over-expressing bacteria. *Eur. J. Biochem.* **194**: 779.

Koski, R. A., Bothwell, A. L. M., and Altman, S. 1976. Identification of a ribonuclease P-like activity from human KB cells. *Cell* **9**: 101.

Kostriken, R., Strathern, J. N., Klar, A. J. S., Hicks, J. B., and Heffron, F. 1983. A site specific endonuclease essential for mating-type switching in *Saccharomyces cerevisiae. Cell* **35**: 167.

Kozarich, J. W., Worth, L., Frank, B. L., Christner, D. F., Vanderwall, D. E., and Stubbe, J. 1989. Sequence-specific DNA-cleaving peptide. *Science* **238**: 1129.

Kramer, W., Kramer, B., and Fritz, H. J. 1984. Different base/base mismatches are corrected with different efficiencies by the methyl-directed DNA mismatch-repair system of *E. coli. Cell* **38**: 879.

Kramer, W., Kramer, B., Williamson, M. S., and Fogel, S. 1989. Cloning and nucleotide sequence of DNA mismatch repair gene PMS1 from *Saccharomyces cerevisiae*: Homology of PMS1 to procaryotic MutL and HexB. *J. Bacteriol.* **171**: 5339.

Krasnow, M. A., and Cozzarelli, N. R. 1982. Catenation of DNA rings by topoisomerases. *J. Biol. Chem.* **257**: 2687.

Kress, M., Giaros, D., Khoury, G., and Jay, G. 1983. Alternative RNA splicing in expression of the H-2K gene. *Nature* **306**: 602.

Kreuzer, K. N., and Cozzarelli, N. R. 1980. Formation and resolution of DNA catenates by DNA gyrase. *Cell* **20**: 245.

Krieser, R. J., and Eastman, A. 1998. The cloning and expression of human Deoxyribonuclease II. A possible role in apoptosis. **273**: 30909–30914.

Krocker, W. D., Hanson, D. M., and Fairley, J. L. 1975. Activity of wheat seedlings nuclease toward single-stranded nucleic acids. *J. Biol. Chem.* **250**: 3767.

Krocker, W., Kowalski, D., and Laskowski, M., Sr. 1976. Mung bean nuclease I. Terminally directed hydrolysis of native DNA. *Biochemistry* **15**: 4463.

Krug, M. C., and Berger, S. L. 1989. Ribonuclease H activities associated with viral reverse transcriptase are endonucleases. *Proc. Natl. Acad. Sci. U.S.A.* **86**: 3539.

Kruger, K., Grabowski, P. J., Zaug, A. J., Sands, J., Gottschling, D. E., and Cech, T. R. 1982. Self-splicing RNA: Autoexcision and autocyclization of the ribosomal RNA intervening sequence of Tetrahymena. *Cell* **31**: 147.

Krupp, G., Cherayil, B., Frendewey, D., Nishikawa, K., and Söll., D. 1986. Two RNA species co-purify with RNase P from the fission yeast *Schizosaccharomyces pombe*. *EMBO J.* **5**: 1697.

Kubo, A., Kusukawa, A., and Komano, T. 1988. Nucleotide sequence of the rci gene encoding shufflon-specific DNA recombinase in the IncI1 plasmid R64: Homology to the site-specific *Mol. Gen. Genet.* **213**(1): 30–5.

Kuhnlein, U., Penhoet, E. E., and Linn, S. 1976. An altered apurinic DNA endonuclease activity in group A and group D *Xeroderma pigmentosum* fibroblasts. *Proc. Natl. Acad. Sci. U.S.A.* **73**: 1169.

Kuhnlein, U., Lee, B., Penhoet, E. E., and Linn, S. 1978. *Xeroderma pigmentosum* fibroblasts of the D group lack an apurinic endonuclease species with a low apparent Km. *Nucleic Acids Res.* **5**: 951.

Kumar, R., Chattopadhyay, D., Bannerjee, A., and Sen, G. C. 1988. Ribonuclease activity is associated with subviral particles isolated from interferon-treated vesicular somatitis virus infected cells. *J. Virol.* **62**: 641.

Kunitz, M. 1940. Crystalline ribonuclease. *J. Gen. Physiol.* **24**: 15.

Kunitz, M. 1950. Crystalline desoxyribonuclease I. Isolation and general properties. Spectrophotometric method for the measurement of deoxyribonuclease activity. *J. Gen. Physiol.* **33**: 349.

Kunkel, T. A. 1988. Exonucleolytic proofreading. *Cell* **53**: 837.

Kunkel, T. A., Eckstein, F., Milbuan, A. S., Koplitz, R. M., and Loeb, L. A. 1981a. Deoxynucleoside [1-thio] triphosphates prevent proofreading during *in vitro* DNA synthesis. *Proc. Natl. Acad. Sci. U.S.A.* **78**: 6734.

Kunkel, T. A., Schaaper, R. M., Beckman, R. A., and Loeb, L. A. 1981b. On the fidelity of DNA replication. *J. Biol. Chem.* **256**: 9883.

Kunkel, T. A., Salatino, R. D., and Bambara, R. A. 1987. Exonucleolytic proofreading by calf thymus DNA polymerase δ. *Proc. Natl. Acad. Sci. U.S.A.* **84**: 4865.

Kunuic, E. B., Ryoji, M., and Worcel, A. 1986. Gyration is required for 5S RNA transcription from a chromatin template. *Proc. Natl. Acad. Sci. U.S.A.* **83**: 1305.

Kurachi, K., Davie, E. W., Strydom, D. J., Riordan, J. R., and Vallee, B. L. 1985. Sequence of the cDNA and gene for angiogenin, a human angiogenesis factor. *Biochemistry* **24**: 5494.

Kurihara, H., Mitsui, Y., Ohgi, K., Irie, M., Mizuno, H., and Nakamura, K. 1992. Crystal and molecular structure of RNase Rh, a new class of microbial ribonucleases from *Rhizopus niveus*. *FEBS Lett.* **306**: 189–192.

Kutsukake, K., and Iino, T. 1980. Inversions of specific DNA segments in flagellar phase variation of *Salmonella* and inversion systems of bacteriophages P1 and Mu. *Proc. Natl. Acad. Sci. U.S.A.* **77**: 4196.

Kuznetsov, G., and Nigam, S. K. 1998. Folding of secretory and membrane proteins. *N. Engl. J. Med.* **339**: 1688–1695.

Labouesse, M., Dujardin, G., and Slonimski, P. P. 1985. The yeast nuclear gene NAM2 is essential for mitochondrial DNA integrity and can cure a mitochondrial RNA-maturase deficiency. *Cell* **41**: 133.

Lachmann, P. J. 1996. The *in vivo* destruction of antigen—A tool for probing and modulating an immune response. *Clin. Exp. Immunol.* **106**: 187–189.

Lacks, S. 1970. Mutants of *Diplococcus pneumoniae* that lack deoxyribonucleases and other activities possibly pertinent to genetic transformation. *J. Bacteriol.* **101**: 373.

Lacks, S. A. 1981. Deoxyribonuclease I in mammalian tissues specificity of inhibition by actin. *J. Biol. Chem.* **256**: 2644.

Lacks, S., and Greenberg, B. 1967. Deoxyribonucleases of *Pneumococcus*. *J. Biol. Chem.* **242**: 3108.

Lacks, S., and Greenberg, B. 1973. Competence for deoxyribonucleic acid uptake and deoxyribonuclease action external to cells in the genetic transformation of *Diplococcus pneumoniae*. *J. Bacteriol.* **114**: 152.

Lacks, S., and Greenberg, B. 1975. A deoxyribonuclease of *Diplococcus pneumoniae* specific for methylated DNA. *J. Biol. Chem.* **250**: 4060–4066.

Lacks, S., and Neuberger, M. 1975. Membrane location of a deoxyribonuclease implicated in the genetic transformation of *Diplococcus pneumoniae*. *J. Bacteriol.* **124**: 1321.

Lacks, S. B., Greenberg, B., and Carson, K. 1967. Fate of donor DNA in pneumococcal transformation. *J. Mol. Biol.* **29**: 327.

Lacks, S., Greenberg, B., and Neuberger, M. 1974. Role of a deoxyribonuclease in the genetic transformation of *Diplococcus pneumoniae*. *Proc. Natl. Acad. Sci. U.S.A.* **71**: 2305.

Lacks, S., Greenberg, B., and Neuberger, M. 1975. Identification of deoxyribonuclease implicated in genetic transformation of *Diplococcus pneumoniae*. *J. Bacteriol.* **123**: 222.

Lacks, S. A., Dunn, J. J., and Greenberg, B. 1982. Identification of base mismatches recognized by the heteroduplex-DNA-repair system of *Streptococcus pneumoniae*. *Cell* **31**: 327.

Laengle-Ranault, F., Maenhaut-Michel, G., and Radman, M. 1986. GATC sequence and mismatch repair in *Escherichia coli*. *EMBO J.* **5**: 2009.

Lahuc, R. S., Su, S. S., and Modrich, P. 1987. Requirement for (GATC) sequences in *Escherichia coli* mutHLS mismatch correction. *Proc. Natl. Acad. Sci. U.S.A.* **84**: 1482.

Lai, M. M. C. 1992. RNA recombination in animal and plant viruses. *Microbiol. Rev.* **56**: 61–79.

Lai, C., and Nathans, D. 1974. Mapping temperature sensitive mutants of simian virus 40: Rescue of mutants by fragments of viral DNA. *Virology* **60**: 466.

Lai C., and Nathans, D. 1975. A map of temperature-sensitive mutants of simian virus 40. *Virology* **66**: 70.

Lambowitz, A. M., and Perlman, P. S. 1990. Involvement of aminacyl-tRNA synthetases and other proteins in group I and group II intron splicing. *Trends Biochem. Sci.* **15**: 440.

Lamond, A. I. 1986. RNA editing and the mysterious undercover genes of trypanosomatid mitochondria. *Trends Biochem. Sci.* **13**: 283.

Lamond, A. I., and Gibson, T. J. 1990. Catalytic RNA and the origin of genetic systems. *Trends Genet.* **6**: 145.

Landweber, L. F., and Gilbert, W. 1994. Phylogenetic analysis of RNA editing: A primitive genetic phenomena. *Proc. Natl. Acad. Sci. U.S.A.* **91**: 918.

Landy, A. 1989. Dynamic, structural, and regulatory aspects of λ site-specific recombination. *Annu. Rev. Biochem.* **58**: 913.

Lang, B. F. 1984. The mitochondrial genome of the fission yeast *Schizosaccharomyces pombe*: Highly homologous introns are inserted at the same position of the otherwise less conserved *coxl* genes in *Schizosaccharomyces pombe* and *Aspergillus nidulans*. *EMBO J.* **3**: 2129.

Lange-Gustafson, B. J., and Nash, H. A. 1984. Purification and properites of Int-h, a variant protein involved in site specific recombination of bacteriophage λ. *J. Biol. Chem.* **259**: 12724.

Langle-Ronault, F., Macnhaut-Michel, G., and Radman, M. 1987. GATC sequences, DNA nicks and the MutH function in *Escherichia coli* mismatch repair. *EMBO J.* **6**: 1121.

Laskowski, M. 1957. Phosphodiesterase from rattle-snake venom. *Nature* **180**: 1181.

Laskowski, M. 1959. Enzymes hydrolyzing DNA. *Ann. N. Y. Acad. Sci.* **81**: 776.

Laskowski, M. 1967. DNases and their use in the studies of primary structure of nucleic acids. *Adv. Enzymol.* **29**: 165.

Laskowski, M. 1982. Nucleases: Historical perspectives. In Linn, S., and R. J. Roberts (eds.). *Nucleases*. Cold Spring Harbor Laboratory Press, Cold Spring Harbor, NY.

Laskowski, M., and Seidel, M. D. 1945. Viscosimetric determination of thymonucleode, polymerase. *Arch. Biochem.* **7**: 465.

Latham, J. A., and Cech, T. R. 1989. Defining the inside and outside of a catalytic molecule. *Science* **245**: 276.

Lawrence, N. P., Ricman, A., Amini, R., and Altman, S. 1987. Heterologous enzyme function in *E. coli* and the selection of genes encoding the catalytic RNA subunit of RNAse P. *Proc. Natl. Acad. Sci. U.S.A.* **84**: 6825.

Lazarides, E., and Lindberg, U. 1974. Actin is the naturally occurring inhibitor of deoxyribonuclease I. *Proc. Natl. Acad. Sci. U.S.A.* **71**: 4742.

Lazowska, J., Jacq, C., and Slonimski, P. P. 1980. Sequence of introns and flanking exons in wild-type and box3 mutants of cytochrome b reveasl an interlaced splicing protein coded by an intron. *Cell* **22**: 333–348.

Lebreton, B., Prasad, P. V., Jayaram, M., and Youderian, P. 1988. Mutations that improve the binding of yeast FLP recombinase to its substrate. *Genetics* **118**: 393.

Lee, D., and Sadowski, P. D. 1981. Genetic recombination of bacteriophage T7 *in vivo* studied by use of a simple physical assay. *J. Virol.* **40**: 839.

Lee, I., Lee, Y.-H., Mikulski, S. M., Lee, J., Covone, K., and Shogen, K. 2000. Tumoricidal effects of onconase on various tumors. *J. Surg. Oncol.* **73**: 164–171.

Lee, K.-H., Kim, D. W., Bae, S.-H., Kim, J.-A., Koo, H.-S., and Seo, Y.-S. 2000. The endonuclease activity of the yeast Dna2 enzyme is essential *in vivo*. *Nuclic Acids Res.* **28**: 2873–2881.

Lee, S. Y., Nakao, Y., and Bock, R. M. 1968. The nuclease of yeast. II., Purification, properties, and specificity of an endonuclease from yeast. *Biochim. Biophys. Acta* **151**: 126.

Leff, S. E., Rosenfeld, M. G., and Evans, R. M. 1986. Complex transcriptional units: Diversity in gene expression by alternative RNA processing. *Annu. Rev. Biochem.* **55**: 1091.

Lehman, I. R., Roussou, G. G., and Prett, E. A. 1962. The deoxyribonucleases of *Escherichia coli. J. Biol. Chem.* **237**: 819.

Lehmann, A. R. 1982. *Xeroderma pigmentosum*, Cockayne syndrome, and ataxia telangiectasia: Disorders relating DNA repair to carcinogenesis. *Cancer Surveys* **1**: 93.

Leong, J. M., Nunes-Düby, S., Lesser, C. F., Youderian, P., Suskind, M. M., and Landy, A. 1985a. The φ80 and P22 attachment sites. *J. Biol. Chem.* **260**: 4468.

Leong, J. M., Nunes-Düby, S. E., and Landy, A. 1985b. Generation of a single base-pair deletions, insertions, and substitutions by a site-specific recombination system. *Proc. Natl. Acad. Sci. U.S.A.* **82**: 6990.

Leong, J. M., Nunesdub, S. E., Oser, A. B., Lesser, C. F., Youderia, P., Susskind, M. M., and Landy, A. 1986. Structural and regulatory divergence among site-specific recombination genes of lambdoid phage. *J. Mol. Biol.* **189**: 603.

Lever, A. M. L., and Goodfellow, P. (eds.). 1995. *Gene Therapy*. Churchill Livingstone, Edinburgh.

Ley, R. D. 1993. Photoreactivation in human. *Proc. Natl. Acad. Sci. U.S.A.* **90**: 4389.

Li, J. J., Geyer, R., and Tan, W. 2000. Using molecular beacons as a sensitive fluorescence assay for enzymatic cleavage of single-stranded DNA. *Nucleic Acid Res.* **28**: 52–61.

Li, J., and Lu, Y. 2000. A highly sensitive and selective catalytic DNA biosensor for lead ions *J. Am. Chem. Soc.* **122**: 10466.

Li, L., and Chandrasegaran, S. 1993. Alteration of the cleavage distance of FokI restriction endonuclease by insertion mutagenesis. *Proc. Natl. Acad. Sci. U.S.A.* **90**: 2764.

Li, Y., and Sen, D. 1998. The modus operandi of a DNA enzyme: Enhancement of substrate basicity. *Chem. Biol.* **5**: 1.

Li, Y. F., Kim, S. T., and Sancar, A. 1994. Evidence for lack of DNA photoreactivating enzymes in humans. *Proc. Natl. Acad. Sci. U.S.A.* **90**: 4389.

Liao, T.-H., Salnikow, J., Moore, S., and Stein, W. H. 1973. Bovine pancreatic deoxyribonuclease A. Isolation of cyanogen bromide peptides: Complete covalent structure of the polypeptide chain. *J. Biol. Chem.* **248**: 1489.

Lichten, M., Goyon, C., Schultes, N. P., Treco, D., Szostak, J. W., Haber, J. E., and Nicholas, A. 1990. Detection of heteroduplex DNA molecules among the product *Saccharomyces cerevisiae* meiosis. *Proc. Natl. Acad. Sci. U.S.A.* **87**: 7653.

Lieber, M. R. 1997. The Fen-1 family of structure-specific nuclease in eukaryotic DNA replication, recombination, and repair. *BioEssays* **19**: 233–240.

Lieber, M. R., Hesse, J. E., Lewis, S., Bosma, G. C., Rosenberg, N., Mizzuchi, K., Bosma, M. J., and Gilbert, M. 1988. The defect in murine severe combined immune deficiency-joining of signal sequences but not coding segments in V(D)J recombination. *Cell* **55**: 7.

Lilley, D. M. 1983. Eukaryotic genes—are they under torsional stress? *Nature* **305**: 276.

Lima, C. D., Wang, J. C., and Mondragon, A. 1994. Three dimensional structure of 67K N-terminal fragment of *E. coli* DNA topoisomerase I. *Nature* **367**: 138.

Lin, W.-C., and Desiderio, S. 1994. Cell cycle regulation of V(D)J recombination-activating protein RAG-2. *Proc. Natl. Acad. Sci. U.S.A.* **91**: 2733.

Lindahl, T. 1979. DNA glycosylases, endonucleases for apurinic/apyrimidinic sites, and base excision repair. *Prog. Nucleic Acid Res.* **22**: 135.

Lindahl, T. 1994. DNA repair. DNA surveillance defect in cancer cells. *Curr. Biol.* **4**, 249–251.

Lindahl, T., Gally, J. A., and Edelman, G. M., 1969. Deoxyribonuclease IV: A new exonuclease from mammalian tissues. *Proc. Natl. Acad. Sci. U.S.A.* **62**: 597–603.

Lindberg, U. 1967a. Purification from calf spleen of two inhibitors of deoxyribonuclease I. Physical and chemical characterization of the inhibitior II. *Biochemistry* **6**: 323.

Lindberg, U. 1967b. Studies on the complex formation between deoxyribonuclease I and spleen inhibitor II. *Biochemistry* **6**: 343.

Lindberg, M. U., and Skoog, L. 1970. Purification fm. calf thymus of an inhibitor of deoxyribonuclease I. *Eur. J. Biochem.* **13**: 326.

Linn, S. 1967. An endonuclease from *Neurospora crassa* specific for polynucleotides lacking an ordered structure. *Methods Enzymol.* **12A**: 247.

Linn, S. 1978. Workshop summary: Enzymology of base excision repair. In Hanawalt, P. C., E. C., Friedberg, and C. F. Fox (eds.), *DNA Repair Mechanisms*. Academic Press, New York.

Linn, S. 1982a. Nucleases involved in DNA repair. In Linn S., and R. J. Roberts (eds.), *Nucleases*. Cold Spring Harbor Laboratory Press, Cold Spring Harbor, NY.

Linn, S. 1982b. The deoxyribonucleases of *Escherichia coli*. In Linn S., and R. J. Roberts (eds.), *Nucleases*. Cold Spring Harbor Laboratory Press, Cold Spring harbor, NY.

Linn, S., and Arber, W. 1968. *In vitro* restriction of phage fd replicative form. *Proc. Natl. Acad. Sci. U.S.A.* **59**: 1300.

Linn, S., and Lehman, I. R. 1965. An endonuclease from *Neurospora crassa* specific for polynucleotides lacking an ordered structure. I. Purification and properties of the enzyme. *J. Biol. Chem.* **240**: 1287.

Linn, S., and Roberts R. J. (eds.). 1982. *Nucleases*. Cold Spring Harbor Laboratory Press, Cold Spring Harbor, NY.

Linn, S., Lloyd, S., and Roberts, R. J. 1993. *Nucleases*, 2nd ed. Cold Spring Harbor Laboratory Press, Cold Spring Harbor, NY.

Linsley, W. S., Penhoet, E. E., and Linn, S. 1977. Human endonuclease specific for apurinic/apyrimidinic sites in DNA. Partial purification and characterization of multiple forms in placenta. *J. Biol. Chem.* **252**: 1235.

Little, J. W., and Mount, D. W. 1982. The SOS regulatory system of *Escherichia coli*. *Cell* **29**: 11.

Liu, L. F., and Wang, J. C. 1979. Interaction between DNA and *Escherichia coli* DNA topoisomerase I. *J. Biol. Chem.* **254**: 11082.

Liu, L. F., and Wang, J. C. 1987. DNA–DNA gyrase complex: The wrapping of the DNA duplex outside the enzyme. *Cell* **15**: 979.

Liu, L. F., Depew, R. E., and Wang, J. C. 1976. Knotted single-stranded DNA rings: A novel topological isomer of circular single-stranded DNA formed by treatment with *Escherichia coli* ω protein. *J. Mol. Biol.* **106**: 439.

Liu, L. F., Liu, C. C., and Alberts, B. M. 1979. T4 DNA topoisomerase; A new ATP-dependent enzyme essential for initiation of T4 bacteriophage DNA replication. *Nature* **281**: 456.

Liu, L. F., Liu, C. C., and Alberts, B. M. 1980. Type II DNA topoisomerases: Enzymes that can unknot a topologically knotted DNA molecule via a reversible double-strand break. *Cell* **19**: 697.

Ljungquist, S. 1977. A new endonuclease from *Escherichia coli* acting at apurinic sites in DNA. *J. Biol. Chem.* **252**: 2808.

Ljungquist, E., and Bertani, L. E. 1983. Properties and products of the cloned int gene of bacteriophage P2. *Mol. Gen. Genet.* **192**: 87.

Ljungquist, S., Lindahl, T., and Howard-Flanders, P. 1976. Methyl methane sulfonate-sensitive mutant of *Escherichia coli,* deficient in an endonuclase specific for apurinic sites in deoxyribonucleic acid. *J. Bacteriol.* **126**: 646.

Lloyd, R. S., and Hanawalt, P. C. 1981. Expression of the denV gene of bacteriophage T4 cloned in *Escherichia coli. Proc. Natl. Acad. Sci. U.S.A.* **78**: 2796.

Lloyd, R. S., and Linn, S. 1993. In Linn, S., S. Lloyd, and R. J. Roberts (eds.), *Nucleases Involved in DNA Repair in Nucleases*, 2nd ed. Cold Spring Harbor Laboratory Press, Cold Spring Harbor, NY.

Loeb, L. A., and Kunkel, T. A. 1982. Fidelity of DNA synthesis. *Annu. Rev. Biochem.* **51**: 429.

Lu, A.-L., and Chang, D. X. 1988. Repair of single base-pair transversion mismatches of *Escherichia coli in vitro:* Correction of certain A:G mismatches is independent of dam methylation and host mut Hls gene functions. *Genetics* **118**: 593.

Lu, A.-L., Clark, S., and Modrich, P. 1983. Methyl-directed repair of DNA base-pair mismatches *in vitro*. *Proc. Natl. Acad. Sci. U.S.A.* **80**: 4639.

Lu, C., Scheuermann, R. H., and Echols, H. 1986. Capacity of *RecA* protein to bind preferentially to U. V. lesions and inhibit the editing subunit (ε) of DNA polymerase III: A possible mechanism for SOS induced targeted mutagenesis. *Proc. Natl. Acad. Sci. U.S.A.* **83**: 619.

Luria, S. E., and Human, M. L. 1952. A non-hereditary, host induced variation of bacterial viruses. *J. Bacteriol.* **64**: 557.

Lutz, M. J., Will, D. W., Breiphol, G., Benner, S. A., and Uhlmann, E. 1999. Synthesis of a monocharged peptide nucleic acid (PNA) analog and its recognition as substrates by DNA polymerases. *Nucleosides Nucleotides* **18**: 393–401.

Lyamichev, V., Brow, M. A., and Dahlberg, J. E. 1993. Structure-specific cleavage of nucleic acids by eubacterial DNA polymerases. *Science* **260**: 778–783.

Lycan, D. E., and Danna, K. J. 1984. S1 mapping of purified nascent transcripts of simian virus 40. *Mol. Cell. Biol.* **4**: 625.

Macanovic, M., Sinicorpi, D., Sach, S., Baughman, S., Thriru, S., and Lachmann, P. J. 1996. The treatment of systemic lupus erythematosus (SLE) in NZB/W FI1 hybrid mice; studies with recombinant murine DNase with dexamethasone. *Clin. Exp. Immunol.* **106**: 254–252.

Machwe, A., Ganunis, R., Bhor, V. A., and Orren, D. K. 2000. Selective blockage of the 3′–5′ exonuclease activity of WRN protein by certain oxidative modifications and bulky lesions in DNA. *Nucleic Acid Res.* **28**: 2762–2770.

Macreadie, I., Scott, G. R. M., Zinn, A. R., and Bustow, R. A. 1985. Transposition of an intron in yeast mitochondria requires a protein encoded by that intron. *Cell* **41**: 395.

Madiraju, M. V. V. S., Templin, A., and Clark, A. J. 1988. Properties of a mutant *RecA*-encoded protein reveal a possible role for *Escherichia coli RecF*-encoded protein in genetic recombination. *Proc. Natl. Acad. Sci. U.S.A.* **85**: 6592.

Mahillion, J., Scurinck, J., Van Rompluy, L., Delcour, J., and Zabeau, M. 1985. Nucleotide sequence and structural organization of an insertion sequence element (IS231) from *Bacillus thuringiensis* strain berliner 1715. *EMBO J.* **4**: 3895.

Mahler, H. R., and Bastos, R. N. Coupling between mitochondrial mutation and energy transduction. *Proc. Natl. Acad. Sci. U.S.A.* **71**: 2241.

Mankovich, J. A., McIntyre, C. A., and Walker, G. C. 1989. Nucleotide sequence of the Salmonella typhimurium mutL gene required for mismatch repair: Homology of MutL to HexB of *Streptococcus pneumoniae* and to PMSI of the yeast *Saccharmyces cerevisiae. J. Bacteriol. Bacteriol.* **171**: 5325.

Mann, M. B., Rao, R. N., and Smith, H. O. 1978. Cloning of restriction and modification genes in *E. coli.* The HhaII system from *Haemophilus haemolyticus. Gene* **3**: 97.

March, P. E., Ahnn, J., and Inouye, M. 1985. The DNA sequence of the gene (rnc) encoding ribonuclease III of *Escherichia coli. Nucleic Acids Res.* **13**: 4677.

Margolin, P., Zumstein, L., Sternganz, R., and Wang, J. C. 1985. The *Escherichia coli* supX locus is topA, the structural gene for DNA topoisomerase I. *Proc. Natl. Acad. Sci. U.S.A.* **82**: 5437.

Marini, J. C., Miller, K. G., and England, P. T. 1980. Decatenation of kinetoplast DNA by topoisomerases. *J. Biol. Chem.* **255**: 4976.

Marisato, D., and Kleckner, N. 1984. Transposase promotes double strand breaks and single strand joints at Tn10 termini *in vivo. Cell* **39**: 181.

Mark, K.-K., and Stuier, F. W. 1981. Purification of the gene 0.3 protein of bacteriophage T7, an inhibitor of the DNA restriction system of *Escherichia coli. J. Biol. Chem.* **256**: 2573.

Mark, J. L. 1987. Animals yield clues to Huntington's disease. *Science* **238**: 1510.

Marshall, L. E., Graham, D. R., Reich, K. A., and Sigman, D. S. 1981. Cleavage of deoxyribonucleic acid by the 1,10-phenanthroline–cuprous complex. Hydrogen peroxide requirement and primary and secondary structure specificity. *Biochemistry* **20**: 244.

Martelmans, K., Friedberg, E. C., Slor, H., Thomas, G., and Cleaver, J. E. 1976. Defective thymine dimer excision by cell-free extracts of *Xeroderma pigmentosum* cells. *Proc. Natl. Acad. Sci. U.S.A.* **73**: 2757.

Martin, J. B. 1987. Molecular genetics: Applications to the clinical neurosciences. *Science* **238**: 765.

Mathias, S. L., Scott, A. F., Kazazian H. H, Jr., Boeke, J. D., and Gabriel, A. 1991. Reverse transcriptase encoded by a human transposable element. *Science* **254**: 1805.

Matsunaga, T., Park, C. H., Bessho, T., Mu, D., and Sancar, A. 1996. Replication protein A confers structure-specific endonuclease activities to the XPF-ERCC1 and XPG sububits of human DNA repair excision nuclease. *J. Biol. Chem.* **271**: 11047–50.

Mauguen, Y., Hartley, R. W., Dodson, G. G., Dodson, E. I., Briscogne, G., Chotia, C., and Jack, A. 1982. Molecular structure of a new family of ribonucleases. *Nature* **297**(5862): 162.

Maxam, A. M., and Gilbert, W. 1977. A new method for sequencing DNA. *Proc. Natl. Acad. Sci. U.S.A.* **74**: 560.

Maylynn, B. A., Blackwell, T. K., Fulop, G. M., Rathbun, G. A., Furley, A. J. W., Ferrier, P., Heinke, L. B., Phillips, R. A., Yancopoulas, G. D., and Alt, F. W. 1988. The SCID defect affects the final step of the Immunoglobulin VDJ recombinase mechanism. *Cell* **54**: 453.

McCardy, S. J., Boyce, J. M., and Cox, B. S. 1987. Excision repair in the yeast *Saccharomyces cerevisiae. J. Cell Sci. Suppl.* **6**: 25.

McCarthy, D. 1979. Gyrase-dependent initiation of bacteriophage T4 DNA replications: Interactions of *Escherichia coli* gyrase with novobiocin, coumermycin and phage DNA-delay gene products. *J. Mol. Biol.* **127**: 265.

McClintock, B. 1951. Chromosome organization and genic expression. *Cold Spring Harbor Symp. Quant. Biol.* **16**: 13.

McClure, B., Haring, V., Ebert, P., Anderson, M., Simpson, R., Sakiyama, F., and Clarke, A. 1989. Style self incompatibility gene product of *Nicotiana alanta* are ribonucleases. *Nature* **342**: 955–957.

McCormick, A. A., Kumagi, M. H., Hanley, K., Turpen, T. H., Hakim, I., Grill, L., Tuse, D., Levy, S., and Levy, R. 1999. Rapid production of specific vaccines for lymphoma by expression of tumor-derived single chain Fv epitopes in tobacco plants. *Proc. Natl. Acad. Sci. U.S.A.* **96**: 703–708.

McCune, J. M., Namikawa, R., Koneshima, H., Shultz, L. D., Lieberman, M., and Weissman, I. L. 1988. The SCID mouse: Murine model for the analysis of human hematolymphoid differentiation and function. *Science* **241**: 1632.

McCutchen T. F., Hansen, J. L., Dame, J. B., and Mullins, A. J. 1984. Mung bean nuclease cleaves plasmidium genomic DNA at sites before and after genes. *Science* **225**: 625.

McEntee, K., Weinstock, G. M., and Lehman, I. R. 1979. Initiation of general recombination catalyzed *in vitro* by the recA protein of *Escherichia coli*. *Proc. Natl. Acad. Sci. U.S.A.* **76**: 2615.

McGrogan, M., Sinsonsen, C., Scott, R., Griffith, J., Ellis, N., Kennedy, J., Campanelli, D., Nathan, C., and Gabay, J. 1988. Isolation of cDNA clone encoding a precursor to human eosinophil major basic protein. *J. Exp. Med.* **169**: 2295.

Mei, H.-Y., and Barton, J. K. 1986. Chiral probe for A-form helices of DNA and RNA: Tris(tetramethylphenanthroline) ruthernium (II). *J. Am. Chem. Soc.* **108**: 7414.

Mei, H.-Y., and Barton, J. K. 1988. Tris(tetramethylphenanthroline) ruthenium (II): A chiral probe that cleaves A-DNA conformations. *Proc. Natl. Acad. Sci. U.S.A.* **85**: 1339.

Mellon, I., Bohr, V. A., Smith, C. A., and Hanawalt, P. C. 1986. Preferential DNA repair of an active gene in human cells. *Proc. Natl. Acad. Sci. U.S.A.* **83**: 8878.

Mendel, D., and Dervan, P. B. 1987. Hoongsteen base pairs proximal and distal to echinomycin binding sites on DNA. *Proc. Natl. Acad. Sci. U.S.A.* **84**: 910.

Merrifield, B. 1986. Solid phase synthesis. *Science* **232**: 341.

Meselson, M. S., and Radding, C. M. 1975. A general model for genetic recombination. *Proc. Natl. Acad. Sci. U.S.A.* **72**: 358.

Meselson, M., Yuan, R., and Heywood, J. 1972. Restriction and modification of DNA. *Annu. Rev. Biochem.* **41**: 447.

Meselson, M., and Yuan, R. 1968. DNA restriction enzyme from *E. coli*. *Nature* **217**: 1110.

Meyer, T. F., and Geider, K. 1979. Bacteriophage fd gene II protein. II. Specific cleavage and relation of supercoiled RF from filamentous phages. *J. Biol. Chem.* **254**: 12642.

Meyer-Leon, L., Gates, C. A., Attwood, J. M., Wood, E. A., and Cox, M. M. 1987. Purification of the FLP site-specific recombinase by affinity chromatography and re-examination of basic properties of the system. *Nucleic Acids Res.* **15**: 6469.

Michaels, M. L., Cruz, C., and Miller, J. H. 1990. MutA and mutC: Two mutator loci in *Escherichia coli* that stimulate transversions. *Proc. Natl. Acad. Sci. U.S.A.* **87**: 9211.

Michel, F., and Dujon, B. 1986. Genetic exchanges between bacteriophage T4 and filamentous fungi. *Cell* **46**: 323.

Michel, F., Umesono, K., and Ozeki, H. 1989. Comparative and functional anatomy of Group II catalytic introns—A review. *Gene* **82**: 5.

Milcarek, C., and Weiss. B. 1972. Mutants of *Escherichia coli* with altered deoxyribo-nucleases, I. Isolation and characterization of mutants for exonuclease III. *J. Mol. Biol.* **68**: 303.

Miller, K. G., Liu, L. F., and Englund, P. T. 1981. A homogenous type II DNA topoisomerase from HeLa cell nuclei. *J. Biol. Chem.* **256**: 9334.

Miller, M., Tanner, J., Aplaugh, M., Benedik, M., and Krause, K. 1994. A structure of *Serratia* endonuclease suggests a mechanism for binding to double stranded DNA. *Nat. Struct. Biol.* **1**: 461–468.

Miller, R. V., and Scurlock, T. R. 1983. DNA gyrase (topoisomerase II) from *Pseudomonas aeruginosa. Biochem. Biophys. Res. Commun.* **110**: 694.

Miller, H. I., Mozola, M. A., and Friedman, D. I. 1980. Int-h and int mutation of phage λ that enhances site-specific recombination. *Cell* **20**: 721.

Miller, C. A., and Cohen, S. N. 1978. Phenotypically cryptic *Eco*RI endonuclease activity specified by the COLE1 plasmid. *Proc. Natl. Acad. Sci. U.S.A.* **75**: 1265.

Mishra, N. C. 1979. DNA-mediated genetic changes in *Neurospora crassa. J. Gen. Microbio.* **113**: 255.

Mishra, N. C. 1985. Gene transfer in fungi. *Adv. Genet.* **23**: 73.

Mishra, N. C. 1986. DNA repair defective mutants of *Neurospora. Mutagenesis–Applied and Basic* **1**: 39.

Mishra, N. C. 1991. Genetics and molecular biology of *Neurospora crassa. Adv. Genet.* **29**: 1.

Mishra, N. C. 1995. *Molecular Biology of Nucleases*. CRC Press, New York, pp. 1–312.

Mishra, N. C. 2001. Fungal cells. *Encyclopedia of Life Sciences*. Macmillan Press, New York, pp. 1–11.

Mishra, N. C., and Forsthoefel, A. M. 1983. Biochemical genetics of *Neurospora nucleases* II. *Genet Res.* **41**: 287.

Misra, T. K., and Apirion, D. 1978. Characterization of an endoribonuclease, RNase. from *Escherichia coli. J. Biol. Chem.* **253**: 5594.

Misra, T. K., and Apirion, D. 1979. RNaseE an RNA processing enzyme from *Escherichia coli. J. Biol. Chem.* **254**: 11154.

Mittler, J. E., and Lenski, R. F. 1990. Excisions of Mu from *E. coli* MCS 2 are not directed mutations. *Nature* **344**: 173.

Mizuuchi, K. 1983. *In vitro* transposition of bacteriophage mu: A biochemical approach to a novel replication reaction. *Cell* **35**: 785.

Mizuuchi, K. 1992. Transpositional recombination: Mechanistic insights form studies of Mu and other elements. *Annu. Rev. Biochem.* **61**: 1011.

Mizuuchi, K., and Craigie, R. 1986. Mechanism of bacteriophage Mu transposition. *Annu. Rev. Genet.* **20**: 385–429.

Mizuuchi, K., Fisher, L. M., O'Dea, M. L., and Gellert, M. 1980. DNA gyrase action involves the introduction of transient double-strand breaks into DNA. *Proc. Natl. Acad. Sci. U.S.A.* **77**: 1847–1851.

Mizuuchi, K., Kemper, B., Hays, J., and Weisberg, R. A. 1982a. T4 endonuclease VII cleaves Holliday structures. *Cell* **29**: 357.

Mizuuchi, K., O'Dea, M. H., and Gellert, M. 1982b. DNA gyrase: Subunit structure and the ATPase activity of the purified enzyme. *Proc. Natl. Acad. Sci. U.S.A.* **75**: 5960.

Modrich, P. 1987. DNA mismatch correction. *Annu. Rev. Biochem.* **56**: 435.

Mohr, G., and Lambowitz, A. M. 1991. Integration of a group I intron into a ribosomal RNA sequence promoted by a tyrosyl-tRNA synthetase. *Nature* **354**: 164.

Molina, H. A., Kierszenbaum, F., Hamann, K., and Gleich, G. K. 1988. Toxic effects produced or mediated by human eosinophil granule components on *Trypanosoma cruzi*. *Am. J. Trop. Med. Hyg.* **37**: 227.

Molloy, P. L., and Symons, R. H. 1980. Cleavage of DNA–RNA hybrids by type II restriction enzymes. *Nucleic Acids Res.* **8**: 2939.

Monaco, A. P., and Kunkel, L. M. 1987. A giant locus for the Duchenne and Becker muscular dystrophy gene. *Trends Genet.* **3**: 33.

Monteilhet, C., Perrin, A., Thierry, A., Colleaux, L., and Dujon, B. 1990. Purification and characterization of the *in vitro* activity of I. SceA novel and highly specific endonuclease encoded by a group I intron. *Nucleic Acids Res.* **18**: 1407.

Mongomery, M. K., and Fire, A. 1998. Double-Stranded RNA as a mediator in sequence-specific genetic silencing and co-suppresion. *Trends Genet.* **14**: 255–258.

Moore, S. 1981. Pancreatic DNase. In *The enzymes*, Vol. 14A. Academic Press, New York.

Moore, S., and Stein, W. H. 1973. Chemical structures of pancreatic ribonuclease and deoxyribonuclease. *Science* **180**: 458.

Morikawa, K., Matsumoto, O., Tsujimote, M., Katayanagi, K., Aniyoshi, M., Doi, T., Ikchara, M., Inaoka, T., and Ohtsuka, E. 1992. X-ray structure of T_4 endonuclease V: An excision repair specific for pyrimidine dimer. *Science* **256**: 523.

Morishima, N., Nakagawa, K., Yamamoto, E., and Shibata, T. 1990. A subunit of yeast site-specific endonuclease SceI is a mitochondrial version of the 70-kDa heat shock protein. *J. Biol. Chem.* **265**: 15189.

Morita, T., Niwata, Y., Ohgi, K., Ogawa, M., and Ine, M. 1986. Distribution of two urinary ribonuclease like enzymes in human organs and body fluids. *J. Biochem.* **99**: 17.

Moroianu, J., and Riordan, J. F. 1994. Nuclear translocation of angiogenin in proliferating endothelial cells is essential to its angiogenic activity. *Proc. Natl. Acad. Sci. U.S.A.* **91**: 1677.

Morrison, A., and Cozzarelli, N. R. 1979. Site-specific cleavage of DNA by *E. coli* DNA gyrase. *Cell* **17**: 175.

Mosbaugh, D. W., and Linn, S. 1980. Further characterization of human fibroblast apurinic/apyrimidinic DNA nucleases: The definition of two mechanistic classes of enzyme. *J. Biol. Chem.* **255**: 11743.

Mosier, D. E., Gulizia, R. J., Baird, S. M., Wilson, D. B. 1988. Transfer of a functional human immune system to mice with severe combined immunodeficiency. *Nature* **335**: 256.

Muegge, K., Vila, M. P., and Durum, S. K. 1993. Interleuken-7: A cofactor for VDJ rearrangement of the T cell receptor β gene. *Science* **261**: 93.

Müller, B. C., Raphael, A. L., and Barton, J. K. 1987. Evidence for altered DNA conformations in the simian virus 40 genome: Site-specific DNA cleavage by the chiral comples Λ-tris(4,7-diphenyl-1,10 phenanthroline) cobalt (III). *Proc. Natl. Acad. Sci. U.S.A.* **84**: 1764.

Mulligan, R. C. 1993. The basic science of gene therapy. *Science* **260**: 926.

Mullis, K. B., and Faloona, F. A. 1987. Specific synthesis of DNA *in vitro* via a polymerase-catalyzed chain reaction. *Methods Enzymol.* **155**: 335.

Murray, N. E., and Murray, K. 1974. Manipulation of restriction targets in phage λ to form receptor chromosomes for DNA fragments. *Nature* **251**: 476.

Murray, A. W., and Szostak, J. W. 1983. Construction of artificial chromosomes in yeast. *Nature* **305**: 189.

Murray, J. M., Tavassoli, M., al-Haithy, R., Sheldrick, K. S., Lehman, A. R., Carr, A. M., and Watts, F. Z. 1994. Structural and functional conservation of the human homolog of the *Schizosacchromyces pombe* rad2 gene, which is required for chromosome segregation and recovery from DNA damage. *Mol. Cell. Biol.* **14**: 4878–4888.

Muscarella, D. E., and Vogt, V. M. 1989. A mobile group I intron in the nuclear rDNA of *Physarum polycephalum. Cell* **10**: 443.

Muskavitch, K. M. T., and Linn, S. 1981. *RecBC*-like enzymes: Exonuclease V deoxyribonucleases. In Boyer, P. D. (ed.), *The Enzymes*, Vol. 14, Academic Press, New York.

Muskavitch, K. M. T. 1982. A unified mechanism for the nuclease and unwinding activities of the *RecBC* enzyme of *Escherichia coli. J. Biol. Chem.* **257**: 2641.

Muster-Nassal, C., and Kolodner, R. 1986. Mismatch correction catalyzed by cell free extract of *Saccharomyces cerevisiae. Proc. Natl. Acad. Sci. U.S.A.* **83**: 7618.

Myler, P. J., Allison, J., Agabian, N., and Stuart, K. 1984. Antigenic variation in African trypanosomes by gene replacement or activation of alternate telomers. *Cell* **39**: 203.

Naas, T., Blot, B., Fitch, W. M., and Arber, W. 1994. Insertion sequence related gentic variation in resting *Escherichia coli* K-12. *Genetics* **136**: 721.

Nakabeppu, V., and Sekiguchi, M. 1981. Physical association of pyrimidine dimer DNA glycosylase and apurinic/apyrimidinic DNA endonuclease essential for repair of ultraviolet-damaged DNA. *Proc. Natl. Acad. Sci. U.S.A.* **78**: 2742.

Nakamura, H., Katayanagi, K., Morikawa, K., and Ikehara, M. 1991. Structural models of ribonuclease H domains in reverse transcriptase from retrovirus. *Nucleic Acids Res.* **19**: 1817.

Namanura, M., Leppert, M., O'Connell, P., Wolff, R., Holm, T., Culver, M., Martin, C., Fujimoto, E., Hoff, M., Kumlin, E., and White, R. 1987. Variable number of tandem repeat (VNTR) markers for human gene mapping. *Science* **235**: 1616.

Nash, H. A. 1981. Integration and excision of bacteriophage λ: The mechanism of conservative site specific recombination. *Annu. Rev. Genet.* **15**: 143.

Nash, H. A., and Robertson, C. A. 1981. Purification and properties of the *Escherichia coli* protein factor required for the integrative recombination. *J. Biol. Chem.* **256**: 9246.

Nash, H. A., Bauer, C. E., and Gardner, J. F. 1987. Role of homology in site-specific recombination of bacteriophage λ: Evidence against joining of cohesive ends. *Proc. Natl. Acad. Sci. U.S.A.* **84**: 4049.

Nelson, E. M., Tewey, K. M., and Liu, L. F. 1984. Mechanism of antitumor drug action: Poisoning of mammalian DNA topoisomerase II on DNA by 4-(9-acridinylamino)-methanesulfur-*m*-anisidide. *Proc. Natl. Acad. Sci. U.S.A.* **81**: 1361.

Nichols, E. K. 1988. *Human Gene Therapy*, Institute of Medicine, National Academy of Sciences. Harvard University Press, Cambridge, MA.

Nichols, M., Söll, D., and Willis, I. 1988. Yeast RNase P: Catalytic activity and substrate binding are separate functions. *Proc. Natl. Acad. Sci. U.S.A.* **85**: 1379.

Nickoloff, J. A., Singer, J. D., and Heffron, F. 1990. *In vivo* analysis of the *Saccharomyces cerevisiae* HO nuclease recognition site by site-directed mutagenesis. *Mol. Cell. Biol.* **10**: 1174.

Nicholson, A. W. 1999. Function, mechanism, and regulation of bacterial ribonucleases. FEMS *Microbiol. Rev.* **23**: 371–390.

Nishikawa, S., Monok, H., Kim, H. J., Fuchimura, K., Tanaka, T., Uesugi, S., Hakoshima, T., Tomita, K., Ohtsuka, E., and Ikehara, M. 1987. Two histidine residues are essential for ribonuclease T1 activity as is the case for ribonuclease A. *Biochemistry* **26**: 8620.

Nishikawa, A., Nanda, A., Gregory, W., Frenz, J., and Kornfeld, S. 1999. Identification of amino acids that modulate mannose phosphorylation of mouse DNaseI, a secretory glycoprotein. *J. Biol. Chem.* **274**: 19309–19315.

Nitta, K., Ozaki, K., Tsukamoto, Y., Furusawa, S., Okhubo, Y., Takimoto, H., Murata, R., and Hakomori, S. 1994. Characterization of a *Rana catesbeina* lectin-resistant mutant of leukemia P388 cells. *Cancer Res.* **54**: 928–934.

Niwata, Y., Ohgi, K., Sanda, A., Takizawa, Y., and Irie, M. 1985. Purification and properties of bovine kidney ribonucleases. *J. Biochem.* **97**: 923.

Nolan, J., Shen, P. B., Park, M. S., and Sklar, L. A. 1996. Kinetic analysis of human flap endonuclease-1 by flow cytometry. *Biochemistry* **35**: 11668–11676.

Noller, H. F., Hoffarth, V., and Zimniak, L. 1992. Unusual resistance of peptidyl transferase to protein extraction procedures. *Science* **256**: 1416.

Notomista, E., Catanzano, F., Graziano, G. Di Gaetano, S., Barne, G., and Di Donato, A. 2001. Contribution of chain termini to the conformational stability and biological activity of onconases. *Biochemistry* **40**: 9097–9103.

Nouspikel, T., and Clarkson, S. G. 1994. Mutations that disable the DNA repair gene XPG in a xeroderma pigmentosum group G patient. *Hum. Mol. Genet.* **3**: 963–967.

Nossal, N. G., and Singer, M. F. 1968. The processive degradation of individual polyribonucletoide chains, I. *Escherichia coli* ribonuclease II. *J. Biol. Chem.* **253**: 913.

Nossal, N. G., and Heppel, L. A. 1966. The release of enzymes by osmotic shock from *Escherichia coli* in exponential phase. *J. Biol. Chem.* **241**: 3055.

Nowak, R. 1994. Mining treasures from "junk DNA." *Science* **263**: 608.

Nunes-Düby, S. E., Matsumoto, L., and Landy, A. 1987. Site-specific recombination intermediates trapped with suicide substrates. *Cell* **50**: 779.

Oakley, M. G., and Dervan, P. B. 1990. Structural motif of the GCN4 DNA binding domain characterized by affinity cleaving. *Science* **248**: 847.

O'Donovan, A., and Wood, R. D. 1993. Identical defects in DNA repair in xeroderma pigmentosum and rodent ERCC group 5. *Nature* **363**: 114–115.

O'Donovan, A., Scherly, D., Clarkson, S. G., and Wood, R. D. 1994. Isolation of active recombinant XPGprotein, a human DNA repair endonuclease. *J. Biol. Chem.* **269**: 15965–15968.

Oettinger, M. A., Baltimore, D., Schatch, D., and Gorkac, G. 1990. Adjacent genes that synergetically activate V(D)J recombination. **248**: 1517.

O'Gorman, S., Fox, D. T., and Wahl, G. M. 1991. Recombinase-mediated gene activation and site specific integration in mammalian cells. *Science* **251**: 1351.

O'Hare, K., and Rubin, G. M. 1983. Structures of P transposable elements and their sites of insertion and excision in the *Drosophila melanogaster* genome. *Cell* **34**: 25.

Ohno, S. 1970. *Evolution by Gene Duplication*, Springer, New York.

Olson, K. A., French, T. C., Vallee, B. L., and Fett, J. W. 1994. A monoclonal antibody to human angiogenin suppresses tumor growth in a thymic mice. *Cancer Res.* **54**: 4576–4579.

Olson, K. A., Fett, J. W., French, T. C., Key, M. E., and Valley, B. L. 1995. Angiogenin antagonists prevent tumor *in vivo*. *Proc. Natl. Acad. Sci. U.S.A.* **92**: 442–446.

Olson, M. 1994. The human genome project. *Proc. Natl. Acad. Sci. U.S.A.* **90**: 4338.

Olsson, I., A.-M., Persson, and Winquist, I. 1986. Biochemical properties of eosinophil cationic protein and demonstration of its biosynthesis *in vitro* in marrow cells from patients with an eosinophila. *Blood* **67**: 498.

Orgel, L. E. 1968. Evolution of genetic apparatus. *J. Mol. Biol.* **38**: 381.

Orgel, L. E. 1986. RNA catalysis and the origin of life. *J. Theor. Biol.* **123**: 127.

Orkin, S. H. 1986. Reverse genetics and human disease. *Cell* **47**: 845.

Overbye, K. M., and Margolin, P. 1981. Role of the *supX* gene in ultraviolet light-induced mutagenesis in *Salmonella typhimurium*. *J. Bacteriol.* **146**: 170.

Palmiter, R. D., and Brinster, R. L. 1986. Germ-line transformation of mice. *Annu. Rev. Genet.* **20**: 465–499.

Pan, C. Q., Dodge, A. H., Baker, D. L., Prince, W. S., Sinicorpi, D. V., and Lazarus, R. A. 1998. Improved potency of hyperactive and actin resistant human DNase I variants for teatment of cystic fibrosis and systemic lupus erythematosus. *J. Biol. Chem.* **273**: 18374–18381.

Pardue, M. L. 1991. Dynamic instability of chromosomes and genome. *Cell* **66**: 427.

Parikh, S. S., Mol, C. D., Hosfiels, D. J., and Trainer, J. A. 1999. Envisioning the molecular choreography of DNA base excision repair. *Curr. Opin. Struct. Biol.* **9**: 37–47.

Peebles, C. L., Ogden, R. C., Knapp, G., and Abelson, J. 1979. Splicing of yeast tRNA precursors: A two-stage reaction. *Cell* **18**: 27.

Peebles, C. L., Gegenheimer, D., and Abelson, J. 1983. Precise excision of intervening sequences from precursor tRNAs by a membrane-associated yeast endonuclease. *Cell* **32**: 525.

Peebles, C. L., Meckenburg, K. L., Perlman, L. S., Tabor, J., and Cheng, H. L. 1986. A self-splicing RNA excises intron lariat. *Cell* **44**: 213.

Penswick, J. R., and Holley, R. W. 1965. Specific cleavage of the yeast alanine RNA into two large fragments. *Proc. Natl. Acad. Sci. U.S.A.* **53**: 543.

Perlman, P. S., Butow, R. A. 1989. Mobile introns and intron-coded proteins. *Science* **246**: 1106.

Pestka, S., Langer, J. A., Zoon, K. C., and Samuels, C. E. 1987. Interferons and their actions. *Annu. Rev. Biochem.* **56**: 727.

Piccoli, R., Tamburrini, M., Piccialli, G., DeDonato, A., Parente, A., and D'Alessio, G. 1992. The dual mode quarternary structure of seminal RNase. *Proc. Natl. Acad. Sci. U.S.A.* **89**: 1870.

Pingoud, A., and Jeltsch, A. 2001. Structure and Function of type II restriction endonucleases. *Nucleic Acids Res.* **15**: 3705–3727.

Piñon, R., and Salts, Y. 1977. Isolation of folded chromosomes from the yeast *Saccharomyces cerevisiae*. *Proc. Natl. Acad. Sci. U.S.A.* **74**: 2850.

Plasterk, R. H. A., Kanaar, R., and van de Putte, P. 1984. A genetic switch *in vitro*. DNA inversion by Gin protein of phage Mu. *Proc. Natl. Acad. Sci. U.S.A.* **9**: 2689.

Plotch, S. J., Bouloy, M., Ulmanen, I., and Krug, R. M. 1981. A unique cap ($m^7GpppXm$)-dependent influenza virion endonuclease cleaves capped RNAs to generate the primers that initiate viral RNA transcription cell. *Cell* **23**: 847.

Plummer, T. H., Jr., and Hirs, C. H. W. 1963. Isolation of ribonuclease B, A glycoprotein from bovine pancreatic juice. *J. Biol. Chem.* **238**: 1396.

Pohl, F. M., and Jovin, T. M. 1972. Salt induced cooperative live conformational change of a synthetic DNA: Equilibrium and kinetic studies with poly (dg-dc). *J. Mol. Biol.* **67**: 375.

Ponticelli, A. S., Scholtz, D. W., Taylor, A. F., and Smith, G. R. 1985. Chi-dependent DNA strand cleavage by *RecBC* enzyme. *Cell* **41**: 145.

Pope, L. E., and Sigman, D. S. 1984. Secondary structure specificity of the nuclease activity of the 1,10-phenanthroline-copper complex. *Proc. Natl. Acad. Sci. U.S.A.* **81**: 3.

Popoff, S. C., Spira, A. S., Johnson, A. W., and Demple, B. 1990. The yeast structural gene (APN1) for the major apurinic endonuclease: Homology to *E. coli* endonuclease IV. *Proc. Natl. Acad. Sci. U.S.A.* **87**: 4193.

Prasad, P. V., Young, L.-J., and Jayaram, M. 1987. Mutation in the 2λm plasmid attaches covalently to DNA via a phosphotyrosyl linkage. *Mol. Cell. Biol.* **5**: 3274.

Prell, B., and Vosberg, H. P. 1980. Analysis of covalent complexes formed between calf thymus DNA topoisomerase and single-stranded DNA. *J. Biochem.* **108**(2): 389.

Priebe, S. D., Hadi, S. M., Greenberg, B., and Lacks, S. A. 1988. Nucleotide sequence of the hexA gene for DNA mismatch repair in *Streptococcus pneumoniae* and homology of HexA to mutS of *Escherichia coli* and *Salmonella typhimurium. J. Bacteriol.* **170**: 190.

Pritchard, A. E., Kowalski, D., and Laskowski, M., Sr. 1977. An endonuclease activity of venom phosphodiesterase specific for single-stranded and superhelical DNA. *J. Biol. Chem.* **252**: 8652.

Privat de Garilhe, M. 1964. *Les Nucleases*. Hermann, Paris.

Privat de Garilhe, M. 1967. *Enzymes in Nucleic Acid Research*. Holden-Day, San Francisco.

Privat de Garilthe, M., and Laskowski, N. 1956. Optical changes occurring during the action of phosphodiesterase on oligonucleotides derived from deoxyribonucleic acid. *J. Biol. Chem.* **223**: 661.

Prolla, T. A., Christie, D. M., and Liskay, R. M. 1994. Dual requirement in yeast DNA mismatch repair for MLH1 and PMS1, two homologs of bacterial mutL gene. *Mol. Cell. Biol.* **14**: 407.

Pruch, J. M., and Laskowski, M., Sr. 1980. Covalently bound ribonucleotides in crab d(A-T) polymer. *J. Biol. Chem.* **255**: 9409.

Prudhomme, M., Matin, B., Mejean, V., and Claverys, J. P. 1989. Nucleotide sequence of the *Streptococcus pneumoniae* hexB mismatch repair gene: Homology of HexB to MutL of *Salmonella typhimurium* and to PMS1 of *Saccharomyces cerevisiae. J. Bacteriol.* **171**: 5332.

Pukkila, P. J., Peterson, J., Herman, G., Modrich, P., and Meselson, M. 1983. Effects of high levels of DNA adenine methylation on methyl-directed mismatch repair in *Escherichia coli. Genetics* **104**: 571.

Putney, S. D., Benkovioc, S. J., and Schimmel, P. R. 1981. A DNA fragment with an α-phosphorothioate nucleotide at one end is asymmetrically blocked from digestion by exonuclease III and can be replicated *in vivo. Proc. Natl. Acad. Sci. U.S.A.* **78**: 7350.

Pyle, A. M., McSwiggen, J. A., and Cech, T. R. 1990. Direct measurement of oligonucleotide substrate binding to wild type and mutant ribozymes from *Tetrahymena. Proc. Natl. Acad. Sci. U.S.A.* **87**: 8187.

Que, B. G., Downey, K. M., and So, A. 1978. Mechanism of selective inhibition of 3 to 5 exonuclease activity of *Escherichia coli* DNA polymerase I by nucleoside 5-monophosphates. *Biochemistry* **17**: 1603.

Quirk, S. M., Bell-Pedersen, D., and Belfort, M. 1989. Intron mobility in the T-even phages: High frequency inheritance of group I introns promoted by intron open reading frames. *Cell* **56**: 455.

Radany, E. H., and Friedberg, E. C. 1980. A pyrimidine dimer-DNA glycosylase activity associated with the gene product of bacteriophage T4. *Nature* **286**: 182.

Radman, M. 1976. An endonuclease from *Escherichia coli* that introduces single polynucleotide chain scissions in ultraviolet-irradiated DNA. *J. Biol. Chem.* **251**: 1438.

Radman, M. 1980. Is there SOS induction in mammalian cells? *Biochem. Photobiol.* **32**: 823.

Raines, R. T. 1998. Ribonuclease A. *Chem. Rev.* **98**: 1045–1066.

Ramotar, R., Auchincloss, A. H., and Fraser, M. J. 1987. Nuclear endo-exonuclease of *Neurospora crassa. J. Biol. Chem.* **262**: 425.

Ramsey, B. W. 1996. Management of pulmonary disease in patients with cystic fibrosis. *N. Engl. J. Med.* **335**: 179–188.

Ratner, V. A., Zabanov, S. A., Kolesnikova, O. V., and Vasilyeva, L. A. 1992. Induction of the mobile genetic element DM-412 transposition in *Drosophila* genome by heat shock treatment. *Proc. Natl. Acad. Sci. U.S.A.* **89**: 5650.

Ray, A., and Apirion, D. 1980. Cloning the gene for ribonuclease E, an RNA processing enzyme. *Gene* **12**: 87.

Ray, B. K., and Apirion, D. 1981. RNase P is dependent on RNase E action in processing monomeric RNA precursors that accumulate in an RNase E^- mutant of *Escherichia coli. J. Mol. Biol.* **149**: 599.

Rayssigvier, C., Thaler, D. S., and Radman, M. 1989. The barrier to recombination between *Escherichia coli* and *Salmonella typhimurium* is disrupted in mismatch-repair mutants. *Nature* **342**: 396.

Reagan, M. S., Pettinger, C., Siede, W., and Friedberg, E. C. 1995. Characterization of a mutant strain of *Saccharomyces cerevisiae* with a deletion of the RAD27 gene, a structural homolog of the RAD2 nucleotide excision repair gene. *J. Bacteriol.* **177**: 364–371.

Reddy, M. K., and Bauer, W. R. 1989. Activation of the vaccinia virus nicking-joining enzyme by trypsinization. *J. Biol. Chem.* **264**: 443.

Reed, R. R. 1981a. Resolution of cointegrates between transposons $\gamma\delta$ and Tn3 define the recombination site. *Proc. Natl. Acad. Sci. U.S.A.* **78**: 3428.

Reed, R. R. 1981b. Transposon-mediated site-specific recombination: A defined *in vitro* system. *Cell* **25**: 713.

Reed, R. R., and Gridley, N. D. F. 1981. Transposon mediated site-specific recombination *in vitro*: DNA cleavage and protein–DNA linkage at the recombination site. *Cell* **25**: 721.

Reed, R., and Maniatis, T. 1985. Intron sequences involved in lariat formation during pre-mRNA splicing. *Cell* **41**: 95.

Reed, R. R., Shibuya, G. I., and Steitz, J. A. 1982. Nucleotide sequence of $\tau\delta$ resolvase gene and demonstration that its gene product acts as a repressor of transcription. *Nature* **300**: 381.

Reich, C., Olsen, G. J., Pace, B., and Pace, N. R. 1988. Role of the protein moiety of ribonuclease ribonucleoprotein P. A enzyme. *Science* **239**: 178.

Reid, L. H., Gregg, R. G., Smithies, O., and Koller, B. H. 1990. Regulatory elements in the introns of the human HPRT gene are necessary for its expression in embryonic stem cells. *Proc. Natl. Acad. Sci. U.S.A.* **87**: 4299.

Reiser, J., and Yuan, R. 1977. Purification and properties of the P15 specific restriction endonuclease from *Escherichia coli*. *J. Biol. Chem.* **252**: 451.

Resnick, M. A., Sugino, A., Nitiss, J., and Chow, T. 1984. DNA polymerases, deoxyribonucleases and recombination during meiosis in *Saccharomyces cerevisiae*. *Mol. Cell. Biol.* **4**: 2811.

Revel, H. R., and Luria, S. E. 1970. DNA-glucosylation in T-even phage: Genetic determination and role in phage-host interaction. *Annu. Rev. Genet.* **4**: 177.

Riazuddin, S., and Grossman, L. 1977. *Micrococcus luteus* correndonuclease II. Mechanism of action of two endonucleases specific for DNA containing pyrimidine dimers. *J. Biol. Chem.* **252**: 6280.

Richards, F. M., and Wyckoff, H. W. 1971. Bovine pancreatic ribonuclease. In *The Enzymes*, Vol. 4, Academic Press, New York.

Rio, D. C. 1990. Molecular mechanisms regulating *Drosophila P* element transposition. *Ann. Rev. Genet.* **24**: 543.

Rio, D. C., Laski, F. A., and Rubin, G. M. 1986. Identification and immunochemical analysis of biologically active *Drosophila* P element transposase. *Cell* **44**: 21.

Roberts, R. J. 1976. Restriction endonucleases. *CRC Crit. Rev. Biochem.* **4**: 123.

Roberts, R. J. 1982. Restriction and modification enzymes and their recognition sequences. *Nucleic Acids Res.* **10**: 117.

Roberts, R. J. 1987. Restriction enzymes and their isoschizomers. *Nucleic Acids Res.* 15 (supplement) r189–r217. (This is a comprehensive list of restriction endonucleases that is updated annually.)

Roberts, R. J., and Macelis, D. 2001. Rebase—restriction enzymes and methylases. *Nucleic Acids Res.* **29**: 268–269.

Robertson, H. D., Webster, R. E., and Zinder, N. D. 1968. Purification and properties of ribonucleases III from *Escherichia coli*. *J. Biol. Chem.* **243**: 82.

Robins, P., Pappin, D., Wood, R. D., and Lindhal, T. 1994. Structural and functional homology between mammalian DNase IV and the 5′-nuclease domain of *E. coli* DNA polymerase I. *J. Biol. Chem.* **269**: 28535–28538.

Robinson, M. A., and Kindt, T. J. 1987. Genetic recombination within the human T-cell receptor α chain gene complex. *Proc. Natl. Acad. Sci. U.S.A.* **84**: 9089.

Roca, A. I., and Cox, M. M. 1990. The *RecA* protein: Structure and function. *CRC. Cri. Rev. Biochem Mol. Biol.* **25**: 415.

Roeder, G. S., and Sadowski, P. D. 1979. Pathways of recombination of bacteriophage T7 DNA *in vitro*. *Cold Spring Harbor Symp. Quant. Biol.* **43**: 1023.

Rogowsky, P., and Schmidt, R. 1984. Resolution of a hybrid cointegrate between transposors Tn 501 and Tn 1721 defines the recombination site. *Mol. Gen. Genet.* **193**: 162.

Rohr, G., and Mannherz, G. 1979. The activation of actin: DNase I complex with rat liver plasma membranes. The possible role of 5 nucleotidase. *FEBS Lett.* **99**: 351.

Rokugawa, K., Fujimoto, M., Kuninaka, A., Yoshino, H. 1979. Immobilization of nuclease P1 on cellulose. *J. Ferment. Techol.* **57**: 570–573.

Rosenberg, H. F., Tenen, D. G., Ackerman, S. J. 1989. Molecular cloning of the human eosinophil-derived neurotoxin: A member of the ribonuclease gene family. *Proc. Natl. Acad. Sci. U.S.A.* **86**: 4460.

Rosenberg, J. M., McClarin, J. A., Frederick, C. A., Wang, B.-C., Grable, J., Boyer, H. W., and Greene, P. 1987. Structure and recognition mechanism of *Eco*RI endonuclease. *Trends Biochem. Sci.* **12**: 395.

Roses, A. D. 2001. Pharmacogenetics. *Hum. Mol. Genet.* **10**: 2261–2267.

Roth, J. S., and Milstein, S. W. 1952. Ribonuclease I, A new method with P^{32} labeled yeast ribonucleic acid. *J. Biol. Chem.* **196**: 489.

Rubin, R. A., Modrich, P., and Vanaman, T. C. 1981. Partial NH_2– and COOH-terminal sequence analyses of *Eco*RI DNA restriction and modification enzymes. *J. Biol. Chem.* **256**: 2140.

Rudd, K. E., and Menzel, R. 1987. His operons of *Escherichia coli* and *Salmonella typhimurium* are regulated by DNA supercoiling. *Proc. Natl. Acad. Sci. U.S.A.* **84**: 517.

Ruddle, F. H. 1984. The Williams Allan Memorial Award Ideas: Reverse genetics and beyond. *Am. J. Human Genet.* **36**: 944.

Rushizky, G. W., Shaternikov, V. A., Mosejko, J. H., and Sober, H. A. 1975. S_1 nuclease hydrolysis of single-stranded nucleic acids with partial double-stranded configuration. *Biochemistry* **14**: 4221.

Ryan, M. J. 1976. Coumermycin A1: A preferential inhibitor of replicative DNA synthesis in *Escherichia coli*. I. *In vivo* characterization. *Biochemistry* **15**: 3769.

Rybak, S. M., Hoogenboom, H. R., Newton, D. L., Raus, J. C., and Youle, J. R. 1992. Rational immunotherapy with ribonuclease chimeras. An approach toward humanizing immunotoxin. *Cell Biophys.* **21**: 121–138.

Sadowski, P. D. 1971. Bacteriophage T7 endonuclease I. Properties of the enzyme purified from T7 phage-infected *Escherichia coli* B. *J. Biol. Chem.* **246**: 209.

Sadowski, P. D. 1977. Genetic recombination of bacteriophage T4 *in vitro* II. Further properties of the *in vitro* recombination packaging reaction. *Virology* **78**: 102.

Sadowski, P. 1986. Site-specific recombinases: Changing partners and doing the twist. *J. Bacteriol.* **165**: 341.

Sadowski, P. D., and Vetter, D. 1976. Genetic recombination of bacteriophage T7 DNA *in vitro*. *Proc. Natl. Acad. Sci. U.S.A.* **73**: 692.

Saenger, W. 1983. *Principles of Nucleic Acid Structure*. Springer-Verlag, New York.

Sakano, H., Maki, R., Kurosawa, Y., Roeder, R., and Tonegawa, S. 1980. Two types of somatic recombination are necessary for the generation of complete immunoglobulin heavy-chain genes. *Nature* **286**: 676.

Salnikow, J., Moore, S., and Stein, W. H. 1970. Comparison of the multiple forms of bovine pancreatic deoxyribonuclease. *J. Biol. Chem.* **245**: 5685.

Salnikow, J., Liao, T.-H., Moore, S., and Stein, W. H. 1973. Bovine pancreatic deoxyribonucleaseIsolation, A., composition, and amino acid sequences of the tryptic and chymotryptic peptides. *J. Biol. Chem.* **248**: 1480.

Saluz, H. P., and Jost, J. P. 1993. Approaches to characterize protein–DNA interactions *in vivo*. *Crit. Rev. Eukaryotic Gene Expression* **3**: 1.

Sambrook, U., Fritch, E. F., and Maniatis, T. 1989. *Molecular Cloning: A Laboratory Manual*, 2nd ed. Cold Spring Harbor Laboratory Press, Cold Spring Harbor, NY.

Sancar, S., and Rupp, W. D. 1983. A novel repair enzyme: *uvrABC* excision nuclease of *Escherichia coli* cuts a DNA strand on both sides of the damaged region. *Cell* **33**: 249.

Sancar, A., and Sancar, G. B. 1988. DNA repair enzymes. *Annu. Rev. Biochem.* **57**: 29.

Sancar, A., Clarke, N. D., Griswold, J., Kennedy, W. J., and Rupp, W. D. 1981a. Identification of the *uvrB* gene product. *J. Mol. Biol.* **148**: 63.

Sancar, A., Wharton, R. P., Selzer, S., Kacinski, B. M., Clarke, N. D., and Rupp, W. D. 1981b. Identification of the *uvrA* gene product. *J. Mol. Biol.* **148**: 45.

Sancar, A., Franklin, K. A., and Sancar, G. 1985. Repair of porsalen and acetylamiofluorene DNA adducts by *uvrABC* exonuclease. *J. Mol. Biol.* **184**: 725.

Sander, M., and Hsieh, T. S. 1985. *Drosophila* topoisomerase II double-strand DNA cleavage: Analysis of DNA sequence homology at cleavage site. *Nucleic Acids Res.* **13**: 1057.

Sander, M., Lowenhaupt, K., and Rich, A. 1991. *Drosophila* RrpI protein: An apurinic endonuclease with homologous recombination activities. *Proc. Natl. Acad. Sci. U.S.A.* **88**: 6780.

Sandmeier, H., Iida, S., Meyer, J., Hiestand-Nauer, R., and Arber, W. 1990. Site-specific DNA recombination system Min of plasmid p15B: A cluster of overlapping invertible DNA segments. *Proc. Natl. Acad. Sci. U.S.A.* **87**: 1109–1113.

Sanger, F., Nicklen, S., and Coulsen, A. R. 1977. DNA sequencing with chain-terminating inhibitors. *Proc. Natl. Acad. Sci. U.S.A.* **74**: 5463.

Santoro, S. W., and Joyce, G. F. 1997. A general purpose RNA-cleaving DNA enzyme. *Proc. Natl. Acad. Sci. U.S.A.* **94**: 4262–4266.

Sanzey, B. 1979. Modulation of gene expression by drugs affecting DNA gyrase. *J. Bacteriol.* **138**: 40.

Sauer, B., and Henderson, N. 1988. Site-specific DNA recombinations in mammalian cells by the Cre recombinase of bacteriophage P1. *Proc. Natl. Acad. Sci. U.S.A.* **85**: 5166–5170.

Schaaper, R. M. 1988. Mechanisms of mutagenesis in the *Escherichia coli* mutator mutD5: Role of DNA mismatch repair. *Proc. Natl. Acad. Sci. U.S.A.* **85**: 8126.

Schaeffer, H. N. 1983. Nucleoside with anti viral activity. In Rideout, J., L. D. Henry, and L. M. Beacham (eds.), *Nucleosides, Nucleotides and Their Biological Application.* Academic Press, New York.

Schatz, D. G., Oettinger, M. A., and Baltimore, D. 1989. The V(D)J recombination acting gene RAG-1. *Cell* **59**: 1035.

Schatz, D. G., Oettinger, M. A., Gorka, C., and Baltimore, D. 1990. RAG-1 and RAG-2: Adjacent genes that synergistically activate V(D)J recombination. *Science* **248**: 1517.

Schatz, O., Cromme, F. V., Gunninger-Leitch, F., and LeGrice, S. F. G. 1989. Point mutations in conserved amino acid residues within the C terminal domain of HIV-1 reverse transcriptase specifically repress RNaseH function. *FEBS Lett.* **257**: 311.

Schein, C. H. 1997. From Housekeeper to microsurgeon: The diagnostic and therapeutic potential of ribonucleases. *Nat. Biotechnol.* **15**: 529–536.

Schein, C. H. 2001. *Methods in Molecular Biology: Nucleases Methods and Protocols*, Vol. 160. Humana Press, Totowa, NJ, pp. 1–525.

Schlessinger, D. 1990. Yeast artificial chromosomes: Tool for mapping and analysis of complex genomes. *Trends Genetics* **6**: 248–258.

Schuler, W., Weiler, I. J., Schuler, A., Phillips, R. A., Rosenberg, N., Mak, T. W., Kearney, J. F., Perry, R., and Bosma, M. J. 1986. Rearrangement of antigen receptor genes is defective in mice with severe combined immune deficiency. *Cell* **46**: 963.

Schleif, R. 1988. DNA binding by proteins. *Science* **241**: 1182.

Schmidt, G. 1955. Nucleases and enzymes attacking nucleic acid components. In Chargaff E., and J. N. Davidson (eds.), *The Nucleic Acids*, Vol. 1. Academic Press, New York.

Schultz, P. G. 1988. The interplay between chemistry and biology in the design of enzymatic catalysts. *Science* **240**: 426.

Schultz, P. G., Taylor, J. S., and Dervan, P. B. 1982. Design and synthesis of a sequence-specific DNA cleaving molecule (Distamycin-EDTA) iron (II). *J. Am. Chem. Soc.* **104**: 6861.

Schultz, P. G., and Dervan, P. B. 1983. Sequence-specific double-strand cleavage of DNA by penta-*N*-methylpyrrolecarboxamide-EDTA-Fe (II). *Proc. Natl. Acad. Sci. U.S.A.* **80**: 6834.

Scurlock, T. T., and Miller, R. V. 1979. Pac Exo IX: A unique deoxyribonuclease from *Pseudomonas aeruginosa* in the presence of EDTA. *Nucleic Acids Res.* **7**: 167.

Selby, C. P., and Sancar, G. B. 1990. Transcription preferentially inhibits nucleotide excision repair of the template DNA strand *in vitro*. *J. Biol. Chem.* **265**: 21330.

Sen, D., and Geyer, C. R. 1998. DNA enzymes. *Curr. Opin. Chem. Biol.* **6**: 680.

Senecoff, J. F., Bruckner, R. C., and Cox, M. M. 1985. The FLP recombinase of the yeast 2-μm plasmid: Characterization of its recombination site. *Proc. Natl. Acad. Sci. U.S.A.* **82**: 7270.

Service, R. F. 1994. Stalking the start of colon cancer. *Science* **263**: 1559.

Sevcik, J., Sanishvili, R. G., Pavlovsky, A. G., and Polyakov, M. 1990. Comparison of active sites of some microbial ribonucleases: Structural basis for guanylic specificity. *Trends Biochem. Sci.* **15**: 185.

Shak, S., Capon, D. J., Hellmiss, R., Marsters, S. A., and Baker, C. C. 1990. Recombinant human DNase I reduces the viscosity of cystic fibrosis sputum. *Proc. Natl. Acad. Sci. U.S.A.* **87**: 9188.

Shamu, C. E., Cox, J. S., and Walter, P. 1994. The unfolded protein response pathway in yeast. *Trends Cell Biol.* **4**: 56–60.

Shapero M. H., Leuther, K. K., Nguyen, A., Scott, M., and Jones, K. W. 2001. SNA genotyping by multiplexed solid-phase amplification and fluorescent minisquencing. *Genome Res.* **11**: 1926–1934.

Shapiro, J. A. 1979. Molecular model for the transposition and replication of bacteriophage Mu and other transposable elements. *Proc. Natl. Acad. Sci. U.S.A.* **76**: 1933.

Shapiro, J. A., and Higgins, N. P. 1989. Differential activity of a transposable element in *E. coli* colonies. *J. Bacteriol.* **171**: 5475.

Shapiro, J., Machattie, L., Eron, L., Ihler, G., Ippen, K., and Beckwith, J. 1969. Isolation of pure *lac* operon DNA. *Nature* **224**: 768.

Shapiro, R., Weremowicz, S., Riordan, J. F., and Vallee, B. L. 1987. Ribonucleolytic activity of angiogenin: Essential histidine, lysine and arginine residues. *Proc. Natl. Acad. Sci. U.S.A.* **84**: 8783.

Sharp, P. A. 1981. Speculations on RNA splicing. *Cell* **23**: 643.

Sharp, P. A. 1985. On the origin of RNA splicing and introns. *Cell* **42**: 397.

Sharp, P. A. 1987. Trans splicing: Variation on a familiar theme. *Cell* **50**: 147.

Sharp, P. A. 1991. Five easy pieces. *Science* **254**: 663.

Sharp, P. A. 1999. RNAi and double stranded RNA. *Genes Dev.* **13**: 139–141.

Shaw, G., and Kamen, R. 1986. A concerved AU sequence from the 3 untranslated region of GM-CSF mRNA mediates selective mRNA degradation. *Cell* **46**: 659.

Shen, B., Nolan, J. P., Sklar, L. A., and Park, M. S. 1995. Functional analysis of a point mutations in human flap endonuclease-1 active site. *Nucleic Acids Res.* **25**: 3332–3338.

Shenk, T. E., Rhodes, C., Rigby, P. W. J., and Berg, P. 1975. Biochemical method for mapping mutational alterations in DNA with S1 nuclease: The location of deletions and temperature-sensitive mutations in simian virus 40. *Proc. Natl. Acad. Sci. U.S.A.* **72**: 989.

Shibahara, D., Mukai, S., Nishibara, T., Inoue, H., and Ohtsuka, E. H. M. 1987. Directed cleavage of RNA. *Nucleic Acids Res.* **15**: 4403.

Shiba, T., and Saigo, K. 1983. Retrovirus-like particles containing RNA homologous to the transposable element copia in *Drosophila melanogaster. Nature* **302**(5904): 119.

Shinagawa, Iwasaki, H. H., Kato, T., and Nakata, A. 1988. *RecA* protein-dependent cleavage of UmuD protein and SOS mutagenesis. *Proc. Natl. Acad. Sci. U.S.A.* **85**: 1806.

Shiomi, T., Harada, Y., Saito, T., Shiomi, N., Okuno, Y., and Yamaizumi, M. M. 1994. An ERCC5 gene homology to yeast RAD2 is involved in group G xeroderma pigmentosum. *Mutat. Res.* **314**: 167–175.

Shirai, T., and Go, M. 1991. RNase-like domain in DNA directed RNA polymerase II. *Proc. Natl. Acad. Sci. U.S.A.* **88**: 9056.

Shortle, D., DiMaio, D., and Nathans, D. 1981. Directed mutagenesis. *Annu. Rev. Genet.* **15**: 265.

Shortle, D. 1983. A genetic system for analysis of staphylococcal nuclease. *Genetics* **22**: 181.

Shub, D. A., Gott, J. M., Xu, M. Q., Lang, B. F., Michel, F., Tomaschewski, J., Lane, J. P., and Belfort, M. 1988. Structural conservation among three homologous introns of bacteriophage T4 and the group I introns of eukaryotes. *Proc. Natl. Acad. Sci. U.S.A.* **85**: 1151.

Shuler, W., Weiler, I. J., and Schiler, A. 1986. Rearrangement of antigen receptor genes is defective in mice with severe combined immune deficiency. *Cell* **46**: 963.

Shuman, S. 1991. Recombination mediated by vaccinia virus DNA topoisomerase I in *Escherichia coli* is sequence specific. *Proc. Natl. Acad. Sci. U.S.A.* **88**: 10104.

Shuster, A. M., Gololobov, G. V., Kvashuk, D. A., Bogomolova, A. E., Smirnov, I. V., and Gabibov, A. G. 1992. DNA hydrolyzing auto antibodies. *Science* **256**: 665.

Sidrauski, C., and Walter, P. 1997. The membrane kinase Ire1p is a site specific endonuclease that initiates mRNAsplicing in a unfolded protein response. *Cell* **90**: 1–20.

Siegel, G., Turchi, J. J., Myers, T. W., and Bambara, R. A. 1992. A 5 to 3 exonuclease functionally interacts with calf DNA polymerase E. *Proc. Natl. Acad. Sci. U.S.A.* **89**: 9377.

Sierakowska, H., and Shugar, D. 1977. Mammalian nucleolytic enzymes and their localization. *Prog. Nucleic Acid Res. Mol. Biol.* **7**: 369.

Sigman, D. S., and Chen, C.-H. B. 1990. Chemical nucleases: New reagents in molecular biology. *Annu. Rev. Biochem.* **59**: 207.

Sigman, D. S., Graham, D. R., D'Aurora, V., and Stern, A. M. 1979. Oxygen-dependent cleavage of DNA by the 1,10-phenanthroline-cuprous complex. *J. Biol. Chem.* **254**: 12269.

Silver, D. P., Spanopoulou, E., Mulligan, R. C., and Baltimore, D. 1993. Dispensable sequence motifs in the RAG-1 and RAG-2 genes for plasmid V(D)J recombination. *Proc. Natl. Acad. Sci. U.S.A.* **90**: 6100.

Silverman, M., and Simon, M. 1983. Phase variation and related systems. In Shapiro, J. A. (ed.), *Mobile Genetic Elements*, Academic Press, New York, pp. 537–557.

Sim, S.-K., and Lown, J. W. 1978. The mechanism of the neocarzinostatin-induced cleavage of DNA. *Biochem. Biophys. Res. Commun.* **81**: 99.

Simon, R. W., and Kleckner, N. 1988. Biological regulation by antisense RNA in prokaryotes. *Annu. Rev. Genet.* **22**: 567.

Simon, M., Zeig, J., Silverman, M., Mandel, G., and Doolittle, R. 1980. Phase variation: Evolution of a controlling element. *Science* **209**: 1370.

Simon, T. J., Smith, C. A., and Friedberg, E. C. 1975. Action of bacteriophage T4 ultraviolet endonuclease on duplex DNA containing one ultraviolet-irradiated strand. *J. Biol. Chem.* **250**: 8748.

Simpson, L. 1990. RNA editing—A novel genetic phenomena. *Science* **250**: 512.

Sinden, R. R., Carlson, J. O., and Pettijohn, D. E. 1980. Torsional tension in the DNA double helix measured with trimethylpsoralen in living *E. coli* cells: Analogous measurements in insect and human cells. *Cell* **21**: 773.

Sinden, R. R., and Pettijohn, D. E. 1981. Chromosomes in living *Escherichia coli* cells are segregated into domains of supercoiling. *Proc. Natl. Acad. Sci. U.S.A.* **78**: 224.

Singwi, S., and Joshi, S. 2000. Potential nuclease based strategies for HIV gene therapy. *Frontiers Biosci.* **5:** 556–579.

Sioud, M., and Leirdal, M. 2000. Therapeutic RNA and DNA enzymes. *Biochem. Pharmacol.* **15**: 1023–1026.

Slifman, N. R., Loegering, D. A., McKean, D. J., and Gleich, G. J. 1986. Ribonuclease activity associated with human eosinophil-derived neurotoxin and eosinophil cationic protein. **137**: 2913.

Sluka, J. P., Horvath, S. J., Bruist, M. F., Simon, M. I., and Dervan, P. B. 1987. Synthesis of a sequence-specific DNA-cleaving peptide. *Science* **238**: 1129.

Smith, C. A., and Hanawalt, P. C. 1978. Phage T4 endonuclease V stimulates DNA repair replication in isolated nuclei from ultraviolet-irradiated human cells including *Xeroderma pigmentosum* fibroblasts. *Proc. Natl. Acad. Sci. U.S.A.* **75**: 295.

Smith, C. L., Kubo, M., and Imamoto, F. 1978. Promoter specific inhibition of transcription by antibiotics which act on DNA gyrase. *Nature* **275**: 420.

Smith, G. R. 1987. Mechanism and control of homologous recombination in *E. coli*. *Annu. Rev. Genet.* **21**: 190.

Smith, G. R. 1989a. Homologous recombination in prokaryotes: Enzymes and controlling sites. *Genome*. **31**: 520.

Smith, G. R. 1989b. Homologous recombination in *E. coli*: Multiple pathways for multiple reasons. *Cell* **58**: 807.

Smith, H. O., and Wilcox, K. W. 1970. A restriction enzyme from *Hemophilus influenzae*. I. Purification and chemical properties. *J. Mol. Biol.* **51**: 371.

Smith, G. R. 1983. Chi hotspots of generalized recombination. *Cell* **34**: 70.

Smith, G. R. 1988. Homologous recombination in prokaryotes. *Microbiol. Rev.* **46**: 1.

Smith, G. R., Comb, M., Schultz, D. W., Daniels, D. L., and Blattner, F. R. 1981. Nucleotide sequence of the Chi recombinational hot spot $X^{+}D$ in bacteriophage λ. *J. Virol.* **37**: 336.

Smith, H. O. 1979. Nucleotide sequence specificity of restriction endonucleases. *Science* **205**: 455.

Smith, H. O., and Nathans, D. 1973. A suggested nomenclature for bacterial host modification and restriction systems and their enzymes. *J. Mol. Biol.* **81**: 419.

Smith, M. 1985. *In vitro* mutagenesis. *Annu. Rev. Genet.* **19**: 423.

Smyth, D. G., Stein, W. H., and Moore, S. 1963. Sequence of amino acid residues in bovine pancreatic ribonuclease-revisions and confirmation. *J. Biol. Chem.* **238**: 227.

Snabes, M. C., Boyd, A. E., Pardue, R. L., and Bryan, J. 1981. A DNase binding-immunoprecipitation assay for actin. *J. Biol. Chem.* **256**: 6291.

Snapk, R. M. 1986. Topoisomerase inhibitors can selectively interfere with different stages of simian virus 40 DNA replication. *Mol. Cell Biol.* **6**: 4221.

Snider, W. D. 1994. Functions of the neurotrophins during nervous system development: What the knockouts are teaching us. *Cell* **77**: 627.

Sogin, M. L., Pace, B., and Pace, N. R. 1977. Partial purification and properties of a ribosomal RNA maturation endonuclease from *B. subtilis. J. Biol. Chem.* **252**: 1350.

Solnick, D. 1985. Alternative splicing caused by RNA secondary structure. *Cell* **43**: 667.

Sorrentino, S., Tucker, G. K., and Glitz, D. G. 1988. Purification and characterization of a ribonuclease from human liver. *Biol. Chem.* **263**: 16125.

Southern, E. M. 1975. Detection of specific sequences of DNA fragments separated by gel electrophoresis. *J. Mol. Biol.* **98**: 503.

Southern, E. M. 1982. New methods for analyzing DNA make genetics simpler. *Biochem. Soc. Trans.* **10**: 1.

Spoerel, N., Herrlich, P., and Bickle, T. A. 1979. A novel bacteriophage defense mechanism: The anti-restriction protein. *Nature* **278**: 30.

Srivenugopal, K. S., Lockshon, D., and Morris, D. R. 1984. *Escherichia coli* DNA topoisomerase III: Purification and characterizations of a new type I enzyme. *Biochemistry* **23**: 1899.

Stahl, F. W. 1979. Specialized sites in generalized recombination. *Annu. Rev. Genet.* **13**: 7.

Stahl, F. W. 1980. *Genetic Recombination: Thinking About It in Phage and Fungi*. Freeman, San Francisco.

Stahl, F. W. 1988. A unicorn in the garden. *Nature* **335**: 112.

Stange, N., Gross, H. J., and Beier, H. 1988. Wheat germ splicing endonuclease is highly specific for plant pre-tRNAs. *EMBO J.* **7**: 3823.

Stark, W. M., Boocock, M. R., and Sherratt, D. J. 1989. Site-specific recombination by Tn3 resolvase. *Trends Genet.* **5**: 304.

Staudenbauer, W. L. 1975. Novobiocin—A specific inhibitor of semiconservative DNA replication in permeabilized *Escherichia coli* cells. *J. Mol. Biol.* **96**: 201.

Staudenbauer, W. L. 1976. Replication of *Escherichia coli* DNA *in vitro:* Inhibition by oxolinic acid. *Eur. J. Biochem.* **62**: 491.

Staudenbauer, W. L., and Orr, E. 1981. DNA gyrase: Affinity chromatography on novobiocin-sepharose and catalytic properties. *Nucleic Acids Res.* **9**: 3589.

Steighner, R. J., and Povirk, L. F. 1990. Bleomycin-induced DNA lesions at mutational hot spots: Implications to the mechanism of double-strand cleavage. *Proc. Natl. Acad. Sci. U.S.A.* **87**: 8350.

Stein, C. A., and Cheng, Y. C. 1993. Antisense oligonucleotides as therapeutic agents—Is the bullet really magical? *Science* **261**: 1004.

Stein, H., and Hausen, P. 1969. Enzyme from calf thymus degrading the RNA moiety of DNA: RNA hybrids effect on DNA dependent RNA polymerase. *Science* **166**: 393.

Sternberg, N., and Hoess, R. 1983. The molecular genetics of bacteriophage P1. *Annu. Rev. Genet.* **17**: 123.

Sternberg, N., Sauer, B., Hoess, R., and Ambremski, K. 1986. Bacteriophage P1 *cre* gene and its regulatory region; evidence for multiple promoters and for regulation by DNA methylation. *J. Mol. Biol.* **187**: 197.

Sternglanz, R., Dinardo, S., Voelkel, K. A., Nishimura, Y., Hirota, Y., Becherer, K., Zumstein, L., and Wang, J. C. 1981. Mutations in the gene coding for *Escherichia coli* DNA topoisomerase I affect transcription and transposition. *Proc. Natl. Acad. Sci. U.S.A.* **78**: 2747.

Stetler, G. L., King, G. J., and Huang, W. W. 1979. T4 DNA-delay proteins, required for specific DNA replication, forms a complex that has ATP-dependent DNA topoisomerase activity. *Proc. Natl. Acad. Sci. U.S.A.* **76**: 3737.

Stirdivant, S. M., Crossland, L. D., and Bogorad, L. 1985. DNA supercoiling affects *in vitro* transcription of two maize chloroplast genes differently. *Proc. Natl. Acad. Sci. U.S.A.* **82**: 4886.

Stevenson, F. K. 1999. DNA vaccines against cancer: From gene to therapy. *Ann. Oncol.* **10**: 1413–1418.

Stonington, G. O., and Pettijohn, D. E. 1971. The folded genome of *Escherichia coli* isolated in a protein DNA–RNA complex. *Proc. Natl. Acad. Sci. U.S.A.* **68**: 6.

Strauss, B. S. 1962. Differential destruction of the transforming activity of damaged deoxyribonucleic acid by a bacterial enzyme. *Proc. Natl. Acad Sci. U.S.A.* **48**: 1670.

Strydom, D. L., Fett, J. W., Lobb, R. R., Alderman, E. M., Bethune, J. L., Riordan, J. F., and Vallee, B. L. 1985. Amino acid sequence of human tumor derived angiognin. *Biochemistry* **24**: 5486.

Stryer, L. 1982. *Biochemistry.* W. H. Freeman, San Francisco.

Su, S. S., and Modrich, P. 1986. *Escherichia coli* mutS-encoded protein binds to mismatched DNA base pairs. *Proc. Natl. Acad. Sci. U.S.A.* **83**: 5057.

Sun, L. Q., Cairns, M. J., Gerlach, W. L., Witherington, C., Wang, L., and King, A. 1999. Suppression of smooth muscle cell proliferation by a c-myc RNA-cleaving deoxyribozyme. *J. Biol. Chem.* **274**: 17236.

Suck, D. 1998. DNA recognition by structure-selective nucleases. *Biopolymers* **44**: 405–421.

Suck, D., Oefner, C., and Kabsch, W. 1984. Three dimensional structure of bovine pancreatic DNase I. *EMBO J.* **3**: 2423.

Südhof, T. C., Goldstein, J. L., Brown, M. S., and Russel, D. W. 1985a. The LDL receptor gene: A mosaic of exons shared with different proteins. *Science* **228**: 815.

Südhof, T. C., Russell, D. W., Goldstein, J. L., and Brown, M. S. 1985b. Cassette of eight exons shared by genes for LDL receptor and EGF precursor. *Science* **228**: 893.

Sugino, A., and Bott, K. 1980. *Bacillus subtilis* deoxyribonucleic acid gyrase. *J. Bacteriol.* **141**: 1331.

Sugino, A., Peebles, G. L., Kreuzer, K. N., and Cozzarelli, N. R. 1977. Mechanism of action of nalidixic acid: Purification of *Escherichia coli* nal-a gene product and its relationship to DNA gyrase and a novel nicking-closing enzyme. *Proc. Natl. Acad. Sci. U.S.A.* **74**: 4767.

Sugino, A., Higgins, N. P., Brown, P. O., Peebles, C. L., and Cozzarelli, N. R. 1978. Energy coupling in DNA gyrase and the mechanism of action of novobiocin. *Proc. Natl. Acad. Sci. U.S.A.* **75**: 4838.

Sugiura, Y., Shiraki, T., Konishi, M., and Oki, T. 1990. DNA intercalation and cleavage of an antitumor antibiotic dynemicin that contains anthracycline and enediyne cores. *Proc. Natl. Acad. Sci. U.S.A.* **87**: 3831.

Sun, X. Z., Hirada, Y. N., Takashi, S., Shiomi, N., and Shiomi, T. 2001. Purkinje cell degeneration in mice lacking the xeroderma pigmentosum group G gene. *J. Neurosci. Res.* **64**: 348–354.

Sung, S.-C., and Laskowski, M., Sr. 1962. A nuclease from mung bean sprouts. *J. Biol. Chem.* **237**: 506.

Symington, L. S., and Kolodner, R. 1985. Partial purification of an enzyme from *Saccharomyces cerevisiae* that cleaves Holliday junctions. *Proc. Natl. Acad. Sci. U.S.A.* **82**: 7247.

Symington, L. S., Fogarty, L. M., and Kolodner, R. 1983. Genetic recombination of homologous plasmids catalyzed by cell-free extracts of *Saccharomyces cerevisiae. Cell* **35**: 805.

Symonds, N. 1989. Anticipatory mutagenesis. *Nature* **337**: 119.

Syvanen, M. 1986. Cross-species gene transfer: A major factor in evolution? *Trends Genet.* **2**: 63.

Szostak, J. W., Orrweaver, T. L., Rothstein, R., and Stahl, F. 1983. The double-strand break repair model for recombination. *Cell* **33**: 25.

Takahashi, K. 1978. Structure and function of ribonuclease T. *J. Biochem.* (*Tokyo*) **67**: 833.

Tamblyn, T. M., and Wells, R. D. 1975. Effect of incubation conditions on the nucleotide sequence of DNA products of unprimed DNA polymerase reactions. *J. Mol. Biol.* **53**: 435.

Tanaka, K., Hayakawa, H., Sekiguchi, M., and Okada, Y. 1977. Specific actions of T4 endonuclease V on damaged DNA in *Xeroderma pigmentosum* cells *in vivo. Proc. Natl. Acad. Sci. U.S.A.* **74**: 2958.

Tanaka, K., Naoyuki, M., Ichiro, S., Iwai, M., Miyomoto, M., Yoshida, M. C., Saloh, Y., Kondo, S., Yascii, A., Drayame, H., and Okada, Y. 1990. Analysis of a human DNA excision repair gene involved in group A *Xeroderma pigmentosum* and containing a zinc-finger domain. *Nature* **348**: 73.

Taylor, A. F., Schultz, D. W., Ponticelli, A. S., and Smith, G. R. 1985. *RecBC* enzyme nicking at Chi sites during DNA unwinding location and orientation-dependence of the cutting. *Cell* **41**: 153.

Templin, A., Kushner, S. R., and Clark, A. J. 1972. Genetic analysis of mutation indirectly suppressing *RecB* and *RecC* mutations. *Genetics* **72**: 205.

Tewey, K. M., Rowe, T. C., Yang, L., Halligan, B. D., and Liu, L. F. 1983. Adriamycin-induced DNA damage mediated by mammalian DNA topoisomerase II. *Science* **226**: 466.

Thierry, A., Perrin, A., Boyer, J., Fairhead, C., Dujon, B., Frey, B., and Schmitz, G. 1991. Cleavage of yeast and bacteriophage T7 genomes at a single site using the rare cutter endonuclease I-Sce I. *Nucleic Acids Res.* **19**: 189.

Thomas, K. R., and Olivera, B. M. 1978. Processivity of DNA exonucleases. *J. Biol. Chem.* **253**: 424.

Thomas, M., and Davis, R. W. 1975. Studies on the cleavage of bacteriophage lambda DNA with *Eco*RI restriction endonuclease. *J. Mol. Biol.* **91**: 315.

Thompson, L. H., Rubin, F. J. S., Cleaver, J. E., Whitmore, G. F., and Brookman, K. 1980. A screening method for isolating DNA repair deficient mutant of CHO cells. *Somat. Cell. Genet.* **6**: 391.

Thompson, L. H., Mooney, C. L., and Brookman, K. W. 1985. Genetic complementation between UV-sensitive CHO mutants and xeroderma pigmentosum fibroblasts. *Mutat. Res.* **150**: 423.

Thompson, L. H. 1998. Chinese hamster cells meet DNA repair: An entirely acceptable affair. *Bioessays* **20**: 589.

Thrash, C., Voekel, K., DiNardo, S., and Sternglanz, R. 1984. Identification of *Saccharomyces cerevisiae* mutants deficient in DNA topoisomerase I activity. *J. Biol. Chem.* **259**: 1375.

Thrash, C., Banker, A. T., Barrell, B. G., and Sternglanz, R. 1985. Cloning, characterization, and sequence of the yeast DNA topoisomerase I gene. *Proc. Natl. Acad. Sci. U.S.A.* **82**: 4374.

Tishkoff, D. X., Filosi, N., Gaida, G. M., and Kolodner, R. D. 1997. A novel mutation avoidance mechanism dependent on *S. cerevisiae* RAD27 is distinct from mismatch repair. *Cell* **88**: 253–263.

Tomlin, N. V., Raveltuchuk, E. B., and Mosevitskay, T. V. 1976. Substrate specificity of the ultraviolet-endonuclease from *Micrococcus luteus*, endonucleolytic cleavage of depurinated DNA. *Eur. J. Biochem.* **69**: 265.

Todd J. A., Livak, K. J., and Marmaro, J. 1995. Towards fully automated genome wide polymorphism screening. *Nat. Genet.* **9**: 341–342.

Tonegawa, S. 1983. Somatic generation of antibody diversity. *Nature* **302**: 575.

Tonegawa, S. 1985. The molecules of the immune system. *Sci. Am.* **253**(4): 122.

Toulmé, J.-J., and C. Héléne, C., 1981. A tryptophan-containing peptide recognizes and cleaves DNA at apurinic sites. *Nature* **292**: 858.

Trash, D. K., DiDonato, J. A., and Muller, M. T. 1984. Rapid detection and isolation of covalent DNA/protein complexes: Application to topoisomerase I and II. *EMBO J.* **3**: 671.

Trilling, D. M., and Aposhian, H. V. 1968. Sequential cleavage of dinucleotides from DNA by phage SP3 DNase. *Proc. Natl. Acad. Sci. U.S.A.* **60**: 214.

Trinks, K., Habermann, P., Begreuther, K., Starlinger, P., and Ehring, R. 1981. An IS4-encoded protein is synthesized in minicells. *Mol. Gen. Genet.* **182**: 183.

Troelstra, C., Van Gool, A., deWitt, J., Vermeulen, W., Bootsma, D., and Hoeijmakers, D. 1992. ERCCB, a member of a subfamily of putative helicases, is involved in Cockayne's syndrome and preferential repair of active genes. *Cell* **71**: 939.

Trubia, M., Sessa, L., and Tarramelli, R. 1997. Mammalian Rh/T2/S-glycoprotein ribonuclease family genes: Cloning of a human member located in a region of chromosome 6(6q27) frequently deleted in human malignancies. *Genomics* **42**: 342–344.

Trucksis, M., and Depew, R. E. 1981. Identification and localization of a gene which specifies production of *Escherichia coli* DNA topoisomerase I. *Proc. Natl. Acad. Sci. U.S.A.* **78**: 2164.

Tsai-Pflugfelder, M., Liu, L. F., Liu, A. A., Tewey, K. M., Whang-Peng, J., Knutsen, T., Huebner, K., Croce, C. M., and Wang, J. C. 1988. Cloning and sequencing of cDNA encoding human DNA topoisomerase II and localization of the gene to chromosome region 17q21-22. *Proc. Natl. Acad. Sci. U.S.A.* **85**: 7177.

Tsai-wu, J.-J., Liu, H.-F., and Lu, A. L. 1992. *Escherichia coli* mut y protein has both *N*-glycosidase and apurinic/apyrimidinic endonuclease activities on AC and AG mispairs. *Proc. Natl. Acad. Sci. U.S.A.* **89**: 8779.

Tse, Y. C., and Wang, J. C. 1980. *E. coli* and *M. luteus* DNA topoisomenase 1 can catalyze catenation or decatenation of double-stranded DNA rings. *Cell* **22**: 269.

Tse-Dinh, Y. C., McCarron, B. G. H., Arentzen, R., and Choudhary, V. 1983. Mechanistic study of *E. coli* DNA topoisomerase 1: Cleavage of oligonucleotides. *Nucleic Acids Res.* **11**: 8691.

Tsuda, Y., and Strauss, B. S. 1964. A deoxyribonuclease reaction requiring nucleoside di- or triphosphates. *Biochemistry* **3**: 1678.

Tsutsumi, S., Aso, T., Nagamachi, Y., Nakajima, T., Yasuda, T., and Kishi, K. 1998. Phenoype 2 of DNaseI may be used as a risk factor for gastric carcinoma. *Cancer* **82**: 1621–1625.

Tucker, P. W., Hazen, E. E., and Cotton, F. A. 1979. Staphylococcal nuclease reviewed: A prototypic study in contemporary enzymology. *Mol. Cell. Biol.* **23**: 131.

Tullius, T. D., and Dombroski, B. A. 1986. Hydroxyl radical "footprinting": High-resolution information about DNA–protein contacts and application to λ repressor and Cro protein. *Proc. Natl. Acad. Sci. U.S.A.* **83**: 5469.

Turcq, B., Dobinson, K. F., Serizawa, N., and Lambowitz, A. 1994. A protein required for RNA processing and splicing in *Neurospora* mitochondria is related to gene products involved in cell cycle phosphatase function. *Proc. Natl. Acad. Sci. U.S.A.* **91**: 1676.

Uemura, T., and Yanagida, M. 1984. Isolation of type I and II DNA topoisomerase mutants from fission yeast: Single and double mutants show different phenotypes in cell growth and chromatin organization. *EMBO J.* **3**: 1737.

Umemura, K., Nagami, F., Okada, T., and Kuroda, R. 2000. AFM characterization of a single strand-specific endonuclease activity on linear DNA. *Nucleic Acids Res.* **28**: 39–45.

Uhlenbeck, O. C. 1987. A small catalytic oligonucleotide. *Nature* **328**: 596.

van de Sande, J. H., Loewen, P. C., and Khorana, H. G. 1972. Studies on polynucleotides CXVIII: A further study of ribonucleotide incorporation into deoxyribonucleic acid chains by deoxyribonucleic acid polymerase I of *E. coli*. *J. Biol. Chem.* **247**: 6140.

van de Putte, P., Plasterk, R., and Kuijpers, A. 1984. A Mu gin complementing function and an invertible DNA region in *Escherichia coli* K-12 are situated on the genetic element e14. *J. Bacteriol.* **158**: 517.

Van der Ploeg, L. H. T., Cornelissen, A. W. C. A., Michels, P. A. M., and Borst, P. 1984. Chromosome rearrangements in *Trypanosoma brucei*. *Cell* **39**: 213.

Van derveen, R., Arnberg, A. C., vander Horst, G., Bonen, L., Tabak, H. F., and Grivell, L. A. 1986. Excised group II introns in yeast mitochondria are lariats and can be formed by self splicing *in vitro*. *Cell* **44**: 225.

VanDyke, M. W., and Dervan, P. B. 1983. Methidium propyl-EDTA. Fe (II) and DNase I footprinting report different small molecule binding site sizes on DNA. *Nucleic Acids Res.* **11**: 5555.

VanDyke, M. W., Hertzberg, R. P., and Dervan, P. B. 1982. Map of distamycin, netropsin, and actinomycin binding sites on heterogeneous DNA: DNA cleavage-inhibition patterns with methidiumpropyl-EDTA-Fe(II). *Proc. Natl. Acad. Sci. U.S.A.* **79**: 5470.

Varlet, I., Radman, M., and Brooks, P. 1990. DNA mismatch repair in *Xenopus* egg extracts: Repair efficiency and DNA repair synthesis for all single base pair mismatches. *Proc. Natl. Acad. Sci. U.S.A.* **87**: 7883.

Varmuza, S. I., and Smiley, J. R. 1985. Signals for site-specific cleavage of HSV DNA: Maturation involves two separate cleavage events at sites distal to the recognition sequence. *Cell* **41**: 793.

Venema, J., Mullenders, L. H. F., Natrajan, A. T., Vanzeeland, A. A., and Mayne, L. V. 1990. The genetic defect in cockayne syndrome is associated with a defect in repair of UV-induced damage in transcriptionally active DNA. *Proc. Natl. Acad. Sci. U.S.A.* **87**: 4707.

Verdine, G. L. 1994. The flip side of DNA methylation. *Cell* **76**: 197.

Vereley, W. G., and Paquette, Y. 1972. An endonuclease for depurinated DNA in *Escherichia coli* B. *Can. J. Biochem.* **50**: 217.

Vereley, W. G., and Rassart, E. 1975. Purification of *Escherichia coli* endonuclease specific for apurinic sites in DNA. *J. Biol. Chem.* **250**: 8214.

Vetter, B., Andrews, B. J., Roberts-Beatty, L., and Sadowski, P. D. 1983. Site-specific recombination of yeast 2-μm DNA *in vitro*. *Proc. Natl. Acad. Sci. U.S.A.* **80**: 7284.

Villeponteau, B., Lundell, M., and Martinson, H. 1984. Torsional stress promotes the DNase I sensitivity of active genes. *Cell* **39**: 469.

Vinograd, J., Lebowitz, J., Radloff, R., Watson, R., and Laipis, P. 1965. The twisted circular form of polyoma viral DNA. *Proc. Natl. Acad. Sci. U.S.A.* **53**: 1104.

Vogt, V. M. 1973. Purification and further properties of single-strand specific nuclease from *Aspergillus oryzae*. *Eur. J. Biochem.* **33**: 192.

Vosberg, H. P. 1985. DNA topoisomerase. Enzymes that control DNA conformation. *Current Topics in Immunology and Microbiology* **114**: 19.

Vovis, G. F. 1973. Adenosine triphosphate-dependent deoxyribonuclease from *Diplococcus pneumoniea*: Fate of transforming deoxyribonucleic acid in a strain deficient in the enzymatic activity. *J. Bacteriol.* **113**: 718.

Waga, S., and Stillman, B. 1998. The DNA replication fork in eukaryotic cells. *Annu. Rev. Biochem.* **67**: 721.

Walder, R. V., and Walder, J. A. 1988. Role of RNase H in hybrid-arrested translation by antisense oligonucleotides. *Proc. Natl. Acad. Sci. U.S.A.* **85**: 5011.

Waldman, A. S., and Liskay, R. M. 1988. Resolution of synthetic Holliday structures by an extract of human cells. *Nucleic Acids Res.* **16**: 10249.

Waldrop, M. M. 1989. Chemzymes mimic biology in miniature. *Science* **245**: 354.

Walker, G. C. 1985. Inducible DNA repair systems. *Annu. Rev. Biochem.* **54**: 425.

Wallace, D. C. 1989. Mitochondrial DNA mutations and neuromuscular diseases. *Trends Genet.* **5**: 9.

Wang, D. 1979. Preparation of the bifunctional enzyme ribonuclease–deoxyribonuclease by cross-linkage. *Biochemistry* **18**: 4449.

Wang, J. C. 1971. Interaction between DNA and an *Escherichia coli* protein. *J. Mol. Biol.* **55**: 523.

Wang, J. C. 1981. Type I DNA topoisomerases. In Boyer, P. (ed.), *The Enzymes*, Vol. XIV. Academic Press, New York.

Wang, J. C., and Liu, L. F. 1979. DNA topoisomerases: Enzymes which catalyze the concerted breaking and rejoining of DNA backbone bonds. In Taylor, J. H. (ed.), *Molecular Genetics*, Part III. Academic Press, New York.

Wang, A. H. J., Quigley, G. J., Kolpack, F. J., Crawford, J. L., Van Bloom, J. H., Vandermarel, G., and Rich, A. 1979. Molecular structure of a left handed double helical DNA fragment at atomic resolution. *Nature* **282**: 680.

Wang, Z. G., X., Wu, and Friedberg, E. C. 1993. Nucleotide excision repair of DNA in cell free extracts of yeast *Saccharomyces cerevisiae*. *Proc. Natl. Acad. Sci. U.S.A.* **90**: 4907.

Wani, A. A., and Hadi, S. M. 1979. Partial purification and characterization of an endonuclease from germinating pea seeds specific for single-stranded DNA. *Arch. Biochem. Biophys.* **196**: 138.

Ward, R., Meagher, A., Tomlinson, I., O'Connor, T., Norrie, M., and Wu, R. 2001. Microsatellite instability and the clinicopathological features of sporadic colorectal cancer. Gut **48**: 821–829.

Warner, H. R., Demple, B. F., Deutsch, W. A., Kane, C. M., and Linn, S. 1980a. Apurinic/apyrimidinic endonucleases in the repair of pyrimidine dimers and other lesions in DNA. *Proc. Natl. Acad. Sci. U.S.A.* **77**: 4602.

Warner, H. R., Persson, M.-L., Benson, R. J., Mosbaugh, D. W., and Linn, S. 1980b. Selective inhibition by harmane of the apurinic/apyrimidinic endonucleases in the repair of pyrimidine dimers and other lesions in DNA. *Proc. Natl. Acad. Sci. U.S.A.* **77**: 4602.

Warner, H. R., Persson, M. L., and Benson, R. J. 1981. Selective inhibition by harmane of the apurinic/apyumidinic endonuclease activity of phage T4-induced UV-endonuclease. *Nucleic Acids Res.* **9**: 6083.

Wasserman, S. A., and Cozzarelli, N. R. 1985. Determination of the stereostructure of the product of Tn3 resolvase by a general method. *Proc. Natl. Acad. Sci. U.S.A.* **82**: 1079.

Wasserman, S. A., and Cozzarelli, N. R. 1986. Biochemical topology: Applications to DNA recombination and replication. *Science* **232**: 951.

Wasserman, S. A., Dungan, J. M., and Cozzarelli, N. R. 1985. Discovery of a predicted DNA knot substantiates a model for site specific recombination. *Science* **229**: 171.

Watabe, H., Iino, T., Kaneko, T., Shibata, T., and Ando, T. 1983. A new class of site-specific endodeoxyribonucleases. *J. Biol. Chem.* **258**: 4663.

Watanabe, T., and Kasai, K. 1978. Studies on sheep kidney nuclease. I. An improved purification method and some properties. *Biochim. Biophys. Acta* **520**: 52.

Waterborg, J., and Kuyper, C. M. A. 1979. Purification of an alkaline nuclease from *Physarum polycephalum. Biochim. Biophys. Acta* **571**: 359.

Watson, J. D., and Crick, F. H. C. 1953a. Molecular structure of nucleic acids: A structure for deoxynucleic acids. *Nature* **171**: 737.

Watson, J. D., and Crick, F. H. C. 1953b. General implications of the structure of deoxyribonucleic acids. *Nature* **171**: 964.

Weatherall, D. J. 1982. *The New Genetics and Clinical Practice*. Nuffield Provincial Hospitals Trust, London.

Weiner, A. M. 1988. Eukaryotic nuclear telomeres: Molecular fossils of the RNP world? *Cell* **52**: 155.

Weickmann, J. L., Elson, M., and Glitz, D. G. 1981. Purification and characterization of human pancreatic ribonuclease. *Biochemistry* **21**: 1272.

Weintraub, H., Cheng, P. F., and Conrad, K. 1986. Expression of transfected DNA depends on DNA topology. *Cell* **46**: 115.

Weintraub, H., and Groudine, M. 1976. Chromosomal subunits in active genes have altered conformation. *Science* **193**: 848.

Weiss, B. 1976. Endonuclease II of *Escherichia coli* is exonuclease III. *J. Biol. Chem.* **251**: 1896.

Weiss, B. 1981. Exodeoxyribonucleases of *Escherichia coli*. In Boyer, P. D. (ed.), *The Enzymes*, Vol. 14. Academic Press, New York.

Weiss, B., Rogers, S. G., and Taylor, A. F. 1978. The endonuclease activity of exonuclease III and the repair of uracil-containing DNA in *E. coli*. In Hanawalt, P. C., E. C. Friedberg, and C. F. Fox (eds.), *DNA Repair Mechanisms*. Academic Press, New York.

Weiss-Brummer, B., Rodel, G., Schweyen, R. J., and Kaudewitz, F. 1982. Expression of the split gene *cob* in yeast: Evidence for a precursor of a "maturase" protein translated from intron 4 and preceding Exons. *Cell* **29**: 527.

Welsh, K. M., Lu, A. L., Clark, S., and Modrich, P. 1987. Isolation and characterization of the *Escherichia coli* mutH gene product. *J. Biol. Chem.* **262**: 15624.

Wenzlau, J. M., Saldanha, R. J., Butow, R. A., and Perlman, P. S. 1989. A latent intron-encoded maturase is also an endonuclease needed for intron mobility. *Cell* **56**: 421.

West, D. K., Belfort, M., Maley, G. F., and Maley, F. 1986. Cloning and expression of an intron-encoded phage T4 td gene. *J. Biol. Chem.* **261**: 13446.

West, D. K., Changchien, L., Maley, G. F., and Maley, F. 1989. Evidence that the intron open reading frame of the phage T4 td gene encodes a specific endonuclease. *J. Biol. Chem.* **264**: 10343.

West, S. 1992. Enzymes and molecular mechanism of genetic recombination. *Annu. Rev. Biochem.* **61**: 603.

West, S. 1994. The processing of recombination intermediate mechanistic insight from studies of bacterial proteins. *Cell* **74**: 9.

Westheimer, F. H. 1987. Why nature chose phosphates? *Science* **235**: 1173.

White, K. A., and Morris, T. J. 1994. Recombination between defective Tombus virus RNAs generates functional hybride genomes. *Proc. Natl. Acad. Sci. U.S.A.* **91**: 3642.

White, J. H., Lusnak, K., and Fogel, S. 1985. Mismatch post meiotic segregation frequency in yeast suggests a heteroduplex recombination intermediate. *Nature* **315**: 350.

White, R., and Lalouel, J.-M. 1988. Chromosome mapping with DNA markers. *Sci. Am.* **258**: 40.

White, R. 1986. The search for the cystic fibrosis gene. *Science* **234**: 1054.

White, H. B. 1976. Coenzymes as fossils of an earlier metabolic state. *J. Mol. Evol.* **7**: 101.

Whitehouse, H. L. K. 1982. *Genetic Recombination*. John Wiley & Sons., New York.

Williams, J. G. K., Shibata, T., and Radding, C. M. 1981. *E. coli RecA* protein protects single-stranded DNA or gapped duplex DNA from degradation by *RecBC* DNase. *J. Biol. Chem.* **256**: 7573.

Wilcox, K. W., and Smith, H. O. 1975. Isolation and characterization of mutants of *Haemophilus influenzae* deficient in an adenosine 5-triphosphate-dependent deoxyribonuclease activity. *J. Bacteriol.* **122**: 443.

Wilcox, K. W., and Smith, H. O. 1976. Mechanism of DNA degradation by the ATP-dependent DNase from *Hemophilus influenzae* Rd. *J. Biol. Chem.* **251**: 6127.

Williamson, M. S., Game, J. C., and Fogel, S. 1985. Meiotic gene conversion mutants in *Saccharomyces cerevisiae:* Isolation and characterization of pmS1-1 and pmS1-2. *Genetics* **110**: 609.

Willis, I., Frendewey, D., Nichols, M., Andrea Hottinger-Werlen, Schaack, J., and Soll, D. 1986. A single base change in the intron of a serine tRNA affects the rate of RNase P cleavage *in vitro* and suppressor activity *in vivo* in *Saccharomyces cerevisiae. J. Biol. Chem.* **261**: 5878.

Wilson, D. S., and Szostak, J. W. 1999. *In vitro* selection of functional nucleic acids. *Annu. Rev. Biochem.* **68**: 611–635.

Wilson, G. G. 1991. Organization of restriction-modification systems. *Nucleic Acids Res.* **19**: 2539.

Winey, M., Edelman, I., and Culbertson, M. R. 1989. A synthetic intron in a naturally intronless yeast pre-tRNA is spliced efficiently *in vivo. Mol. Cell Biol.* **9**: 329.

Winkler, G. S., Sugasawa, K., Eker A. P., deLaatand, W. I., and Hoeijmakers, J. H. 2001. Novel functional interactions between nucleotide excision DNA repair proteins influencing the enzymatic activities of TFIIH, XPG, and ERCC1-XPF. *Biochemistry* **40**: 160–165.

Wintersberger, U., and Wintersberger, E. 1987. RNA makes DNA: A speculative view of the evolution of DNA replication mechanisms. *Trends Genet.* **3**: 198.

Witkin, E. M. 1976. Ultraviolet mutagenesis and inducible DNA repair in *Escherichia coli. Bacteriol. Rev.* **40**: 869.

Woodhead, J. L., and Malcolm, A. D. B. 1980. Non-specific binding of restriction endonuclease *Eco*RI to DNA. *Nucleic Acids Res.* **8**: 389.

Worcel, A., and Burgi, E. 1972. On the structure of the folded chromosome of *Escherichia coli*. *J. Mol. Biol.* **71**: 127.

Woron, R. G., and Thompson, M. W. 1988. Genetics of Duchenne muscular dystrophy. *Annu. Rev. Genet.* **22**: 601.

Wu, H., Lima, W. F., and Crooke, S. T. 2001. Investigating the structure of human RNaseH1 by site directed mutagenesis. *J. Biol. Chem.* **276**: 23547–23553.

Wu, J. C., Kozarich, J. W., and Stubbe, J. 1983. The mechanism of free base formation from DNA by bleomycin. *J. Biol. Chem.* **258**: 4694.

Wu, T. C., and Lichten, M. 1994. Meiosis-induced double strand break sites determined by yeast chromatic structure. *Science* **263**: 515.

Wu, Y., Yu, L., McMahon, R., Rossi, J. J., Forman, S. J., and Snyder, D. S. 1999. Inhibition of bcr-abl oncogene expression by novel deoxyribozymes (DNAzymes). *Hum. Gene. Ther.* **20**: 2847.

Wycoff, E., and Hsieh, T. 1988. Functional expression of a *Drosophila* gene in yeast: Genetic complementation of DNA topoisomerase II. *Proc. Natl. Acad. Sci. U.S.A.* **85**: 6272.

Yagima, H., and Fujii, N. 1980. Chemical synthesis of bovine pancreatic ribonuclease A. *J. Chem. Soc. Chem. Commun.* pp 115–116.

Yajko, D. M., Valentine, M. C., and Weiss, B. 1974. Mutants of *Escherichia coli* with altered deoxyribonucleases, II. Isolation and characterization of mutants for exonuclease I. *J. Mol. Biol.* **85**: 323.

Yajko, D. M., and Weiss, B. 1975. Mutations simultaneously affecting endonuclease II and endonuclease III in *Escherichia coli* (Xth mutations/alkylation/DNA repair). *Proc. Natl. Acad. Sci. U.S.A.* **72**: 688.

Yamage, M., Debrabant, A., and Dwyer, D. M. 2000. Molecular characterization of a hyperinducible, surface membrane anchored, Class I nuclease of a trypanosomatid parasite. *J. Biol. Chem.* **275**: 36369–36379.

Yamamoto, N., and Droffner, M. L. 1985. Mechanisms determining aerobic or anoerobic growth in the facultative anaerobic *Salmonella typhimurium*. *Proc. Natl. Acad. Sci. U.S.A.* **82**: 2077.

Yancopoulos, G. D., and Alt, F. W. 1985. Developmentally controlled and tissue-specific expressions of unrearranged V_H gene segments. *Cell* **40**: 271.

Yancopoulos, G. D., and Alt, F. W. 1988. Reconstruction of an immune system. *Science* **241**: 1632.

Yancopoulos, G. D., Blackwell, T. K., Son, H., Hood, L., and Alt, F. W. 1986. Introduced T cell receptor variable region gene segments recombine in pre-B cells: Evidence that B and T cells use a common recombinase. *Cell* **44**: 251.

Yang, D., Lu, H., and Erickson, J. W. 2000. Evidence that processed small dsRNAs may mediate sequence-specific mRNA degradation during RNAi in *Drosophila* embryos. *Curr. Biol.* **10**: 1191–1200.

Yang, L., Heller, K., Gellert, M., and Zubay, G. 1979. Differential sensitivity of gene expression *in vitro* to inhibitors of DNA gyrase. *Proc. Natl. Acad. Sci. U.S.A.* **76**: 3304.

Yang, L., Rowe, T. C., Nelson, E. M., and Liu, L. F. 1985. *In vitro* mapping of DNA topoisomerase 11-specific cleavage sites on SV40 chromatin. *Cell* **41**: 127.

Yang, L., Wold, M. S., Li, J. J., Kelly, T. J., and Liu, L. F. 1987. Roles of DNA topoisomerases in simian virus 40 DNA replications *in vitro*. *Proc. Natl. Acad. Sci. U.S.A.* **84**: 950.

Yang, W., Hendrickson, W. A., Crouch, R. J., and Satow, Y. 1990. Structure of ribonuclease H phased at 2 Å resolution by MAD analysis of the selenomethionyl protein. *Science* **249**: 1398.

Yang, J. H., Cedergren, R., and Nadal-Ginard, B. 1994. Catalytic activity of an RNA domain derived from U6-U4 RNA complex. *Science* **263**: 77.

Yen, L., Strittmatter, S. M., and Kalb, R. G. 1999. Sequence-specific cleavage of Huntingtin mRNA by catalytic DNA. *Am. Neurol.* **46**: 366.

Yeung, A. T., Mattes, W. B., Oh, E. Y., and Grossman, L. 1983. Enzymatic properties of purified *Escherichia coli uvrABC* proteins. *Proc. Natl. Acad. Sci. U.S.A.* **80**: 6157.

Youle, R. J., and D'Alessio, G. 1997. Antitumor ribonucleases. In D'Alessio, G., and J. F. Riordan (eds.), *Ribonucleases: Structures and Functions*. Academic Press, New York.

Youle, R. J., Wu, Y.-N., Mikulski, S. M., Shogen, K., Hamilton, R. S., Newton, D., D'Alessio, G., and Gravell, M. 1994. RNase inhibition of human immunodeficiency by virus infection of H9 cells. *Proc. Natl. Acad. Sci. U.S.A.* **91**: 6012.

Young, E. T., and Sinsheimer, R. L. 1965. Comparison of initial actions of spleen deoxyribonuclease and pancreatic deoxyribonuclease. *J. Biol. Chem.* **240**: 1274.

Youngquist, R. S., and Dervan, P. B. 1985. Sequence specific recognition of B-DNA by oligo *N*-methylpyrrolecarboxamides. *Proc. Natl. Acad. Sci. U.S.A.* **82**: 2565.

Yu, C.-E., Oshima, J., Fu, Y.-H., Wijsma, E. M., Hisama, F., Aliesch, R.S., Matthews, Nakura, J., Miki, T., Ouais, S., Martin, G. M., Mulligan, J., and Schellenberg, G. D. 1996. Positional cloning of the Werner's syndrome gene. *Science* **271**: 258–262.

Yuan, R. 1981. Structure and mechanism of multifunctional restriction endonucleases. *Annu. Rev. Biochem.* **50**: 285.

Zafeiriou, D., Thorel, I. F., Andreou, A., Kleijer, W. J., Raams, A., Garristen, V. H., Gombakis, N., Zasper, N. G., and Clarkson, S. G. 2001. Xeroderma group G with severe neurological involvement and features of Cockayne syndrome in infancy. *Pediatr. Res.* **49**: 407–412.

Zajac-Kaye, M., and Tso, P. O. P. 1984. DNase 1 encapsulated in liposomes can induce neoplastic transformation of syrian hamster embryo cells in culture. *Cell* **39**: 427.

Zein, N., Sinha, A. M., McGahren, W. J., and Ellestad, G. A. 1988. Calicheamicin $\tau 1^1$: An antitumor antibiotic that cleaves double-stranded DNA site specifically. *Science* **240**: 1198.

Zieg, J., and Simon, M. 1980. Analysis of the nucleotide sequence of an invertible controling element. *Proc. Natl. Acad. Sci. U.S.A.* **77**: 4196.

Zieg, J., Silverman, M., Hilmen, M., and Simon, M. 1978. Recombinational switch for gene expression. *Science* **196**: 170.

Zinn, A. R., and Butow, R. A. 1985. Nonreciprocal exchange between alleles of the yeast mitochondrial 21S rRNA gene: Kinetics and the involvement of a double-strand break. *Cell* **40**: 887.

Zambrano, M. M., Sieglele, D. A., Almiron, M., Tormo, A., and Kolter, R. 1993. Microbiol compettition: *Escherichia coli* mutants that take over stationary phase culture. *Science* **259**: 1757.

Zhang, X., Xu, Y., Ling, H., and Hattori, T. 1999. Inhibition of infection of incoming HIV-1 virus by RNA-cleaving DNA enzyme. *FEBS Lett.* **458**: 151.

Zollner, E. J., Helm, W., Zahn, R. K., Beck, J., and Reitz, M. 1974. Different deoxyribonucleases in human lymphocytes. *Nucleic Acids Res.* **1**: 1069.

Zuag, A. J., Been, M. O., and Cech, T. R. 1986. The tetrahymena ribozyme acts like an RNA restriction endonuclease. *Nature* **324**: 429.

Zuag, A. J., and Cech, T. R. 1985. Oligomenzation of intervening sequence RNA molecules in the absence of proteins. *Science* **229**: 1060.

Zwelling, L. A., Kerrigan, D., and Marton, L. J. 1985. Effect of difluoromethylornithine, and inhibitor of polyamine biosynthesis, on the topoisomerase II-mediated DNA scission produced by 4-(9-acridiylamino) methanesulfon-*m*-anisidide in L 1210 murine leukemia cells. *Cancer Res.* **45**: 1122.

INDEX